SUSTAINABLE ENGINEERING AND AGRO-FOOD PROCESSING

Renewable and Nonrenewable Resources

SUSTAINABLE ENGINEERING AND AGRO-FOOD PROCESSING

Renewable and Nonrenewable Resources

José Juan Buenrostro-Figueroa, PhD
Julio César Tafolla-Arellano, PhD
Mónica Chávez-González, PhD
Pedro Aguilar-Zárate, PhD
A. K. Haghi, PhD
Editors

AAP | APPLE ACADEMIC PRESS

First edition published 2025

Apple Academic Press Inc.
1265 Goldenrod Circle, NE,
Palm Bay, FL 32905 USA

760 Laurentian Drive, Unit 19,
Burlington, ON L7N 0A4, CANADA

CRC Press
2385 NW Executive Center Drive,
Suite 320, Boca Raton FL 33431

4 Park Square, Milton Park,
Abingdon, Oxon, OX14 4RN UK

Apple Academic Press exclusively co-publishes with CRC Press, an imprint of Taylor & Francis Group, LLC

Library and Archives Canada Cataloguing in Publication

CIP data on file with Canada Library and Archives

Library of Congress Cataloging-in-Publication Data

CIP data on file with US Library of Congress

ISBN: 978-1-77491-774-9 (hbk)
ISBN: 978-1-77491-775-6 (pbk)
ISBN: 978-1-00356-104-0 (ebk)

About the Editors

José Juan Buenrostro-Figueroa, PhD
Researcher, Research Center for Food and Development, A.C., Mexico

José Juan Buenrostro-Figueroa, PhD, is Titular Researcher at the Research Center for Food and Development, A.C., Mexico. He is a food engineer. His MSc and PhD degrees were Food Science and Technology. Dr. Buenrostro has experience in bioprocess development, including microbial processes for enzyme production and recovery of bioactive compounds, valorization of agroindustrial byproducts, and extraction and characterization of bioactive compounds from plants and agroindustrial wastes. He has published 40 papers in indexed journals, 25 book chapters, six books as editor, and more than 70 contributions at scientific meetings, with over 1160 citations and h-index of 18 (Google Scholar database). He is member of the Mexican Association of Food Science (AMECA), Mexican Society for Food Protection affiliate of the International Association for Food Protection, and Food Safety Thematic Network: Agroindustrial By-products Valorization and Reduction of Food Losses and Waste (PDA). Dr. Buenrostro is a member of SNII (National System of Researches, Mexico), since 2014.

Julio César Tafolla-Arellano, PhD
Research Professor, Department of Basic Sciences,
Autonomous Agrarian University Antonio Narro, Coahuila, Mexico

Julio César Tafolla-Arellano, PhD, is a full-time Research Professor in the Department of Basic Sciences of the Autonomous Agrarian University Antonio Narro in Coahuila, Mexico. He is the Head of the Biotechnology and Molecular Biology Laboratory and a member of the Mexican Association of Food Engineering and Biotechnology. He is the editor in-chief of the journals *Universitas Agri* and *Revista Agraria.* Dr. Tafolla-Arellano is a member of the Mexican National System of Researchers (level I), and his research is focused on the study of the biotechnological use of plants through an interdisciplinary omics approach, biocatalysis, bioinformatics, and food biotechnology. He studied engineering in food science and technology at the

Autonomous Agrarian University Antonio Narro. He earned his PhD degree in Physiology and Plant Molecular Biology at the Center for Research and Food and Development A.C. of Sonora, Mexico. Also, he performed post-doctoral research on the postharvest shelf life of fruit and transcriptomics analysis. He has published several books chapters and articles in indexed journals and developed eight research projects. He has been advisor of several PhD, MSc, and BSc theses.

Mónica Chávez-González, PhD
Professor, School of Chemistry of the Universidad Autónoma de Coahuila, Mexico

Mónica Chávez-González, PhD, is a Research Professor at the School of Chemical Sciences of the Universidad Autónoma de Coahuila. She is a pharmacobiological chemist; her master's and doctoral programs were in Food Science and Technology. She is the General Coordinator and founder of the Bioprocessing Mx-LATAM Network, whose objective is to promote R&D in the area of bioprocessing through interaction between universities, institutes, research centers, and relevant industries in Mexico and Latin America. She has been awarded the Latin American Women in Chemistry Awards in the category of "Emerging Leader," granted by the American Chemical Society and the Latin American Federation of Chemical Associations (FLAQ). Monica's expertise is in the areas of fermentation processes, microbial biotransformation, enzyme production, industrial food waste valorization, extraction of bioactive compounds (antioxidants) and chemical characterization.

Pedro Aguilar-Zárate, PhD
Professor, National Technological Institute of Mexico, Technological Institute of Ciudad Valles, Mexico

Pedro Aguilar-Zárate, PhD, is Professor at the National Technological Institute of Mexico, Technological Institute of Ciudad Valles, Mexico. He earned his PhD degree in Food Science and Technology at Autonomous University of Coahuila, Mexico. His research focuses on the development of bioprocesses for the extraction and purification of secondary plant metabolites, the development of microbial fermentations in solid and liquid media to produce metabolites of industrial interest, and the study of protein-polyphenols interactions. Dr. Aguilar has published over 40 articles in indexed journals, over 20 book chapters, and nine articles in refereed journals. He has directed and co-directed 24 undergraduate theses and two MSc theses and has made several presentations at scientific and academic events. He has been in charge of four research projects funded by the National Technological Institute of Mexico and collaborated in two projects funded by the sectoral funds SEP-CONACYT and CONAFOR-CONACYT. He is guest editor of *Frontiers in Food Science and Technology, Polymers, Plant Science Today*, and the *International Journal of Research and Technological Innovation* (RIIT). This has earned him recognition as a member of the National System of Researchers level 1 (Mexico). His research and scientific career have been recognized by the Mexican Association of Food Science (2014), the Mexican Society of Biotechnology and Bioengineering (2015), the Mexican Association of Food Engineering and Biotechnology (2020), and the Potosino Council of Science and Technology (2021).

A. K. Haghi, PhD
Research Associate, Department of Chemistry,
University of Coimbra, Portugal

A. K. Haghi, PhD, is a retired professor and has written, cowritten, edited, and coedited more than 1000 publications, including books, book chapters, and papers in refereed journals with over 3800 citations and h-index of 32, according to the Google Scholar database. He is currently a research associate at university of Coimbra, Portugal. Professor Haghi has received several grants, consulted for several major corporations, and is a frequent speaker to national and international audiences. He is the founder and former editor-in-chief of the *International Journal of Chemoinformatics and Chemical Engineering* and *Polymers Research Journal*. Professor Haghi has acted as an editorial board member of many international journals. He has served as a member of the Canadian Research and Development Center of Sciences & Cultures. He has supervised several PhD and MSc theses at the University of Guilan (UG) and co-supervised international doctoral projects. Professor Haghi holds a BSc in urban and environmental engineering from the University of North Carolina (USA) and holds two MSc degrees, one in mechanical engineering from North Carolina State University (USA) and another one in applied mechanics, acoustics, and materials from the Université de Technologie de Compiègne (France). He was awarded a PhD in engineering sciences at the Université de *Franche-Comté* (France). He is a persistent reviewer for leading international journals.

Contents

Contributors .... *xi*

Preface .... *xv*

PART 1: SUSTAINABLE WASTE MANAGEMENT STRATEGIES 1

1. *Litchi chinensis*: Challenges and Opportunities in Agro-Industrial Waste Biorefineries 3

Christian Iván Cano-Gómez, Diana Beatriz Muñiz-Marquez, Abigail Reyes-Munguía, and Jorge Enrique Wong-Paz

2. Bioactive Compounds from Coffee Agro-Industrial Waste: An Opportunity Area within the Circular Economy 23

María de los A. Vázquez-Núñez, Nimbe Cuervo-Gama, Mariela R. Michel-Michel, Candy Carranza-Álvarez, Pedro Aguilar-Zárate, Jorge Enrique Wong-Paz, and Abigaíl Reyes-Munguía

3. Sustainable Technologies for Resource Recovery from Waste: A Case Study for Better Integrating Environmental and Economic Issues 49

Alfredo Ávila-Galarza, Claudia Yazmín Ortega-Montoya, Norma Lilia López-Díaz, and Any Lucely Sierra-Negrette

PART 2: SUSTAINABLE SPRAY DRYING TECHNOLOGY FOR PRESERVATION OF BIOACTIVE COMPOUNDS 69

4. Spray Drying: An Alternative for Seasonal Fruit Peservation 71

Gisela Rodríguez-Montes, Diana Beatriz Muñiz-Márquez, Abigail Reyes-Munguía, and Jorge Enrique Wong-Paz

5. Optimization of Spray-Drying Microencapsulation of Curcumin as a Strategy for Enhancing Water Solubility and Evaluation of Antioxidant Activities 91

Mayra Aguilar-Zárate, Alma Yolanda Vázquez-Sánchez, Dulce María Hernández-Medina, Romeo Rojas, Guillermo Cristian Guadalupe Martínez-Ávila, Ricardo Gómez-García, Mariela R. Michel-Michel, and Pedro Aguilar-Zárate

PART 3: SUSTAINABLE TECHNOLOGIES AND ENGINEERING 111

6. Potentialities from Neglected and Underutilized Species for Sustainable Fermented Beverages Production 113
Flor Edith Calderón-Martínez, Jorge Luis González-Escobar, Rubén F. González-Laredo, Anahí Jobeth Borrás-Enriquez, Pedro Aguilar-Zárate, and Fabiola Veana

7. *Trichoderma* Roles in Sustainable Agriculture 147
Erika Sireni Rodríguez-Martínez, Edel Torres-Torres, César Guigón-López, and Mónica Alvarado-Gonzalez

8. Use of Oregano as a Sustainable Perspective 189
Yahaira J. Navarro-Melgarejo, Mónica L. Chávez-Gonzalez, Cristóbal N. Aguilar, Cristian Torres-Léon, Leonardo Sepulveda-Torre, and Anna Iliná

9. Oleogels: Healthy and Sustainable Replacement for Saturated and Hydrogenated Fats 209
Mayra Aguilar-Zárate, Laura Araceli López-Martínez, and Jorge Fernando Toro-Vazquez

10. Production of Fungi for Biological Control in Solid-State Culture Bioreactors: A Sustainable Approach 227
Fernando Méndez-González, José Miguel Castillo-Minjarez, María Esmeralda Bibián-León, Ebandro Uscanga-Mortera, Javier Barrios-González, and José Juan Buenrostro-Figueroa

11. Utilization of Genome Edition to Design Tropical Plants with Increased Resistance to Biotic Stress for Agricultural Sustainability....247
José Alberto Valenzuela-Avilés, Eduardo Antonio Trillo Hernández, Miguel Angel Hernández Oñate, and Martín Ernesto Tiznado Hernández

12. Sustainable Use of Chihuahua Desert Endemic Plants: An Overview of Traditional Uses, Phytochemistry and Biological Properties 263
J. Omar Díaz-Rivas, O. Miriam Rutiaga-Quiñones, M. Humberto Chairez-Ramirez, and Julio César Tafolla-Arellano

13. Biostimulation of Grains and Plant Seeds using Probiotic Microorganisms: Shaping the Production of Sustainable and Healthy Foods for the Future 281
Stephany Nefertari Chávez-García, Dora Elisa Cruz-Casas, Erik De León-Juárez, Álvaro Javier Pastrana-Pastrana, Steffanny Sánchez-Portillo, and Adriana Carolina Flores-Gallegos

Index 295

Contributors

Mayra Aguilar-Zárate
Universidad Autónoma de San Luis Potosí, Facultad de Ciencias Químicas, San Luis Potosí, SLP, México

Pedro Aguilar-Zárate
Departamento de Ingenierías. Tecnológico Nacional de México/I. T. de Ciudad Valles, Ciudad Valles, San Luis Potosí, México

Cristóbal N. Aguilar
Food Research Department, School of Chemistry, Universidad Autónoma de Coahuila, Saltillo, Coahuila, México

Javier Barrios-González
Departamento de Biotecnología, Universidad Autónoma Metropolitana, Iztapalapa, Mexico City, Mexico

Anahí Jobeth Borrás-Enriquez
División de Ingeniería en Industrias Alimentarias, Tecnológico Nacional de México/TES San Felipe del Progreso. Estado de México, México

José Juan Buenrostro-Figueroa
Laboratorio de Biotecnología y Bioingeniería, Centro de Investigación en Alimentación y Desarrollo, Delicias, Chihuahua, Mexico

Flor Edith Calderón-Martínez
Tecnológico Nacional de México/Instituto Tecnológico de Durango. Durango, Durango, México

Christian Iván Cano-Gómez
Universidad Autónoma de San Luis Potosí, Facultad de Estudios Profesionales Zona Huasteca, Cd. Valles, San Luis Potosi, México

Candy Carranza-Álvarez
Facultad de Estudios Profesionales Zona Huasteca. Universidad Autónoma de San Luis Potosí, Ciudad Valles, San Luis Potosí, México

José Miguel Castillo-Minjarez
División de Electromecánica Industrial, Universidad Tecnológica de Tecámac, Tecámac, Mexico

M. Humberto Chairez-Ramirez
Unidad de investigación Biomédica de Cáncer, Instituto de investigaciones biomédicas, Universidad Nacional Autónoma de México & Instituto Nacional de Cancerología, Secretaria de Salud, Tlalpan, CDMX, México

Stephany Nefertari Chávez-García
Food Research Department, School of Chemistry, Saltillo, Coahuila, México

Dora Elisa Cruz-Casas
Food Research Department, School of Chemistry, Saltillo, Coahuila, México

Nimbe Cuervo-Gama
Facultad de Estudios Profesionales Zona Huasteca. Universidad Autónoma de San Luis Potosí, Ciudad Valles, San Luis Potosí, México

J. Omar Díaz-Rivas
Laboratorio de química analítica, Instituto de Investigación y Transferencia de Tecnología de Zonas Parras de la Fuente, Coahuila, México

Norma Lilia López Díaz
Universidad Autónoma de San Luis Potosí. Programa Multidisciplinario de Posgrado en Ciencias Ambientales, San Luis Potosí, México

Adriana Carolina Flores-Gallegos
Food Research Department, School of Chemistry, Saltillo, Coahuila, México

Alfredo Ávila Galarza
Universidad Autónoma de San Luis Potosí. Programa Multidisciplinario de Posgrado en Ciencias Ambientales, San Luis Potosí, México

Ricardo Gómez-García
Universidade Católica Portuguesa, CBQF - Centro de Biotecnologia e Química Fina – Laboratório Associado, Escola Superior de Biotecnologia, Porto, Portugal

Jorge Luis González-Escobar
Tecnológico Nacional de México/Instituto Tecnológico de Ciudad Valles. Ciudad Valles, San Luis Potosí, México

Rubén F. González-Laredo
Tecnológico Nacional de México/Instituto Tecnológico de Durango. Durango, Durango, México

Sandra Mónica Alvarado González
Centro de Investigación en Alimentación y Desarrollo, A.C. Coordinación Delicias, Fracc, Vencedores del desierto. Delicias, Chihuahua, México, C. P. (Mónica Alvarado-González)

Dulce María Hernández-Medina
Engineering Department, Tecnológico Nacional de México/ Instituto Tecnológico de Ciudad Valles, Ciudad Valles, San Luis Potosí, México

Miguel Ángel Hernández-Oñate
Estancias Posdoctorales-Consejo Nacional de Ciencia y Tecnología, Coordinación de Apoyos a Becarios e Investigadores, Dirección de Posgrado, Ciudad de México
CONACYT-Coordinación de Tecnología de Alimentos de Origen Vegetal, Centro de Investigación en Alimentación y Desarrollo, A.C., Victoria. Hermosillo, Sonora, México

Martín Ernesto Tiznado Hernández
Coordinación de Tecnología de Alimentos de Origen Vegetal, Centro de Investigación en Alimentación y Desarrollo A.C., Hermosillo, Sonora, México

Anna Iliná
Bioprocesses and Bioproducts Department, School of Chemistry, Universidad Autónoma de Coahuila, Saltillo, Coahuila, México

María Esmeralda Bibían León
Posgrado en Botánica, Colegio de Posgraduados-Montecillo, Texcoco, Mexico State, Mexico

Erik De León-Juárez
Food Research Department, School of Chemistry, Saltillo, Coahuila, México

Laura Araceli López-Martínez
Universidad Autónoma de San Luis Potosí, Coordinación Académica Región Altiplano Oeste, Salinas de Hidalgo, SLP, México

César Guigón López
Universidad Autónoma de Chihuahua. Facultad de Ciencias Agrícolas y Forestales. Delicias, Chihuahua, México

Guillermo Cristian Guadalupe Martínez-Ávila
School of Agronomy, General Escobedo, Universidad Autónoma de Nuevo León, Monterrey, México

Fernando Méndez-González
Laboratorio de Biotecnología y Bioingeniería, Centro de Investigación en Alimentación y Desarrollo, Delicias, Chihuahua, Mexico

Mariela R. Michel-Michel
Department de Ingenierías. Tecnológico Nacional de México/ I. T. de Ciudad Valles, Ciudad Valles, San Luis Potosí, México

Claudia Yazmín Ortega Montoya
Universidad Autónoma de Coahuila. Facultad de Ciencias Biológicas. Torreón, México

Diana Beatriz Muñiz-Marquez
Universidad Autónoma de San Luis Potosí, Facultad de Estudios Profesionales Zona Huasteca, Cd. Valles, San Luis Potosi, México

Yahaira J. Navarro-Melgarejo
Food Research Department, School of Chemistry, Universidad Autónoma de Coahuila, Saltillo, Coahuila, México

Any Lucely Sierra Negrette
Universidad Autónoma de San Luis Potosí. Programa Multidisciplinario de Posgrado en Ciencias Ambientales, San Luis Potosí, México

Álvaro Javier Pastrana Pastrana
Food Research Department, School of Chemistry, Saltillo, Coahuila, México

Steffanny Sánchez Portillo
Food Research Department, School of Chemistry, Saltillo, Coahuila, México

Abigail Reyes-Munguía
Universidad Autónoma de San Luis Potosí, Facultad de Estudios Profesionales Zona Huasteca, Cd. Valles, San Luis Potosi, México

Erika Sireni Rodríguez-Martínez
Centro de Investigación en Alimentación y Desarrollo, A.C. Coordinación Delicias, Fracc, Delicias, Chihuahua, México

Gisela Rodríguez-Montes
Tecnológico Nacional de México, Campus Instituto Tecnológico de Ciudad Valles, Ciudad Valles, San Luis Potosí, México

Romeo Rojas
School of Agronomy, General Escobedo, Universidad Autónoma de Nuevo León, Monterrey, México

O. Miriam Rutiaga-Quiñones
Departamento de Ings. Química-Bioquímica, Laboratorio de tecnología postcosecha. Tecnológico Nacional de México Instituto Tecnológico de Durango. Durango, Dgo, México

Leonardo Sepulveda-Torre
Bioprocesses and Bioproducts Department, School of Chemistry, Universidad Autónoma de Coahuila, Saltillo, Coahuila, México

Julio César Tafolla-Arellano
Laboratorio de Biotecnología y Biología Molecular.Departamento de Ciencias Básicas. Universidad Autónoma Agraria Antonio Narro. Saltillo, Coahuila, México

Jorge Fernando Toro-Vazquez
Universidad Autónoma de San Luis Potosí, Facultad de Ciencias Químicas, San Luis Potosí, SLP, México

Cristian Torres-Léon
Bioprocesses and Bioproducts Department, School of Chemistry, Universidad Autónoma de Coahuila, Saltillo, Coahuila, México

Edel Torres-Torres
Universidad Autónoma de Chihuahua. Facultad de Ciencias Agrícolas y Forestales. Poniente, Delicias, Chihuahua, México

Eduardo Antonio Trillo-Hernández
Unidad de Tecnología de Alimentos-Secretaría de Investigación y Posgrado, Universidad Autónoma de Nayarit, Tepic, Nayarit, Mexico
Estancias Posdoctorales-Consejo Nacional de Ciencia y Tecnología, Coordinación de Apoyos a Becarios e Investigadores, Dirección de Posgrado, Ciudad de México

Ebrando Uscanga-Mortero
Posgrado en Botánica, Colegio de Posgraduados-Montecillo, Texcoco, Estado de México, Mexico

José Alberto Valenzuela-Avilés
Estudiante de Doctorado. Coordinación de Tecnología de Alimentos de Origen Vegetal, Centro de Investigación en Alimentación y Desarrollo A.C., Hermosillo, Sonora. México

María de los A. Vázquez-Núñez
Facultad de Estudios Profesionales Zona Huasteca. Universidad Autónoma de San Luis Potosí, Ciudad Valles, San Luis Potosí, México

Alma Yolanda Vázquez-Sánchez
Área Agroindustrial Alimentaria, Universidad Tecnológica de Xicotepec de Juárez, Xicotepec de Juárez, Puebla, México

Fabiola Veana
Tecnológico Nacional de México/Instituto Tecnológico de Ciudad Valles. Vista Hermosa, Ciudad Valles, San Luis Potosí, México

Jorge Enrique Wong-Paz
Universidad Autónoma de San Luis Potosí, Facultad de Estudios Profesionales Zona Huasteca, Cd. Valles, San Luis Potosi, México

Preface

This book provides the most recent research progress, innovations, and applications in the agro-food processing for sustainable agriculture and food production. This book also focuses on the challenges of implementing sustainability in different contexts of biodiversity and ecosystem for shaping future engineering.

The chapters in this research-oriented book address the sustainable applications of renewable and nonrenewable resources. It also provides a practical multidisciplinary overview of solutions to help achieve sustainable production.

This book highlights the engineering advancements and processing solutions for agricultural and food engineering and cover chapters that discuss agro-industrial waste biorefineries, bioactive compounds from agro-industrial waste, sustainable technologies for resource recovery from waste, spray drying, healthy and sustainable replacement for saturated and hydrogenated fats, and healthy foods for the future.

This new volume is designed to provide a balanced overview of agro-processing methods from the basic concepts to the case studies related to agro-processing and food engineering.

This book:

- Addresses different aspects of agro-food processing levels of the food chain.
- Suggests practical solutions for the food industry to control and reduce environmental impact.
- Provides ideas to reduce environmental footprint and securing our future through biodiversity.

PART 1

Sustainable Waste Management Strategies

CHAPTER 1

Litchi chinensis: Challenges and Opportunities in Agro-Industrial Waste Biorefineries

CHRISTIAN IVÁN CANO-GÓMEZ, DIANA BEATRIZ MUÑIZ-MARQUEZ, ABIGAIL REYES-MUNGUÍA, and JORGE ENRIQUE WONG-PAZ*

Universidad Autónoma de San Luis Potosí, Facultad de Estudios Profesionales Zona Huasteca, San Luis Potosi, México

ABSTRACT

Litchi chinensis (litchi) is increasingly in demand due to its high nutritional value and health benefits. It is widely distributed and consumed worldwide, mainly in Southeast Asia, followed by Africa and South America. Today, litchi production and consumption involve the significant generation of agro-industrial wastes, such as peel, seeds, and leaves, which are a huge pollution problem. These wastes contain valuable compounds, which can be exploited through different applications. Some studies have already evaluated the bioactivities of these compounds; however, only a general vision of waste valorization pathways with a focus on a linear economy has been studied. The circular economy is a model that involves the full use of existing materials and products to create added value and extend the life cycle of the products. The biorefinery concept involves the integration of multiple processes to recover and utilize the maximum value of plant-based waste materials, trying to reach a circular economy. This not only reduces waste and saves resources but also generates valuable products such as food ingredients, functional foods, cosmetic ingredients, and biofuels. Despite the many benefits of biorefineries, several challenges remain to be overcome, such as technical and economic feasibility, scalability, and sustainability; however, there are also numerous

Sustainable Engineering and Agro-Food Processing: Renewable and Nonrenewable Resources.
José Juan Buenrostro-Figueroa, Julio César Tafolla-Arellano, Mónica Chávez-González, Pedro Aguilar-Zárate, A. K. Haghi (Eds.)

growth opportunities, particularly in the development of new technologies and processes to optimize the use of bioactive compounds from plant-based waste materials. This chapter aims to discuss the feasibility that exists in the agro-industry, focusing on the production of litchi, with the objective of taking advantage of the waste generated through the implementation of biorefineries to achieve a circular economy in the future.

1.1 INTRODUCTION

Litchi chinensis, commonly known as litchi or lychee, is a tropical fruit that has gained significant popularity worldwide due to its unique taste, nutritional value, and versatility in culinary applications (Menzel and Waite, 2005). With the increasing global demand for litchi and the expansion of its cultivation areas, there is a growing interest in the sustainable management of agro-industrial waste generated during litchi production and processing.

Litchi is consumed and appreciated in many regions worldwide, with Asia being the largest producer and consumer. The global distribution and consumption patterns of litchi have evolved, with significant variations in production and demand trends across different regions (Huang et al., 2005). As litchi production continues to grow, there is a pressing need to address the challenges associated with the generation and management of agro-industrial wastes.

The processing of litchi generates a substantial amount of waste, primarily in the form of peels, seeds, and pulp. If not managed properly, these wastes can pose environmental and economic challenges. However, they also represent an untapped resource as they contain valuable components such as phenolic compounds, dietary fiber, and essential oils (Chen et al., 2017).

The concept of biorefineries offers a promising approach to transforming litchi wastes into value-added products, promoting a more sustainable and circular bioeconomy (Cherubini, 2010). Various technologies have been explored for the extraction and utilization of valuable components from litchi wastes, including enzymatic hydrolysis, solvent extraction, and membrane separation (Zhao et al., 2021). Despite the potential benefits of biorefineries, there are several limitations and challenges that need to be addressed for their widespread implementation. These include the availability and sustainability of feedstocks, the development of efficient and cost-effective conversion technologies, and the integration of different processes within a single biorefinery (Jönsson and Martín, 2016a; Kumar et al., 2009).

Looking ahead, future trends in biorefineries will likely focus on the diversification of biomass feedstocks, advances in conversion technologies, and process integration and waste valorization (Hallenbeck, 2009; Lynd et al., 2008). By exploring these trends and understanding their implications, we can better develop the biorefinery sector and ensure its success in promoting the sustainable development of the bioeconomy.

1.2 LITCHI: PRODUCTION AND CONSUMPTION

The origin of litchi can be traced back to the subtropical region of southern China and northern Vietnam, between latitudes 23 and 27°N. China has been cultivating litchi for over 2000 years. Eventually, litchi spread to Myanmar, India, Thailand, Madagascar, Mauritius, Hawaii, South Africa, and Australia. The main export markets for litchi are the United States, the European Union, and other Asian countries. China is the largest exporter of litchi, followed by Thailand and Vietnam (FAOSTAT, 2021).

In 2018, the global production of litchi was approximately 3.5 million tons, with China contributing 80% (2.8 million tons) of this total (see Fig. 1.1). Other significant producers included India (0.67 million tons), Vietnam (0.38 million tons), Madagascar (0.10 million tons), and Thailand (0.048 million tons). In Mexico, litchi cultivation in 2019 was 4040.20 hectares harvested, obtaining a production of 0.025 million tons. Litchis are also cultivated in countries like Bangladesh, Australia, South Africa, Reunion, Brazil, and Israel. The primary markets for China's fresh litchi exports are Hong Kong, the United States, and Malaysia. In 2018, India exported a modest amount of litchi (around 150 tons). South Africa produced about 11,000 tons of litchi in the 2017/18 season, with 56.4% primarily exported to Europe (Menzel and Waite, 2005; Mitra and Pan, 2020; SIAP, 2019).

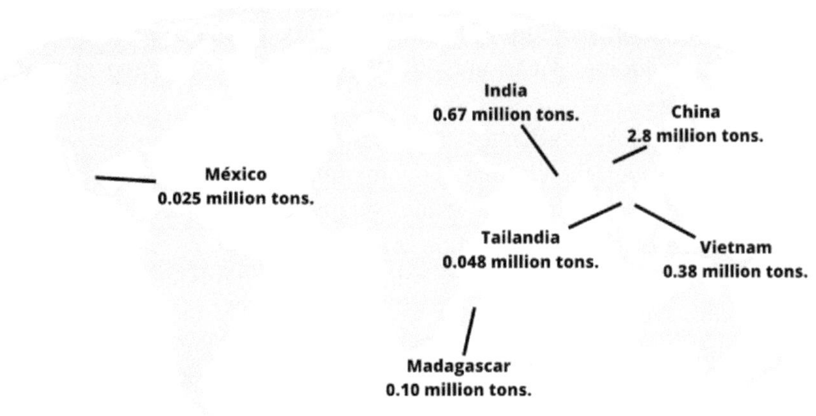

FIGURE 1.1 Production of litchi in the world.

Litchi, as a fresh fruit, has a unique taste and texture, characterized by a thin, leathery skin covering a juicy and aromatic white pulp and a single large seed. Its flavor is often described as a mix of grape, pear, and watermelon with a floral aroma (Menzel and Waite, 2005). Litchi is rich in vitamin C, potassium, and other nutrients, making it a popular choice for health-conscious consumers. In addition to fresh fruit consumption, litchi is used in various food products such as juices, jams, jellies, syrups, and ice creams. The fruit is also processed and canned to be sold in international markets, extending its shelf life and availability.

In most litchi-producing countries, a significant proportion of the fruit is consumed domestically. For instance, in China, around 90% of the litchi produced is consumed locally (Menzel and Waite, 2005). Similarly, in India, the greatest amount of litchi is consumed within the country (APEDA, 2021). In other countries, litchi is also appreciated for its sweet taste and distinctive aroma. In the United States and Europe, litchi is mainly consumed as a fresh fruit and is also used in beverages and desserts (Blancke, 2016). In South Africa, litchi fruit is used in the production of wines and spirits (Blancke, 2016).

Despite the popularity and economic value of litchi, producers face multiple challenges, including rising production costs, climate change, and market fluctuations (Menzel and Waite, 2005). Inflation, unpredictable weather patterns, and reduced yields are putting pressure on the litchi industry to maintain profitability. In response to these challenges, the industry is looking for alternatives within the litchi production system to offset the negative impacts. One such approach is the circular economy, which focuses on utilizing the entire biomass generated from litchi production (Menzel and Waite, 2005). Wastes such as litchi peel, seed, and pulp can be used for various applications, including food ingredients, animal feed, and bioenergy production, reducing waste and creating additional revenue streams for producers.

To mitigate these challenges, the industry is exploring innovative uses for litchi waste through the circular economy approach. Revalorization of litchi wastes can create additional revenue streams for producers while reducing waste. The global litchi production and trade volume have been increasing steadily over the years, with China being the leading producer and the United States being the leading importer. Overall, the litchi industry has a promising future, and embracing sustainable practices such as circular economy approaches can help ensure its long-term success. Litchi has a rich history and is a popular fruit worldwide because the fruit offers unique organoleptic properties and has been used in a variety of food products. However, the industry faces challenges due to rising production costs and climate change. Adopting the circular economy concept and exploring innovative uses for litchi waste can help mitigate these challenges and ensure a sustainable and profitable future for the industry.

1.3 GENERATION OF AGRO-INDUSTRIAL WASTES FROM LITCHI

While litchi is highly valued for its flavor and nutritional properties, its production generates significant agro-industrial waste, such as husks and seeds. Proper management of these wastes is crucial for environmental sustainability and offers opportunities to create value-added products.

The seeds make up approximately 10–15% of the total fruit weight (see Fig. 1.2). The seeds contain a high percentage of fatty acids, proteins, and bioactive compounds, such as saponins and flavonoids, and are discarded as waste during consumption (Menzel and Waite, 2005; Mühlbauer and Müller, 2020), resulting in the loss of potentially highly valuable materials. In recent years, there have been some reports describing the use of litchi seed as an economical and effective carbon source (Yao et al., 2021).

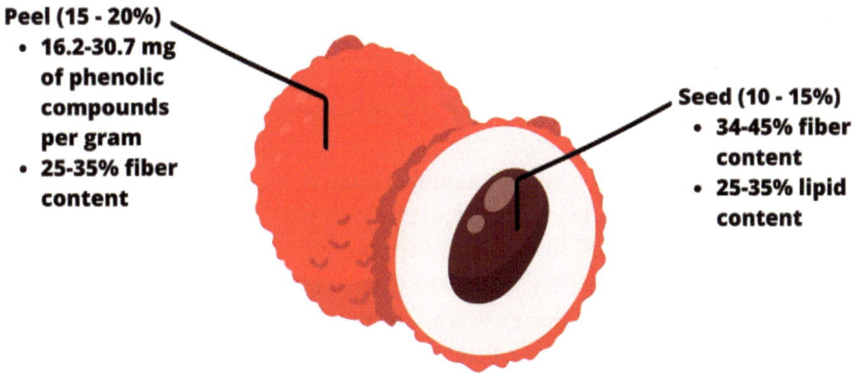

FIGURE 1.2 Generation of litchi agro-industrial wastes and potential uses of these wastes.

The litchi peel accounts for about 15–20% of the total fruit weight (see Fig. 3.1) and is a significant source of waste. The peels, which are rich in bioactive compounds such as polyphenols and flavonoids, which have potential applications in various industries (Menzel and Waite, 2005; Mühlbauer and Müller, 2020), are often simply discarded in the garbage or incinerated

(Nur Hanani et al., 2018). Litchi pericarp, as an affordable and environmentally friendly material rich in active ingredients, was developed and applied in many fields, such as the fabrication of porous carbon nanosheets/particle composite materials, medical diagnosis, environmental protection, food, and healthcare product development (Yao et al., 2021).

The litchi production generates significant amounts of agro-industrial wastes from different parts of the fruit, including peels, seeds, stems, and leaves. These wastes are rich in nutrients and bioactive compounds, but their disposal can lead to environmental pollution and health hazards. Therefore, it is important to explore potential uses for these wastes, such as the production of value-added products and organic fertilizers, to contribute to sustainable agriculture and reduce environmental pollution.

1.4 VALUABLE COMPONENTS FROM LITCHI WASTES

The processing of litchi fruits generates significant amounts of waste, such as peels, seeds, and pulp residues, which are often discarded and lead to environmental pollution and waste of resources. However, recent studies have revealed that these wastes contain valuable components with potential applications in the food, pharmaceutical, and cosmetic industries (Dahiya et al., 2018; Duan et al., 2007; Huang et al., 2016; Liu et al., 2013).

Phenolic compounds are abundant in litchi wastes, especially in the peels, which have been reported to contain 16.2–30.7 mg of phenolic compounds per gram of dry weight (see Fig. 3.1) (Li et al., 2014). These compounds possess antioxidant, anti-inflammatory, and antimicrobial properties, with the main phenolic compounds found in litchi peels including epicatechin, procyanidin B2, and proanthocyanidins (Zhao et al., 2021).

Litchi seeds and peels are particularly rich in dietary fiber, with the seed fiber content ranging from 35–45% (dry weight) and the peel fiber content ranging from 25–35% (dry weight) (Chen et al., 2017). Litchi seed and peel fiber exhibit good water-holding capacity, oil-binding capacity, and swelling capacity, suggesting that these fibers could be used as functional food ingredients to improve the textural properties and nutritional profile of food products.

Lipids and fatty acids in litchi seeds have potential applications in the food, pharmaceutical, and cosmetic industries (Sui et al., 2016). The lipid content in litchi seeds ranges from 25–35% (dry weight), with primary fatty acids found in litchi seed oil including oleic acid (43–50%), linoleic acid (33–37%), and palmitic acid (12–15%).

Litchi wastes, particularly the peels, are rich in flavonoids, which are a group of bioactive compounds with antioxidant, anti-inflammatory, and anticancer properties (Zhao et al., 2020). The flavonoid content in litchi peels has been reported to range from 2.1 to 6.3 mg/g of dry weight.

Litchi wastes can also be used as a source of natural colorants due to the presence of anthocyanins, specifically cyanidin-3-rutinoside and cyanidin-3-glucoside, which are responsible for the red color of litchi peels (Zhang et al., 2017). These natural colorants can be utilized in the food, beverage, and cosmetic industries as safe and ecofriendly alternatives to synthetic colorants.

Given the potential applications of litchi waste in various industries, it is crucial to develop sustainable and efficient methods to extract and utilize these valuable components. This would not only help reduce environmental pollution and waste of resources but also contribute to the development of ecofriendly and high-value products. Some of the extraction techniques being explored for litchi wastes include solvent extraction, enzyme-assisted extraction, ultrasonic-assisted extraction, and microwave-assisted extraction (Zhang et al., 2013).

Litchi wastes, such as peels and seeds, are rich in valuable components with potential applications in the food, pharmaceutical, and cosmetic industries. By developing sustainable and efficient extraction methods, these wastes can be transformed into high-value products, contributing to environmental protection and resource utilization.

1.5 BIOREFINERY CONCEPT IN AGRO-INDUSTRIAL WASTE

The biorefinery concept has emerged as a promising approach for the sustainable utilization of agro-industrial waste, converting it into valuable products such as biofuels, chemicals, and materials. This concept is analogous to the traditional petroleum refinery, but instead of using fossil resources, it relies on renewable biomass, reducing the environmental impact and dependence on nonrenewable resources (Cherubini, 2010).

1.5.1 AGRO-INDUSTRIAL WASTE AS FEEDSTOCK

Agro-industrial waste, generated from agricultural and food processing activities, is an abundant and renewable source of biomass that can be used as feedstock in biorefineries. These wastes include crop residues, food processing wastes, and waste from animal production (Kamm and Kamm,

2004). Utilizing agro-industrial waste not only provides a valuable resource for biorefinery operations but also helps to reduce environmental pollution and waste disposal issues (Pandey et al., 2015).

1.5.2 INTEGRATED BIOREFINERY SYSTEMS

Integrated biorefinery systems are designed to convert agro-industrial waste into various value-added products using a combination of biochemical and thermochemical processes. These systems (see Fig. 1.3) can produce a wide range of products, such as biofuels (e.g., bioethanol, biodiesel), chemicals (e.g., organic acids, biopolymers), and materials (e.g., fibers, composites) (Venkata Mohan et al., 2016).

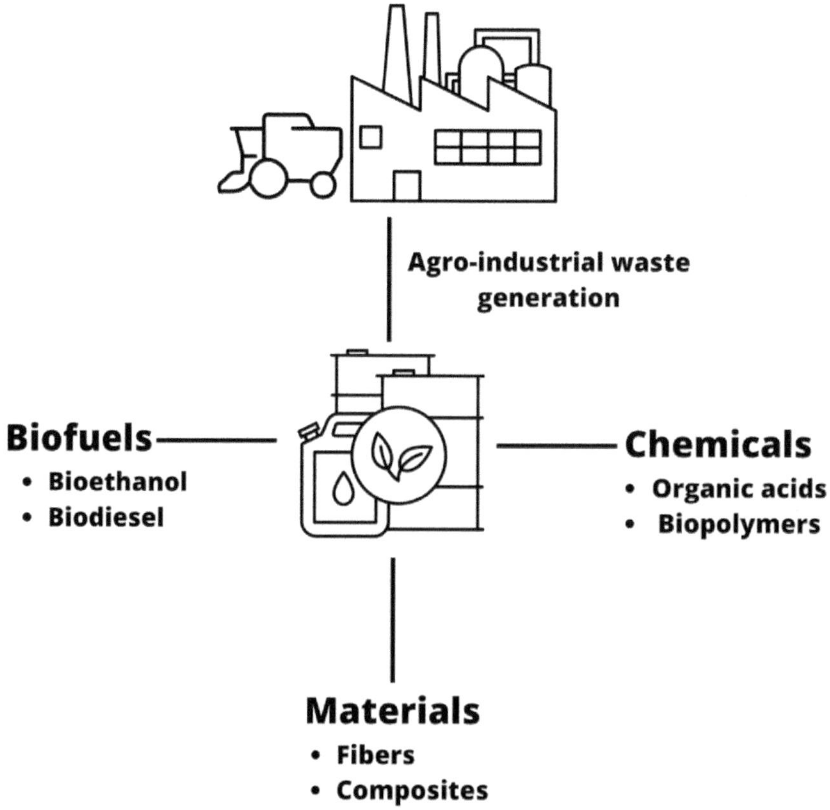

FIGURE 1.3 Production of products from agro-industrial wastes.

1.5.3 BENEFITS OF BIOREFINERIES

The implementation of biorefineries offers several benefits, including waste reduction, resource efficiency, and the development of a circular economy. By transforming waste into valuable products, biorefineries can reduce the environmental impact of waste disposal and contribute to more sustainable use of resources (Carvalheiro et al., 2008). The biorefinery concept provides an innovative approach to the sustainable utilization of agro-industrial waste, offering environmental and economic benefits. By converting waste into valuable products, biorefineries contribute to the development of a more sustainable and circular economy.

The biorefinery concept is a promising approach to the sustainable management of agro-industrial waste. The concept aims to convert various types of waste into value-added products through integrated processes that utilize all components of the waste stream. Biorefinery can contribute to the development of a circular economy and reduce environmental pollution caused by the disposal of waste.

1.6 BIOREFINERY FOR THE TRANSFORMATION OF LITCHI WASTES

Litchi generates a significant amount of waste during processing, such as peels, seeds, and pulp residues. These wastes are rich in valuable components, making them ideal candidates for biorefinery processes. Litchi seeds have been reported to contain valuable components such as proteins, polysaccharides, and lipids (Chen et al., 2011). The development of biorefineries for the transformation of litchi seeds into valuable products can contribute to the sustainability of the litchi industry and the circular economy. The biorefinery concept, inspired by the petroleum industry, aims to convert biomass or waste materials into valuable products, such as biofuels, chemicals, and materials, in a sustainable and environmentally friendly manner (Cherubini, 2010). In this context, the transformation of litchi wastes can contribute to the circular economy, resource efficiency, and waste minimization (see Fig. 1.4).

1.6.1 EXTRACTION OF PHENOLIC COMPOUNDS

Litchi peels are rich in phenolic compounds, which possess strong antioxidant, anti-inflammatory, and antimicrobial properties (Zhao et al., 2006). Solvent extraction is a traditional method for the extraction of phenolic compounds from litchi wastes, involving the use of organic solvents such

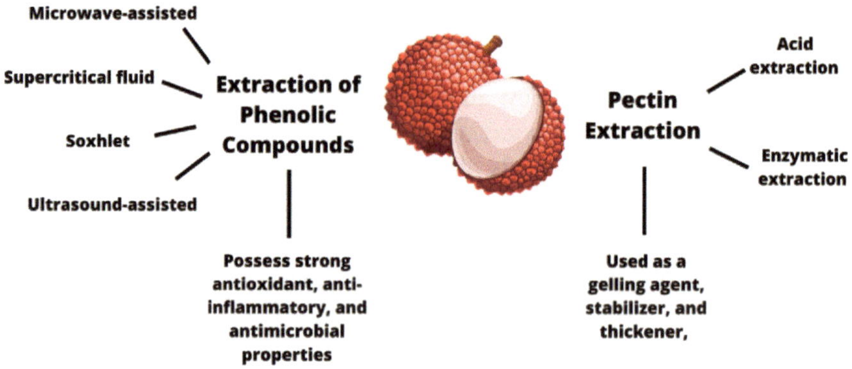

FIGURE 1.4 Technologies for the extraction of valuable compounds from agro-industrial wastes.

as ethanol, methanol, and acetone. This method is widely used due to its simplicity and low cost. However, it can be time-consuming and may require a considerable amount of solvent, which can have negative environmental impacts (Gallo et al., 2010).

Soxhlet extraction is another conventional method used for the extraction of phenolic compounds from litchi waste. This technique involves continuous extraction using a solvent in a closed system, which can improve the extraction efficiency and reduce solvent consumption compared to the solvent extraction method (Gallo et al., 2010).

Ultrasound-assisted extraction (UAE) is an advanced extraction technique that has gained popularity in recent years due to its effectiveness and ecofriendliness. The UAE uses ultrasonic waves to facilitate the release of phenolic compounds from plant materials by disrupting cell walls and enhancing solvent penetration. This technique has been successfully applied to extract phenolic compounds from litchi pericarp, resulting in higher yields and shorter extraction times compared to conventional methods (Rao, 2010).

Microwave-assisted extraction (MAE) is another advanced extraction technique that uses microwave energy to heat solvents and accelerate the extraction of phenolic compounds from plant materials. MAE offers several advantages over traditional extraction methods, such as shorter extraction times, reduced solvent consumption, and higher extraction yields (Gallo et al., 2010).

Supercritical fluid extraction (SFE) is an advanced extraction technique that uses supercritical fluids, such as supercritical carbon dioxide ($scCO_2$), as the extraction solvent. SFE offers several advantages over conventional extraction methods, including reduced solvent consumption, shorter extraction times, and higher selectivity for the target compounds. However, SFE has some limitations, such as high equipment costs and the requirement for specialized knowledge for its operation (Da Porto et al., 2013).

1.6.2 PECTIN EXTRACTION

Pectin is a naturally occurring polysaccharide found in the cell walls of various fruits and vegetables. It is widely used in the food industry as a gelling agent, stabilizer, and thickener and has potential applications in the pharmaceutical and cosmetic industries as well (Picot-Allain et al., 2020). These wastes are rich in pectin and can be utilized as a sustainable source for pectin extraction (Prakash Maran et al., 2013).

Acid extraction is the most conventional method used for pectin extraction from plant materials. The process involves the use of diluted acids, such as citric, hydrochloric, or sulfuric acid, to solubilize the pectin from the plant matrix, followed by precipitation using alcohol (Wang et al., 2015). Acid extraction has been employed to extract pectin from orange peel, with citric acid being the most commonly used acid (Prakash Maran et al., 2013).

Enzymatic extraction is an alternative method that utilizes enzymes, such as pectinase and cellulase, to break down the plant cell wall and release pectin. This method has several advantages over acid extraction, including milder extraction conditions, higher pectin quality, and reduced environmental impact (Picot-Allain et al., 2020).

1.6.3 POLYSACCHARIDE EXTRACTION

Litchi seeds also contain water-soluble polysaccharides that can be extracted using various methods. Chen et al. (2011) reported the optimization of extraction conditions for water-soluble polysaccharides from litchi seeds using response surface methodology. The extracted polysaccharides demonstrated antioxidant and immunomodulatory activities, indicating their potential applications in the pharmaceutical and nutraceutical industries (Chen et al., 2011).

The transformation of litchi seeds into valuable products through biorefinery processes presents an opportunity to add value to the litchi industry and promote the sustainable use of agro-industrial waste. By extracting and converting proteins, polysaccharides, and lipids from litchi seeds, high-value products with potential applications in the food, pharmaceutical, and renewable energy sectors can be obtained.

1.7 LIMITATIONS AND CHALLENGES OF SUSTAINABLE BIOREFINERIES

The transition toward a sustainable, low-carbon economy has led to the emergence of biorefineries as a promising approach for the production of biobased products, including biofuels, biomaterials, and biochemicals (Clark and Deswarte, 2008; Kamm and Kamm, 2004). Biorefineries employ a variety of biological and chemical processes to convert biomass feedstocks, such as agricultural residues, forestry wastes, and energy crops, into value-added products (Piotrowski et al., 2008). These biorefineries have the potential to address global challenges associated with energy security, climate change, and resource depletion while promoting rural development and green jobs (Cherubini, 2010; Sims et al., 2010).

However, the successful implementation of sustainable biorefineries faces numerous limitations and challenges across different stages of their development, ranging from feedstock production and supply to conversion technologies, process integration, and market dynamics (Venkata Mohan et al., 2016). Additionally, the broader environmental, social, and economic implications of biorefineries warrant thorough investigation and assessment to ensure the long-term sustainability of this emerging sector (Tilman et al., 2009; Zilberman et al., 2013).

1.7.1 FEEDSTOCK AVAILABILITY AND SUPPLY

One of the primary challenges in establishing sustainable biorefineries is the availability and supply of feedstock. The seasonal and regional variability of biomass resources can impact the consistent supply of feedstock to biorefineries (Kumar et al., 2009). Moreover, competition for resources with other industries, such as agriculture and forestry, can lead to higher feedstock costs (Cherubini, 2010). To address this challenge, biorefineries need to develop a diverse portfolio of feedstocks, including lignocellulosic biomass, agricultural residues, and municipal solid waste (Balat, 2011).

1.7.2 CONVERSION TECHNOLOGIES

The development and optimization of efficient conversion technologies are crucial for the success of sustainable biorefineries. Some of the current biomass conversion technologies, such as biochemical and thermochemical processes, are still not fully mature or cost-effective (Mohr and Raman, 2013). Furthermore, the complex nature of biomass feedstock requires the development of flexible and robust conversion technologies capable of handling a wide range of feedstocks (Bozell and Petersen, 2010).

1.7.3 INTEGRATION AND SCALE-UP

The integration of multiple processes within a biorefinery is a critical aspect that needs to be addressed to achieve economic viability (Carvalheiro et al., 2008). Biorefineries should consider integrating multiple conversion technologies, product lines, and waste management strategies to maximize resource utilization and reduce environmental impacts (Fernando et al., 2006). Additionally, scaling up biorefinery processes from the laboratory to commercial-scale operations is a significant challenge as it requires substantial investments and infrastructure development (Clark et al., 2009).

1.7.4 ENVIRONMENTAL AND SOCIAL IMPACTS

The environmental and social impacts of biorefineries must be carefully considered during their planning and operation. Biorefineries can lead to land use change, which may have adverse effects on biodiversity and ecosystem services (Zilberman et al., 2013). Moreover, the cultivation of energy crops for biorefineries can compete with food production, potentially leading to food security issues (Tilman et al., 2009). To address these concerns, biorefineries should focus on utilizing waste and residue streams as feedstocks, implementing sustainable agricultural practices, and conducting life cycle assessments to evaluate and minimize their environmental impacts (Cherubini, 2010).

1.7.5 POLICY AND MARKET SUPPORT

The development and expansion of sustainable biorefineries require supportive policies and market incentives. Currently, policies and financial incentives for bioenergy and bioproducts are not as well-established as those for fossil-based industries (Sims et al., 2010). Governments should implement policies that support the development of biorefineries and the market for their products, such as tax credits, grants, and renewable energy mandates (Balan et al., 2013). Additionally, creating awareness and demand for bioproducts among consumers is essential for the long-term success of biorefineries (Piotrowski et al., 2008).

Sustainable biorefineries face several challenges and limitations, including feedstock availability and supply, conversion technology development, process integration and scale-up, environmental and social impacts, and policy and market support. Overcoming these challenges requires a combination of technological advancements, strategic planning, supportive policies, and increased public awareness. Researchers, industry leaders, and policymakers must work together to develop and implement innovative solutions that ensure the long-term success and sustainability of biorefineries.

1.8 FUTURE TRENDS OF BIOREFINERIES

Biorefineries have been increasingly recognized as a key component in the transition toward a more sustainable and circular bioeconomy, offering potential solutions to the challenges of resource depletion, climate change, and energy security (Cherubini, 2010; Kamm and Kamm, 2004). As the field of biorefineries continues to evolve, several emerging trends are shaping the future landscape of this sector, including advancements in feedstock development, breakthroughs in conversion technologies, and new approaches to process integration and waste valorization (Ragauskas et al., 2006; Sims et al., 2010).

1.8.1 DIVERSIFICATION AND SUSTAINABILITY OF BIOMASS FEEDSTOCKS

One of the critical factors influencing the future of biorefineries is the development of more diverse and sustainable biomass feedstocks that can reduce competition with food production, minimize land-use change and environmental impacts, and enhance resource-use efficiency (Cherubini, 2010; Sims et al., 2010). Examples of such feedstocks include lignocellulosic materials from agricultural and forestry residues, dedicated energy crops grown on marginal lands, and various waste streams, such as municipal solid waste, food waste, and industrial wastes (Bentsen and Felby, 2012; Ragauskas et al., 2006). Furthermore, advances in plant breeding and biotechnology are enabling the development of new feedstock varieties with improved traits, such as higher biomass yield, enhanced digestibility, and increased resistance to pests and diseases (Pauly and Keegstra, 2008; Vanholme et al., 2010).

1.8.2 ADVANCES IN CONVERSION TECHNOLOGIES

The development of more efficient and versatile conversion technologies is another key trend shaping the future of biorefineries. This includes advances in both biochemical and thermochemical conversion processes, as well as the emergence of novel hybrid and integrated systems (Huber et al., 2006; Wyman, 2007). For instance, progress in enzymatic hydrolysis and fermentation technologies is enabling the cost-effective production of cellulosic ethanol and other advanced biofuels, while improvements in gasification, pyrolysis, and catalytic processes are facilitating the conversion of biomass into a wide range of platform chemicals and materials (Bridgwater, 2012; Lynd et al., 2008). Additionally, the growing interest in the production of biogas and biohydrogen from biomass and waste streams highlights the potential for expanding the portfolio of bioenergy carriers and integrating biorefineries with other energy systems (Angelidaki et al., 2018; Hallenbeck, 2009).

1.8.3 PROCESS INTEGRATION AND WASTE VALORIZATION

Another emerging trend in biorefineries is the increasing focus on process integration and waste valorization, aiming to maximize the value derived from biomass feedstocks and minimize waste generation (Jönsson and Martín, 2016b; Kumar et al., 2009). This involves the development of novel biorefinery configurations that integrate multiple conversion pathways and coproduce a diverse range of bio-based products, such as fuels, chemicals, materials, and energy (Kamm and Kamm, 2004; Ragauskas et al., 2006). Furthermore, the utilization of waste streams and wastes generated within the biorefinery, as well as external sources of waste, can contribute to the circularity of the bioeconomy and enhance the overall sustainability of biorefinery operations (Cherubini, 2010; Jönsson and Martín, 2016a).

All contribute to the ongoing development and maturation of this sector. These trends also underscore the importance of a holistic and system-oriented approach to biorefinery research and development, which is essential for realizing the full potential of biorefineries as a cornerstone of the sustainable bioeconomy.

The litchi industry plays a significant role in the global fruit market, with substantial production and demand across various regions. The increasing popularity of litchi has led to a considerable generation of agro-industrial wastes, including peels, seeds, and pulp. Biorefineries for the transformation of litchi wastes offer a promising approach to address the environmental and economic challenges associated with agro-industrial waste management; however, there are limitations and challenges in the development and implementation of sustainable biorefineries, such as the optimization of extraction techniques, scale-up, and commercial viability. Future trends in biorefineries involve the integration of innovative technologies, the development of new applications for litchi-derived components, and the combination of waste valorization processes to improve the overall sustainability and economic performance of litchi biorefineries. By addressing these challenges and exploring new opportunities, the litchi industry can contribute to the transition toward a more sustainable and circular economy.

KEYWORDS

- **litchi**
- **waste valorization**
- **biorefinery**
- **sustainable development goals**
- **circular economy**

REFERENCES

Angelidaki, I.; Treu, L.; Tsapekos, P.; Luo, G.; Campanaro, S.; Wenzel, H.; Kougias, P. G. Biogas Upgrading and Utilization: Current Status and Perspectives. *Biotechnol. Adv.* **2018,** *36* (2), 452–466. https://doi.org/10.1016/J.BIOTECHADV.2018.01.011

APEDA. 2021, https://agriexchange.apeda.gov.in/inttrade/inttrade.aspx

Balan, V.; Chiaramonti, D.; Kumar, S. Review of US and EU Initiatives Toward Development, Demonstration, and Commercialization of Lignocellulosic Biofuels. *Biofuels Bioprod. Biorefin.* **2013,** *7* (6), 732–759. https://doi.org/10.1002/BBB.1436

Balat, M.Production of Bioethanol from Lignocellulosic Materials via the Biochemical Pathway: A Review. *Energy Convers. Manag.* **2011,** *52* (2), 858–875. https://doi.org/10.1016/J.ENCONMAN.2010.08.013

Bentsen, N. S.; Felby, C. Biomass for Energy in the European Union—A Review of Bioenergy Resource Assessments. *Biotechnol. Biofuels* **2012,** *5* (1), 1–10. https://doi.org/10.1186/1754-6834-5-25/FIGURES/6

Blancke, R. *Tropical Fruits and Other Edible Plants of the World: An Illustrated Guide - Rolf Blancke - Google Libros* 2016.

Bozell, J. J.; Petersen, G. R. Technology Development for the Production of Biobased Products from Biorefinery Carbohydrates—the US Department of Energy's "Top 10" Revisited. *Green Chem.* **2010,** *12* (4), 539–554. https://doi.org/10.1039/B922014C

Bridgwater, A. V. Review of Fast Pyrolysis of Biomass and Product Upgrading. *Biomass Bioenergy* **2012,** *38,* 68–94. https://doi.org/10.1016/J.BIOMBIOE.2011.01.048

Carvalheiro, F.; Duarte, L. C.; Gírio, F. M.; Gírio, F. M. Hemicellulose Biorefineries: A Review on Biomass Pretreatments. *J. Sci. Indus. Res.* **2008,** *67.* https://www.researchgate.net/publication/236878322

Chen, L. G.; Chang, C. W.; Tsay, J. G.; Weng, B. B. C. Hepatoprotective Effects of Litchi (*Litchi chinensis*) Procyanidin A2 on Carbon Tetrachloride-Induced Liver Injury in ICR Mice. *Exp. Ther. Med.* **2017,** *13* (6), 2839–2847. https://doi.org/10.3892/etm.2017.4358

Chen, Y.; Luo, H.; Gao, A.; Zhu, M. Ultrasound-assisted Extraction of Polysaccharides from Litchi (*Litchi chinensis* Sonn.) Seed by Response Surface Methodology and Their Structural Characteristics. *Innov. Food Sci. Emerg. Technol.* **2011,** *12* (3), 305–309. https://doi.org/10.1016/J.IFSET.2011.03.003

Cherubini, F. The Biorefinery Concept: Using Biomass Instead of Oil for Producing Energy and Chemicals. *Energy Convers. Manag.* **2010,** *51* (7), 1412–1421. https://doi.org/10.1016/J.ENCONMAN.2010.01.015

Clark, J. H.; Deswarte, F. E. I. Introduction to Chemicals from Biomass. In *Wiley Series in Renewable Resources,* Chichester: Wiley-VCH;**2008.** https://www.wiley.com/en-ie/Introduction+to+Chemicals+from+Biomass%2C+2nd+Edition-p-9781118714485

Clark, J. H.; Deswarte, F. E. I.; Farmer, T. J. The Integration of Green Chemistry Into Future Biorefineries. *Biofuels, Bioprod. Bior.* **2009,** *3* (1), 72–90. https://doi.org/10.1002/BBB.119

Da Porto, C.; Porretto, E.; Decorti, D. Comparison of Ultrasound-Assisted Extraction with Conventional Extraction Methods of Oil and Polyphenols from Grape (*Vitis vinifera* L.) Seeds. *Ultrason. Sonochem.* **2013,** *20* (4), 1076–1080. https://doi.org/10.1016/J.ULTSONCH.2012.12.002

Dahiya, S.; Kumar, A. N.; Shanthi Sravan, J.; Chatterjee, S.; Sarkar, O.; Mohan, S. V. Food Waste Biorefinery: Sustainable Strategy for Circular Bioeconomy. In *Bioresource Technology;* Elsevier Ltd, 2018; vol. 248, pp 2–12. https://doi.org/10.1016/j.biortech.2017.07.176

Duan, X.; Wu, G.; Jiang, Y. Evaluation of the Antioxidant Properties of Litchi Fruit Phenolics in Relation to Pericarp Browning Prevention. *Molecules* **2007,** *12* (4), 759–771. https://doi.org/10.3390/12040759

FAOSTAT. 2021. https://www.fao.org/faostat/en/#data/QCL

Fernando, S.; Adhikari, S.; Chandrapal, C.; Murali, N. Biorefineries: Current Status, Challenges, and Future Direction. *Energ. Fuels* **2006,** *20* (4), 1727–1737. https://doi.org/10.1021/EF060097W

Gallo, M.; Ferracane, R.; Graziani, G.; Ritieni, A.; Fogliano, V. Microwave Assisted Extraction of Phenolic Compounds from Four Different Spices. *Molecules* **2010,** *15* (9), 6365. https://doi.org/10.3390/MOLECULES15096365

Hallenbeck, P. C. Fermentative Hydrogen Production: Principles, Progress, and Prognosis. *Int. J. Hydrogen Energy* **2009,** *34* (17), 7379–7389. https://doi.org/10.1016/J.IJHYDENE.2008.12.080

Huang, F.; Zhang, R.; Liu, Y.; Xiao, J.; Liu, L.; Wei, Z.; Yi, Y.; Zhang, M.; Liu, D. Dietary Litchi Pulp Polysaccharides Could Enhance Immunomodulatory and Antioxidant Effects in Mice. *Int. J. Biol. Macromol.* **2016,** *92,* 1067–1073. https://doi.org/10.1016/j.ijbiomac.2016.08.021

Huang, X.; Subhadrabandhu, S.; Mitra, S. K.; Ben-Arie, R.; Stern, R. A. *Origin, History, Production and Processing* [Online], 2005. https://www.researchgate.net/publication/294688844

Huber, G. W.; Iborra, S.; Corma, A. Synthesis of Transportation Fuels from Biomass: Chemistry, Catalysts, and Engineering. *Chem. Rev.* **2006,** *106* (9), 4044–4098. https://doi.org/10.1021/CR068360D/ASSET/CR068360D.FP.PNG_V03

Jönsson, L. J.; Martín, C. Pretreatment of Lignocellulose: Formation of inhibitory Wastes and Strategies for Minimizing Their Effects. *Bioresour. Technol.* **2016a,** *199,* 103–112. https://doi.org/10.1016/J.BIORTECH.2015.10.009

Jönsson, L. J.; Martín, C. Pretreatment of Lignocellulose: Formation of inhibitory Wastes and Strategies for Minimizing Their Effects. *Bioresour. Technol.* **2016b,** *199,* 103–112. https://doi.org/10.1016/J.BIORTECH.2015.10.009

Kamm, B.; Kamm, M. Principles of Biorefineries. *Appl. Microbiol. Biotechnol.* **2004,** *64* (2), 137–145. https://doi.org/10.1007/S00253-003-1537-7

Kumar, P.; Barrett, D. M.; Delwiche, M. J.; Stroeve, P. Methods for Pretreatment of Lignocellulosic Biomass for Efficient Hydrolysis and Biofuel Production. *Ind. Eng. Chem. Res.* **2009,** *48* (8), 3713–3729. https://doi.org/10.1021/IE801542G

Li, H.; Deng, Z.; Liu, R.; Loewen, S.; Tsao, R. Bioaccessibility, *In Vitro* Antioxidant Activities and *In Vivo* Anti-Inflammatory Activities of a Purple Tomato (*Solanum lycopersicum* L.). *Food Chem.* **2014,** *159,* 353–360. https://doi.org/10.1016/J.FOODCHEM.2014.03.023

Liu, J. R.; Ye, Y. L.; Lin, T. Y.; Wang, Y. W.; Peng, C. C. Effect of Floral Sources on the Antioxidant, Antimicrobial, and Anti-Inflammatory Activities of Honeys in Taiwan. *Food Chem.* **2013,** *139* (1–4), 938–943. https://doi.org/10.1016/j.foodchem.2013.02.015

Lynd, L. R.; Laser, M. S.; Bransby, D.; Dale, B. E.; Davison, B.; Hamilton, R.; Himmel, M.; Keller, M.; McMillan, J. D.; Sheehan, J.; Wyman, C. E. How Biotech can Transform Biofuels. *Nat. Biotechnol.* **2008,** *26* (2), 169–172. https://doi.org/10.1038/nbt0208-169

Menzel, C.; Waite, G. K. *Litchi and Longan Botany, Cultivation and Uses*; 2005.

Mitra, S. K.; Pan, J. Litchi and longan production and trade in the world. *Acta Hortic.* **2020,** *1293,* 1–6. https://doi.org/10.17660/ACTAHORTIC.2020.1293.1

Mohr, A.; Raman, S. Lessons from First Generation Biofuels and Implications for the Sustainability Appraisal of Second Generation Biofuels. *Energy Policy* **2013,** *63,* 114–122. https://doi.org/10.1016/J.ENPOL.2013.08.033

Mühlbauer, W.; Müller, J. Litchi (*Litchi chinensis* Sonn.). In *Drying Atlas*; Elsevier, 2020; pp 305–313. https://doi.org/10.1016/b978-0-12-818162-1.00035-3

Nur Hanani, Z. A.; Aelma Husna, A. B.; Nurul Syahida, S.; Nor Khaizura, M. A. B.; Jamilah, B. Effect of Different Fruit Peels on the Functional Properties of Gelatin/Polyethylene Bilayer Films for Active Packaging. *Food Packag. Shelf Life* **2018,** *18,* 201–211. https://doi.org/10.1016/j.fpsl.2018.11.004

Pandey, A.; Höfer, R.; Larroche, C.; Taherzadeh, M.; Nampoothiri, K. M. *Industrial Biorefineries and White Biotechnology*; 2015.

Pauly, M.; Keegstra, K. Cell-wall Carbohydrates and Their Modification as a Resource for Biofuels. *Plant J.* **2008,** *54* (4), 559–568. https://doi.org/10.1111/J.1365-313X.2008.03463.X

Picot-Allain, M. C. N.; Ramasawmy, B.; Emmambux, M. N. Extraction, Characterisation, and Application of Pectin from Tropical and Sub-Tropical Fruits: A Review. **2020,** *38* (3), 282–312. https://doi.org/10.1080/87559129.2020.1733008

Piotrowski, S.; Carus, M.; Carrez, D. *European Bioeconomy in Figures2008–2015* [Online]. 2008. www.biconsortium.eu

Prakash Maran, J.; Sivakumar, V.; Thirugnanasambandham, K.; Sridhar, R. Optimization of Microwave Assisted Extraction of Pectin from Orange Peel. *Carbohydr. Polym.* **2013,** *97* (2), 703–709. https://doi.org/10.1016/J.CARBPOL.2013.05.052

Ragauskas, A. J.; Williams, C. K.; Davison, B. H.; Britovsek, G.; Cairney, J.; Eckert, C. A.; Frederick, W. J.; Hallett, J. P.; Leak, D. J.; Liotta, C. L.; Mielenz, J. R.; Murphy, R.; Templer, R.; Tschaplinski, T. The Path Forward for Biofuels and Biomaterials. *Science* **2006,** *311* (5760), 484–489. https://doi.org/10.1126/SCIENCE.1114736

Rao, G. Optimization of Ultrasound-Assisted Extraction of cyanidin 3-Rutinoside from litchi (*Lichi chinensis* Sonn.) Fruit Pericarp. *Anal. Methods* **2010,** *2* (8), 1166–1170. https://doi.org/10.1039/C0AY00203H

SIAP. 2019, https://nube.siap.gob.mx/cierreagricola/

Sims, R. E. H.; Mabee, W.; Saddler, J. N.; Taylor, M. An Overview of Second Generation Biofuel Technologies. *Bioresour. Technol.* **2010,** *101* (6), 1570–1580. https://doi.org/10.1016/J.BIORTECH.2009.11.046

Sui, Y.; Zheng, Y.; Li, X.; Li, S.; Xie, B.; Sun, Z. Characterization and Preparation of Oligomeric Procyanidins from *Litchi chinensis* Pericarp. *Fitoterapia* **2016,** *112,* 168–174. https://doi.org/10.1016/j.fitote.2016.06.001

Tilman, D.; Socolow, R.; Foley, J. A.; Hill, J.; Larson, E.; Lynd, L.; Pacala, S.; Reilly, J.; Searchinger, T.; Somerville, C.; Williams, R. Beneficial Biofuels—The Food, Energy, and Environment Trilemma. *Science* **2009,** *325* (5938), 270–271. https://doi.org/10.1126/SCIENCE.1177970

Vanholme, R.; Demedts, B.; Morreel, K.; Ralph, J.; Boerjan, W. Lignin Biosynthesis and Structure. *Plant Physiol.* **2010,** *153* (3), 895–905. https://doi.org/10.1104/PP.110.155119

Venkata Mohan, S.; Nikhil, G. N.; Chiranjeevi, P.; Nagendranatha Reddy, C.; Rohit, M. V.; Kumar, A. N.; Sarkar, O. Waste Biorefinery Models Towards Sustainable Circular Bioeconomy: Critical Review and Future Perspectives. *Bioresour. Technol.* **2016,** *215,* 2–12. https://doi.org/10.1016/J.BIORTECH.2016.03.130

Wang, W.; Ma, X.; Xu, Y.; Cao, Y.; Jiang, Z.; Ding, T.; Ye, X.; Liu, D. Ultrasound-Assisted Heating Extraction of Pectin From Grapefruit Peel: Optimization and Comparison with the Conventional Method. *Food Chem.* **2015,** *178,* 106–114. https://doi.org/10.1016/J.FOODCHEM.2015.01.080

Wyman, C. E. What is (and is not) Vital to Advancing Cellulosic Ethanol. *Trends Biotechnol.* **2007,** *25* (4), 153–157. https://doi.org/10.1016/J.TIBTECH.2007.02.009

Yao, P.; Gao, Y.; Simal-Gandara, J.; Farag, M. A.; Chen, W.; Yao, D.; Delmas, D.; Chen, Z.; Liu, K.; Hu, H.; Xiao, J.; Rong, X.; Wang, S.; Hu, Y.; Wang, Y. Litchi (*Litchi chinensis* Sonn.): A Comprehensive Review of Phytochemistry, Medicinal Properties, and Product Development. In *Food and Function;* Royal Society of Chemistry; 2021; vol. 12 (20), pp 9527–9548. https://doi.org/10.1039/d1fo01148k

Zhang, R.; Su, D.; Hou, F.; Liu, L.; Huang, F.; Dong, L.; Deng, Y.; Zhang, Y.; Wei, Z.; Zhang, M. Optimized Ultra-High-Pressure-Assisted Extraction of Procyanidins from Lychee Pericarp Improves the Antioxidant Activity of Extracts. *Biosci. Biotechnol. Biochem.* **2017,** *81* (8), 1576–1585. https://doi.org/10.1080/09168451.2017.1321953

Zhang, R.; Zeng, Q.; Deng, Y.; Zhang, M.; Wei, Z.; Zhang, Y.; Tang, X. Phenolic Profiles and Antioxidant Activity of Litchi Pulp of Different Cultivars Cultivated in Southern China. *Food Chem.* **2013,** *136* (3–4), 1169–1176. https://doi.org/10.1016/J.FOODCHEM.2012.09.085

Zhao, L.; Ruan, S.; Yang, X.; Chen, Q.; Shi, K.; Lu, K.; He, L.; Liu, S.; Song, Y. Characterization of Volatile Aroma Compounds in Litchi (Heiye) Wine and Distilled Spirit. *Food Sci. Nutrit.* **2021,** *9* (11), 5914. https://doi.org/10.1002/FSN3.2361

Zhao, L.; Wang, K.; Wang, K.; Zhu, J.; Hu, Z. Nutrient Components, Health Benefits, and Safety of Litchi (*Litchi chinensis* Sonn.): A Review. *Compr. Rev. Food Sci. Food Saf.* **2020,** *19* (4), 2139–2163. https://doi.org/10.1111/1541-4337.12590

Zhao, M.; Yang, B.; Wang, J.; Li, B.; Jiang, Y. Identification of the Major Flavonoids from Pericarp Tissues of Lychee Fruit in Relation to Their Antioxidant Activities. *Food Chem.* **2006,** *98* (3), 539–544. https://doi.org/10.1016/J.FOODCHEM.2005.06.028

Zilberman, D.; Hochman, G.; Rajagopal, D.; Sexton, S.; Timilsina, G.; Zilberman, D.; Hochman, G.; Rajagopal, D.; Sexton, S.; Timilsina, G. The Impact of Biofuels on Commodity Food Prices: Assessment of Findings. *Am. J. Agric. Econ.* **2013,** *95* (2), 275–281. https://doi.org/10.1093/AJAE/AAS037

CHAPTER 2

Bioactive Compounds from Coffee Agro-industrial Waste: An Opportunity Area within the Circular Economy

MARÍA DE LOS A. VÁZQUEZ-NÚÑEZ[1], NIMBE CUERVO-GAMA[1], MARIELA R. MICHEL-MICHEL[2], CANDY CARRANZA-ÁLVAREZ[1], PEDRO AGUILAR-ZÁRATE[2], JORGE ENRIQUE WONG-PAZ[1], and ABIGAÍL REYES-MUNGUÍA[1*]

[1]*Facultad de Estudios Profesionales Zona Huasteca. Universidad Autónoma de San Luis Potosí, Ciudad Valles, San Luis Potosí, México*

[2]*Departamento de Ingenierías. Tecnológico Nacional de México/I. T. de Ciudad Valles, Ciudad Valles, San Luis Potosí, México*

ABSTRACT

The coffee plant is originally from East Africa and is the most cultivated plant in the world; in Mexico, it occupies the tenth place with the highest production. The high rate of growth of the coffee agroindustry generates large quantities of waste, such as coffee pulp, whose accumulation has a negative impact on economic and environmental aspects. Therefore, for several years, the solution has leaned toward what is known as circular economy and the chemical revaluation of its compounds, among which are compounds such as polyphenols, alkaloids, and diterpenes, which are of great interest for the food, pharmaceutical, and cosmetic industries, mainly due to their bioactive properties, which demonstrates the great potential of this type of waste.

Sustainable Engineering and Agro-Food Processing: Renewable and Nonrenewable Resources.
José Juan Buenrostro-Figueroa, Julio César Tafolla-Arellano, Mónica Chávez-González, Pedro Aguilar-Zárate, A. K. Haghi (Eds.)

2.1 INTRODUCTION

The origin of the name "coffee", which means strength, comes from the Arabic word "qahwah" (Siedentopp, 2009), from where it was later transferred to Europe and America, initially to the region of Brazil (Barbosa and Ferrão, 2020), becoming the largest producer worldwide (Figueroa Hernández et al., 2015).

In Mexico, coffee was introduced, and its cultivation began in 1790 (Clark de Lara and Speckman Guerra, 2005). It has become an important source of employment for more than 500,000 producers in 14 states of the Mexican Republic and in a total of 480 municipalities, positioning Chiapas as the main producing state nationwide due to its topography, altitude, climate, and soil.

2.1.1 CHARACTERISTICS OF COFFEA ARABICA

Coffea arabica, classified as one of the best worldwide (SEDARH, 2018), is the most cultivated coffee species, with a production of 60% of the total coffee grown worldwide. Arabica trees grow at an altitude between 1000 and 1300 meters above sea level; exceeding them would freeze them, and if it is less than 1200, it must be under other trees to survive and avoid the presence of pests. The ideal temperature is between 13 and 26°C; if it is lower, the leaves and fruits freeze, and in terms of rainfall, the best average is between 1200 and 2500 milliliters per year (Romero and Camilo, 2015). Harvesting begins between 18 and 20 months; however, it is not until 3 years of age that the tree reaches its maximum reproductive peak (SADER, 2018).

2.1.2 CHARACTERISTICS OF COFFEA CANEPHORA

Coffea canephora is also known as Robusta coffee. This species has its origins in West Africa, the African continent, and Brazil, the places with the highest production of this species. In comparison with Coffea arabica, it requires less demanding cultivation conditions. *Robusta* beans are used for the production of instant coffee due to the fact that the quality is considered to be lower and the caffeine content is 3 times higher (Figueroa Hernández et al., 2015). It has a high yield, and it is more resistant to pests and diseases; however, the very bitter and acid taste is a major disadvantage (Chadnurn and Davis, 2017).

The optimal characteristics for the cultivation of this species, according to ECOCROP (2014), are the following: a minimum temperature of 20°C, a maximum of 30°C, annual rainfall of at least 1700 mm, and at most 3000 mm. This species is self-pollinated and self-sterile, so its propagation is by asexual methods to avoid problems of self-incompatibility and production.

2.2 COFFEE INDUSTRIALIZATION

The USDA (2021) points out that the top ten coffee producers worldwide are the following: Brazil, Vietnam, Colombia, Indonesia, Ethiopia, Honduras, India, Peru, Uganda, and Mexico, which rank 10th with a production of 3.3 million bags, contributing 5% of world production (León Carrasco, 2021). The states of Chiapas (40%), Veracruz (24%), and Puebla (17%) generate the most production (SIAP, 2022).

The USDA (2016) (SAGARPA, 2017) reports that 51.95% of the national production of green coffee is used for domestic use and the rest is exported, mainly to the United States (53.85%) and countries that are part of the European Union, as well as Japan, Cuba, and Canada (SAGARPA, 2017). Of the coffee that is not exported, 54.2% is consumed as soluble coffee, 40.5% as ground coffee, and 5.3% corresponds to the consumption of roasted coffee beans (Fig. 2.1), consuming annually an average of between 1.3 and 1.6 kg of coffee per person, satisfying this demand with national production according to SADER (2020).

However, it is worth mentioning that, during the last decade, production has decreased by 43.51% and increases in imports from Brazil, Vietnam, and Honduras have been identified (SAGARPA, 2017). However, by 2030, it is estimated that consumption at the national level will increase from 0.80 to 0.94 Mt (millions of tons) and that production will increase to 4.70 Mt, with the aim of diversifying exports.

The state of San Luis Potosi has about 14,500 hectares with approximately 17,500 producers (SEDARH, 2018) in the municipalities of Aquismón, Huehuetlán, Matlapa, Tamazunchale, Axtla de Terrazas, Xilitla, and Coxcatlán dedicated to coffee cultivation (SEDARH, 2022), which are favored by a humid climate and adequate altitudes in highland areas for obtaining successful harvests, with an approximate production of 46 kg/ha.

2.3 AGRO-INDUSTRIAL WASTE

2.3.1 GENERATION OF AGRO-INDUSTRIAL WASTE

Agroindustry is a set of economic activities that integrate agricultural, livestock, or forestry primary production processes and industrial activities with the purpose of obtaining food or raw materials for their transformation or

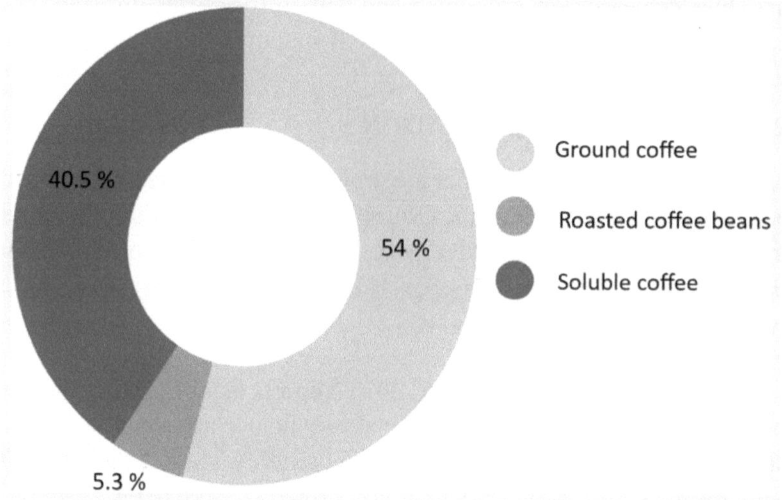

FIGURE 2.1 Coffee consumption according to its type.

Source: Own elaboration with information collected by AMECAFE 2016.

destined for the broad market for their commercialization (Corredor Vargas and Pérez Pérez, 2018; Cury et al., 2017; Peñaranda Gonzalez et al., 2017; Saval Bohórquez, 2012). In addition, it constitutes a part of the industrial sector that is responsible for producing, transforming, storing, and marketing products that come from the field (Saval Bohórquez, 2012).

Agroindustry is currently recognized due to the high demand for production worldwide, promoting development in various areas, such as economic, social, and global environmental areas (Corredor Vargas and Pérez Pérez, 2018; Cury et al., 2017), as long as it keeps environmental protection activities under development (Corredor Vargas and Pérez Pérez, 2018). Currently, in addition to their productive and economic role, agroindustries are valued for their relationship with the environment (Cury et al., 2017; Peñaranda González et al., 2017).

Agro-industrial wastes are produced from these, which can be found in a solid or liquid state and are not of value or utility in the process in which they were generated. However, it is considered that they are susceptible to transformation with the aim of generating different products with economic, commercial, or social value (Freitas et al., 2021; Saval Bohórquez, 2012).

In 2018, it was reported that worldwide, the production of agro-industrial waste represented approximately one-third of the food produced for human consumption, which is estimated to be equivalent to 1300 million tons of agro-industrial waste produced (Peñaranda González et al., 2017; Ravindran et al., 2018). However, by the year 2020, it was estimated that the production would rise to approximately 146 billion tons of waste (Perez Sandoval, 2020). The generation of these wastes depends mainly on each country due to its average income and population consumption (Peñaranda González et al., 2017) as they are generated by the harvesting and processing of high-demand products of the region. Countries such as Mexico, Colombia, Ecuador, Paraguay, and Panama produce waste mainly from livestock or agriculture (Peñaranda González et al., 2017).

Commonly, these agro-industrial wastes are grouped under this concept to differentiate them from other wastes; however, in general, the characteristics of this type of materials can be very varied since they depend largely on the raw material they come from and the processes they receive to generate the waste (Cabrera Rodríguez et al., 2016; Ravindran et al., 2018; SAGARPA, 2015; Saval Bohórquez, 2012). Due to the large differences, the development of techniques for the utilization of these wastes requires knowledge of the different characteristics, properties, and different chemical compositions of each material with the aim of developing effective techniques (Perez Sandoval, 2020).

Due to the high rate of growth of agroindustries, increasing amounts of waste are produced, which, when released into the environment, can cause problems and harmful effects on the environment and human health (Leite et al., 2019; Ravindran et al., 2018; Sadh et al., 2018), representing a great challenge for society, for which the development of new techniques for the use of these residues are issues of great importance that must be addressed.

2.3.2 CLASSIFICATION

All agro-industrial processes, regardless of scale, generate liquid, gaseous, and mainly solid waste (Corredor Vargas and Pérez Pérez, 2018; Cury et al., 2017; Riera et al., 2019), which is considered the main classification; however, each subsector of agroindustry can also produce different residues at different levels (Corredor Vargas and Pérez Pérez, 2018) and can therefore be differentiated into primary and secondary wastes (Riera et al., 2019; Sadh et al., 2018).

Primary residues are generated in the agricultural sector from the cultivation and harvest of raw materials; in turn, these can also be divided into field residues and process residues. Field residues are produced from the cultivation and harvest process; for example, leaves, stems, seeds, and process residues are produced after harvest when they are transformed into resources (Sadh et al., 2018; Saval Bohórquez, 2012).

Secondary wastes are obtained in the industrial sector with the subsequent processing of raw materials, where we can find animal and vegetable wastes such as pulp, peel, and bagasse from different products (Godoy Padilla et al., 2020; Ríos-Pérez et al., 2020; Saval Bohórquez, 2012). Table 2.1 shows some examples of primary and secondary solid, liquid, and gaseous wastes from different agro-industrial sectors, where the coffee industry is one of those that produce large amounts of waste, mainly due to the high worldwide demand for this product (Valencia-Hernandez et al., 2021).

TABLE 2.1 Primary and Secondary Wastes from Different Agroindustries.

Primary wastes			
Sector	**Type**	**Example**	**Source**
Livestock	Solid	Manure	(Ríos-Pérez et al., 2020)
Crops		Peels, leaves, and stems	
Secondary wastes			
Cocoa industry	Solid	Grain husk and mucilage	(Ríos-Pérez et al., 2020)
Coffee industry		Pulp	(Kc et al., 2021; Loukri et al., 2020; Wong-Paz et al., 2021a)
		Coffee grounds	(Afolabi et al., 2020a; Schmidt Rivera et al., 2020)
Wine industry		Grape seeds and skin	(Valencia-Avilés et al., 2017; Venkitasamy et al., 2019; Zou et al., 2019)
Citrus or juice industry		Orange and lemon peel	(Imene Ousaadi et al., 2021)
Meat industry	Liquid	Wastewater	(Ríos-Pérez et al., 2020)
Agroinsdustries in general	Gaseous	Carbon dioxide emissions	(Riera et al., 2018)

2.4 AGRO-INDUSTRIAL WASTE FROM COFFEE

The coffee beverage is one of the most consumed worldwide (Valencia-Hernandez et al., 2021) and is obtained from the coffee cherry, which is mainly made up of a seed or bean covered by a silvery skin, parchment, mucilage, pulp, and the outer husk or skin (Fig. 2.2) (Hendroko Setyobudi et al., 2018).

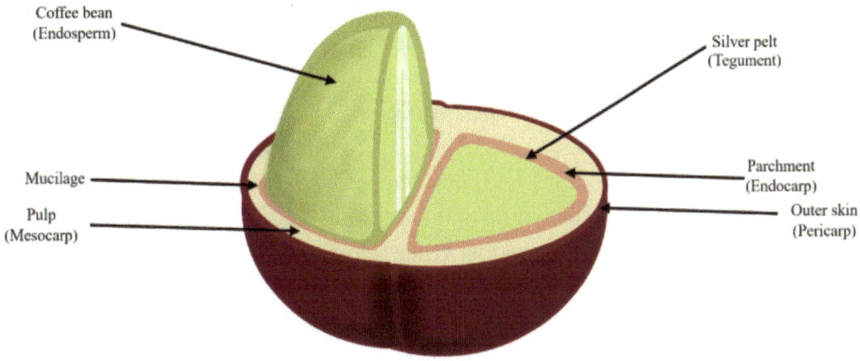

FIGURE 2.2 Coffee bean structure.

Source: Own elaboration with information collected by Hendroko Setyobudi et al. (2018).

The processing of cherries to obtain coffee beans and the consumption of these are processes that generate different agro-industrial wastes, always depending on the type of process (Loukri et al., 2020; Magoni et al., 2018). In the coffee industry, the processing of cherries is commonly carried out by three methods: the dry method, the wet method (Echeverria and Nuti, 2017; Kieu Tran et al., 2020; Magoni et al., 2018; Marín Tello et al., 2020), and the semidry method (Magoni et al., 2018).

The dry method is the one that is mainly adopted in the industry, regardless of the coffee species and the region of production (Echeverria and Nuti, 2017). It consists of drying the cherries (Buck et al., 2021; Echeverria and Nuti, 2017) and generates the husk as residue, which consists of the cherry skin, the pulp, the mucilage, and the parchment (Echeverria and Nuti, 2017; Magoni et al., 2018). In the wet method, drying is not necessary (Echeverria and Nuti, 2017), but it is helped by a fermentation process where mucilage and pulp residues are produced (Buck et al., 2021; Echeverria and Nuti, 2017; Magoni et al., 2018).

During this processing, the losses generated are approximately 40–45% (Sangta et al., 2021), with the production of a large amount of various agro-industrial wastes (Buck et al., 2021; Loukri et al., 2020; Munguía Ameca et al., 2018; Sangta et al., 2021), which annually generates more than 10 million tons of waste worldwide (Atila, 2019). Where the main wastes generated are coffee pulp and coffee grounds (Buck et al., 2021; Echeverria and Nuti, 2017) as coffee pulp reaches up to 29% of the total weight of waste (Buck et al., 2021; Sangta et al., 2021), and coffee grounds represent almost the total processed coffee bean as it is generated after the preparation of the beverage (Afolabi et al., 2020b).

2.4.1 COFFEE PULP

Coffee pulp is the main waste from the coffee industry (Fig. 2.3) (Kc et al., 2021; Loukri et al., 2020; Marín Tello et al., 2020) that is generated during wet and semidry processes (Kc et al., 2021; Magoni et al., 2018; Munguía Ameca et al., 2018; Rosales Delgado et al., 2019).

In general, the coffee pulp includes the husk (skin, epidermis, and exocarp) as well as the pulp proper (mesocarp) (Marín Tello et al., 2020) and has been reported to represent from 29 to 55% of the dry weight of the cherry (Hendroko Setyobudi et al., 2018; Kc et al., 2021; Marín Tello et al., 2020; Munguía Ameca et al., 2018; Rosales Delgado et al., 2019; Wong-Paz et al., 2021b), and its composition can be varied depending on various factors (Marín Tello et al., 2020).

In 2019, there was a global production of 15 million tons of agro-industrial waste produced by the coffee industry (Loukri et al., 2020; Magoni et al., 2018; Rosales Delgado et al., 2019), of which 56% (8.5 million) were coffee pulp (Rosales Delgado et al., 2019), and by 2020, 9.4 million tons were registered (Marín Tello et al., 2020).

Currently, this residue is considered an environmental pollutant (Hendroko Setyobudi et al., 2018; Magoni et al., 2018; Valencia-Hernandez et al., 2021) that generates a high management cost (Magoni et al., 2018), and it is given very limited uses due to the presence of caffeine and tannins that are considered antinutrients (Atila, 2019; Valencia-Hernandez et al., 2021). As a result, they have not currently been given a technical or sustainable use (Munguía Ameca et al., 2018; Rosales Delgado et al., 2019), causing them to be burned to eliminate their compounds (Atila, 2019) or disposed of in landfills without any treatment (Hendroko Setyobudi et al., 2018; Munguía Ameca et al., 2018).

2.4.2 COFFEE GROUNDS

The beverage prepared from roasted coffee beans is the second most consumed beverage worldwide, after tea (Atabani et al., 2019; Zhang et al., 2020). It has been assessed that about 50% of the coffee produced worldwide is destined for the preparation of this beverage (Karmee, 2018; Taleb et al., 2020) and coffee as such is a clear example of products containing a high rate of waste that are unavoidable during consumption (Schmidt Rivera et al., 2020), as millions of tons of waste in the form of coffee grounds,

FIGURE 2.3 Coffee pulp (Own elaboration).

also known as spent coffee grounds (SCGs), are also produced from coffee (Afolabi et al., 2020; Karmee, 2018; Schmidt Rivera et al., 2020).

This inedible residue generally has a fine particle size, high levels of organic load and moisture, and results from the solid and insoluble leftover portion of the coffee after brewing by a hot water percolation process (Afolabi et al., 2020b; Atabani et al., 2019; Ganesh et al., 2020; Hellström et al., 2009; Karmee, 2018; McNutt and He, 2019; Saberian et al., 2021; Schmidt Rivera et al., 2020); therefore, this residue is commonly recovered from coffee pot filters or any filter used for brewing coffee.

It has been recorded that this coffee consumption generates a worldwide production of approximately 6 million tons of coffee grounds (Afolabi et al.,

2020b; Atabani et al., 2019; Taleb et al., 2020). Due to the fact that currently, these residues do not have much commercial value (Taleb et al., 2020), in addition to the lack of specific guidelines for the management of these residues (Schmidt Rivera et al., 2020).

However, as these residues are derived from coffee, it is expected that they present similar properties or compounds with bioactivities that are present in coffee beans, which would represent a possible strategy to take advantage of these residues in different industrial applications, giving them an added value, which is why there is currently an increase in research to evaluate the possible benefits that can be obtained from these residues (Taleb et al., 2020).

FIGURE 2.4 Coffee grounds (Own elaboration).

2.4.3 THE PROBLEM OF ACCUMULATION AND POOR DISPOSAL OF AGRO-INDUSTRIAL WASTE

Despite the great contributions made by agroindustries to the economic sector, there is another reality, which is the uncontrolled generation of excessive amounts of agro-industrial waste (Corredor Vargas and Pérez Pérez, 2018; Riera et al., 2019). These wastes can cause negative economic impacts due to the need for adequate techniques for their disposal, which generates large costs for the companies since they are directly responsible for the final disposition.

However, agro-industrial waste can cause environmental impacts under inadequate management and poor disposal, which commonly happens since many of the companies do not have an adequate management plan, mainly due to the high costs involved in the treatment and disposal processes (Corredor Vargas and Pérez Pérez, 2018; Ríos-Pérez et al., 2020). The main problem derives from the lack of environmental awareness for waste management, coupled with the lack of technology and economic resources for adequate treatment, as well as specific legislation for the management of this type of waste (Cury et al., 2017; Ríos-Pérez et al., 2020).

Environmental pollution is defined as a change in the physical, chemical, and biological quality of some resource caused by human activities such as agroindustry (Corredor Vargas and Pérez Pérez, 2018), which is the direct effect caused by improper disposal of agro-industrial waste. The most visible adverse effects are the loss of forests and biomes resulting from fires caused by the accumulation of waste in dry regions not suitable for disposal, as well as the production of greenhouse gases related to the release of CO_2, toxic products, and unpleasant odors due to the decomposition of organic matter (Corredor Vargas and Pérez Pérez, 2018; Freitas et al., 2021; Sadh et al., 2018), in addition to the proliferation of pests such as rats or insects (Corredor Vargas and Pérez Pérez, 2018) and pathogenic microorganisms (Freitas et al., 2021).

However, the negative effects they cause go beyond those that can be assessed with the naked eye, such as the environmental alterations they cause negatively harming the development of living beings (Corredor Vargas and Pérez Pérez, 2018) in addition to the existence and potential use of resources (Corredor Vargas and Pérez Pérez, 2018), seriously affecting the quality of life of the population (Freitas et al., 2021; Sadh et al., 2018). The main resources affected are soils and water, both at the surface and subway levels (Corredor Vargas and Pérez Pérez, 2018) due to the presence of toxic compounds such as chromium, chlorine, and tannins, which cause the deterioration of these resources. The atmosphere is also affected directly by the production of gaseous-type waste in agro-industrial processes or indirectly by the burning of different solid-type waste such as coffee pulp, which can cause the release of compounds such as CO_2, sulfur oxides, hydrocarbons, and nitrous oxides, among others (Corredor Vargas and Pérez Pérez, 2018; Sadh et al., 2018).

However, in recent years, this has caused companies and researchers to rethink technology and technical allocations to minimize and/or reverse the negative effects caused by the accumulation of these wastes.

2.4.4 REVALORIZATION OF AGRO-INDUSTRIAL WASTE: CIRCULAR ECONOMY

The protection of the environment has taken on great importance for the agro-industrial sector; therefore, the use of the waste generated by these activities is also an aspect that has taken on great importance in recent years. Due to the current environmental concerns of society, agroindustries must seek the development of a social conscience of not producing at the expense of the environment and the planet. For this reason, the need arises to reduce the exploitation of natural resources by taking advantage of agro-industrial waste, generating useful and value-added products, and making use of the circular economy.

The circular economy has been defined as a system that focuses on slowing and closing resource flows, waste generation, and energy leakage and extending product life cycles through maintenance, reuse, repair, remanufacturing, and recycling (Corvellec et al., 2022; Mies and Gold, 2021) with the creation of circular resource flow loops (Fig. 2.4) implemented at the microlevel with consumers and businesses or industries such as agroindustries, at the mesolevel with integrated agents, and at the macrolevel of cities, regions, and governments (Grafström and Aasma, 2021).

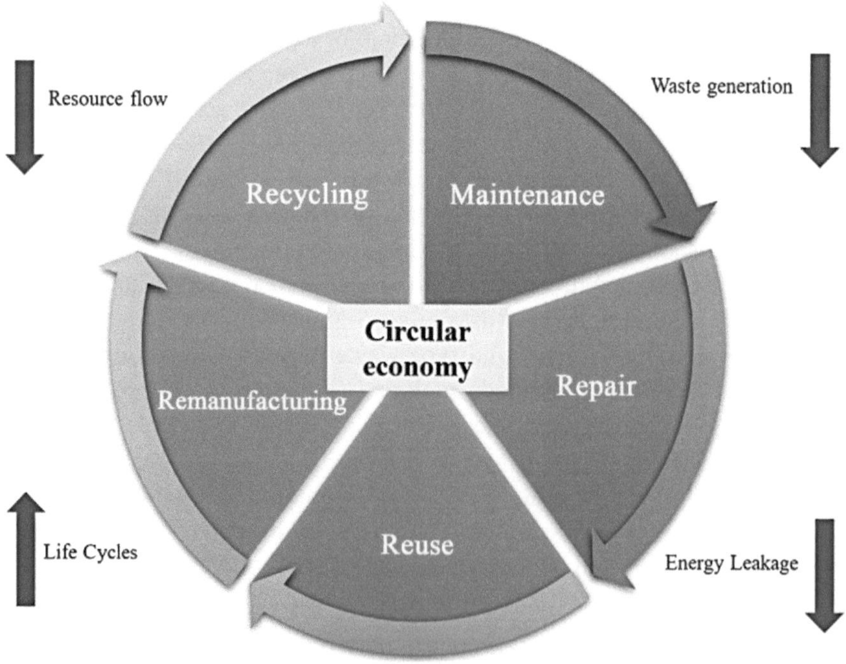

FIGURE 2.5 Components of the circular economy.

Source: Own elaboration information collected by Corvellec et al. (2022); Mies and Gold (2021).

This economy promotes a change to reduce the use of resources and the environmental impact, contributing to a transition toward sustainability as the main objective (Corvellec et al., 2021; Mies and Gold, 2021). Under the understanding of the circular economy, agro-industrial waste is not considered the end of a chain but the beginning of a new one, from which a benefit can be obtained in more than one area, allowing us to make great progress in sustainable production since revaluation processes should be prioritized before the final disposal of waste, ensuring the protection of the environment and the health of the population (Ravindran et al., 2018).

Therefore, great benefits can be obtained from agro-industrial wastes since due to their great diversity of chemical and biological characteristics, they present many possibilities for utilization (Corredor Vargas and Pérez Pérez, 2018; Freitas et al., 2021; Peñaranda González et al., 2017; Riera et al., 2019), always considering that the valorization potential of these wastes depends largely on the type of waste.

The utilization of first-generation waste generally ranges from composting (Corredor Vargas and Pérez Pérez, 2018; Cury et al., 2017) to the production of renewable energy or bioenergy such as biofuels (Cabrera Rodríguez et al., 2016; Corredor Vargas and Pérez Pérez, 2018; Cury et al., 2017; Godoy Padilla et al., 2020; Riera et al., 2019; Ríos-Pérez et al., 2020; Rojas-González et al., 2019) and animal feed (Corredor Vargas and Pérez Pérez, 2018; Cury et al., 2017; Godoy Padilla et al., 2020; Ravindran et al., 2018; Riera et al., 2018), which have been the most used strategies for many decades.

The chemical valorization of second-generation agro-industrial wastes turns out to be a much more complex process; however, they have had a great boom in recent years (Corredor Vargas and Pérez Pérez, 2018) since it allows obtaining a large number of marketable compounds that allow giving an added value to the wastes (Corredor Vargas and Pérez Pérez, 2018; Rojas-González et al., 2019), for the manufacture of chemical specialties, new products, or adding value to original products (Ravindran et al., 2018; Riera et al., 2018; Sangta et al., 2021) trend that originates from the low or no cost, high abundance, and easy disposal of these materials worldwide (Cabrera Rodríguez et al., 2016).

For research, it is especially of great importance to obtain secondary metabolites with bioactivities that can be exploited in other products and processes (Ravindran et al., 2018; Rojas-González et al., 2019) since it is possible to think that these compounds preserve the properties of the products of origin. A comparative approach has shown that obtaining bioactive compounds from agro-industrial residues has evolved significantly as there are numerous investigations that demonstrate the presence of a huge spectrum of secondary metabolites with diverse biological activities present in different agro-industrial residues of plant origin (Freitas et al., 2021).

It is important to mention that this can help to encourage the use of technologies for the sustainable transformation of natural resources and decrease the use of these resources as raw materials for certain products and generate economic benefit (Corredor Vargas and Pérez Pérez, 2018; Freitas et al., 2021; Ravindran et al., 2018; Rojas-González et al., 2019; Sadh et al., 2018), prevent pollution (Peñaranda González et al., 2017) in addition to decreasing the sources of pollution from processes in the industry by avoiding the improper disposal of waste (Peñaranda González et al., 2017; Ravindran et al., 2018).

2.5 PHYTOCHEMICALS FROM COFFEE AGRO-INDUSTRIAL RESIDUES

Coffee, apart from being one of the most popular beverages worldwide, is a source of residues rich in phytochemicals of great interest in pure research, which is responsible for the chemical study of natural products and the relationship between their structure/activity and the use of bioactive compounds to develop or elaborate products such as functional foods or nutraceuticals (Arias Lamos et al., 2018).

Currently, the use of aqueous extracts or pure compounds obtained from plants and their residues has increased because they can be obtained in the laboratory and their broad biological value has been demonstrated. Despite this, few of the products obtained have reached the boom they deserve in industrialized markets (Martínez Alemán et al., 2019). Some of the components of coffee are byproducts of plant metabolism and in recent years the use of coffee byproducts, as a source of bioactive compounds (Pacheco Machicado and Almanza Vega, 2018), which provide them with important properties to human health, such as antioxidant capacity that contributes to protection against chronic diseases, such as cancer (Khochapong et al., 2021; Méndez dos Santos et al., 2021; Myo et al., 2021), in addition to anti-inflammatory, antihypoglycemic (Magoni et al., 2018; Myo et al., 2021), antimicrobial and antiallergic (Pacheco et al., 2018), making these residues attractive materials for application in areas such as scientific research and industries such as food, pharmaceutical (Kieu Tran et al., 2020; Méndez dos Santos et al., 2021; Taleb et al., 2020; Wen et al., 2019) and cosmetics (Pacheco et al., 2018).

2.5.1 ALKALOIDS

They are secondary metabolites, mainly of plants, and they are used as a mechanism of adaptation to the environment. They are derived from amino acids and are organic nitrogenous; most of these compounds have a bitter taste. Their uses are diverse since they have analgesic, anesthetic, and healing properties, as well as antiviral, antimicrobial, antitumor, etc. activities (Chavarro et al., 2020). This group includes caffeine and trigonelline (Cid Carda and Concheiro Nine, 2013).

2.5.1.1 CAFFEINE

This compound is the main alkaloid in coffee and is characterized by being thermostable; in the *Arabica* variety, it is present up to 1.3%, while in the *Robusta* variety, it represents 2.5%, indicating that its concentration will depend on the variety (Vega et al., 2017). One of its best-known properties is that it stimulates the central nervous system of the striated central nervous muscle, thus causing a state of alertness (Cid Carda and Concheiro Nine, 2013).

2.5.1.2 TRIGONELLINE

This compound is derived from vitamin B6 and influences the bitterness of coffee. It is produced by the enzymatic methylation of nicotinic acid. Trigonelline is related to its capacity to inhibit the invasion of cancer cells as well as its capacity to regenerate dendrites and axons, which could have a positive impact on memory. Additionally, the decrease of oxidative compounds in the pancreas is a corroborated benefit (Romualdo, 2020).

2.5.1.3 METHYLXANTHINES

Methylxanthines are psychostimulant compounds but also exert their function in the gastrointestinal tract and cardiovascular system skeletal muscle apparatus, among others; however, their main function is to keep the central nervous system alert by blocking specific adenosine A1 and A2a receptors. Its ability to enhance dopaminergic neurotransmission is also important, exerting a sensation of well-being. It is also hypotensive due to the release of nitric oxide (Torres Ugalde et al., 2018).

2.5.2 DITERPENES

2.5.2.1 CAFESTOL AND KAHWEOL

Diterpenic alcohols are important bioactive compounds in coffee residues, such as kahweol and cafestol (Torres Ugalde et al., 2018). They have the effect of increasing blood lipids; however, in different in vitro assays, their pharmacological actions have been shown to have potential. The benefits are anti-inflammatory activities by significantly inhibiting PGE2 and NO synthesis in RAW 264 macrophages due to decreased COS-2 and iNOS protein expression (Ji Young et al., 2006), hepatoprotective, and anticarcinogenic because it induces prevention of tumorigenesis, inhibits tumor cell proliferation and metastasis (Kotowski et al., 2015), and antidiabetic and antiosteoclastogenic (Cárdenas et al., 2015; Ren et al., 2019). There is evidence of the accumulation of cafestol and its metabolites in the liver and gastrointestinal tract through the enterohepatic cycle, causing changes in its metabolism (Ren et al., 2019). Cafestol and kahweol are the main diterpenes in coffee and constitute 20% of its lipid fraction (Wuerges et al., 2020).

2.5.3 POLYPHENOLS

Polyphenolic compounds are a group of metabolites widely distributed in the plant kingdom that appear to be omnipresent, so their presence in coffee agro-industrial residues is not extraordinary. This group is divided into subgroups, such as phenolic acids and stilbenes, which are nonflavonoid polyphenols, and flavonols, flavones, flavanols, and anthocyanins, which are flavonoid polyphenols. Within these subgroups there are many different compounds, of which several are present in the pulp and other coffee residues in appreciable quantities, suggesting that they could be used as a source of bioactive compounds (Araújo Teles and Tavares Cavalcanti Liberato, 2021; Muzaifa et al., 2021; Sharma et al., 2020). This property is mainly due to their reducing capacity and chemical structure as polyphenols can reduce oxidative stress by playing an important role in neutralizing free radicals and providing health benefits to coffee residues (Araújo Teles and Tavares Cavalcanti Liberato, 2021). In addition, they happen to be the main group of phytochemicals to which the properties of coffee residues are attributed, some of which are described in the following sections.

2.5.3.1 NONFLAVONOID POLYPHENOLS

Phenolic Acids

Caffeic Acid

Caffeic acid is esterified with quinic acid as chlorogenic, isochlorogenic, neclorogenic, and cryptochlorogenic acids. It is considered a potent antioxidant and cancer inhibitor; it is also used to treat diseases such as HIV and herpes, as well as to boost the immune system. It is widely used in the medical and food industries (Morales et al., 2016).

Ferulic Acid

As for ferulic acid, it has been shown that sodium ferulate salt has the ability to prevent the oxidation of low-density lipoproteins known as LDL (Perea Ramírez, 2014). It is a weak organic acid and possesses two ionizable groups: carboxylic and phenolic (Valladares Juárez, 2006).

Chlorogenic Acid

Chlorogenic acids were isolated from coffee in the XIX century, and their name derives from the pigmentation obtained by the reaction that occurs between caffeic acid and ferric chloride, as well as from the color obtained in the precipitate formed in the chlorogenic acids thanks to the oxidation reactions. Chlorogenic acids are part of the esters and are formed by the union of a hydroxynamic acid and a quinic acid; they are the majority in the green bean, and the acids found in coffee in order of presence are caffeic acid, ferulic acid, and p-coumaric acid (Chaves-Ulate and Esquivel-Rodíguez, 2019).

Ellagic Acid

This bioactive compound has beneficial effects for treating patients with Diabetes Mellitus Type 2 (DMT2) because its molecular structure is very similar to that of insulin, thus acting on the pancreatic receptors. It also has antioxidant action and prevents various types of cancer (Martínez Alemán et al., 2019; Ruiz-Pompa and González-Hernández, 2021).

2.5.3.2 FLAVONOID POLYPHENOLS

Many authors have generally identified flavonoid polyphenols (Kieu Tran et al., 2020; Rosales Delgado et al., 2019; Sangta et al., 2021); however, compounds from their subgroups, mainly flavonols and anthocyanins, have also been specifically identified.

Flavanols

Compounds such as tannins are the main flavanol-type polyphenols that have been identified and extensively characterized by different authors in coffee residues, mainly in the pulp (Khochapong et al., 2021; Kieu Tran et al., 2020; Marín Tello et al., 2020; Muzaifa et al., 2021; Rosales Delgado et al., 2019; Sharma et al., 2020). Although the presence of these compounds may limit their use as animal feed (Munguía Ameca et al., 2018; Wong-Paz et al., 2021), it has been reported that condensed tannins provide a high antioxidant capacity to these residues, so they have come to be considered as a possible functional food (Araújo Teles and Tavares Cavalcanti Liberato, 2021; Rosales Delgado et al., 2019), as well as a possible natural source of functional ingredients for the pharmaceutical and cosmetic industry (Kieu Tran et al., 2020) due to their antimicrobial properties (Araújo Teles and Tavares Cavalcanti Liberato, 2021).

Proanthocyanidins are another example of flavanols that are also present in coffee residues (Kieu Tran et al., 2020), and authors such as Marín Tello et al. (2020) mention their application for the control of plant pathogens. Proanthocyanidins are derived from proanthocyanidins; although there is no precise structural characterization, several studies have reported their presence and that of their subunits (Sangta et al., 2021; Valencia-Hernandez et al., 2021). Studies such as those conducted by Wong-Paz et al. (2021) have identified A, B, and C-type procyanidins with polymerization degrees of trimers and pentamers. These, in addition to antioxidant activity, possess properties such as cell signaling (Wong-Paz et al., 2021), anticancer and antihyperglycemic capacity, and cardioprotective properties, among many others (Araújo Teles and Tavares Cavalcanti Liberato, 2021; Hou and Wang, 2022; Rue et al., 2018).

Anthocyanins

Anthocyanins and their derivatives have been identified in coffee residues by authors such as Rosales Delgado et al. (2019). Specifically, anthocyanins have also been identified in this type of residue, which are responsible for pigmentation in plants, so due to their properties, they have potential applications in the food industry as natural colorants (Méndez dos Santos et al., 2021; Munguía Ameca et al., 2018; Muzaifa et al., 2021), in addition to the antioxidant, anticarcinogenic, anti-inflammatory, and antihypoglycemic effects that have been proven in anthocyanins such as cyanadin-3-ruthinoside (Thaiphanit et al., 2020).

Likewise, compounds have been identified that have an important healing capacity such as flavanones, flavonols, and xanthones (Araújo Teles and Tavares Cavalcanti Liberato, 2021). In addition, bioactive compounds in coffee residues have been shown to have antimicrobial activity (Khochapong et al., 2021), and specifically, polyphenolic compounds have the capacity to inhibit fungal activity in vitro, a capacity that increases with increasing concentration. However, it has been difficult to identify the exact compounds that contribute to this property; despite this, they can be considered an anti-fungal agent (Sangta et al., 2021).

2.6 CONCLUSIONS

Currently, the incorrect disposal of agro-industrial wastes has generated a negative economic and environmental impact; therefore, it is of great importance to design and apply production and consumption models such as the circular economy that allow the revaluation of agro-industrial wastes, avoiding their accumulation. Agro-industrial wastes are very useful because they are rich in bioactive compounds; specifically, agro-industrial wastes from the coffee industry are sources of many bioactive compounds that have a wide range of properties, which demonstrates the great potential for application in areas such as the pharmaceutical and food industries that these wastes have, which until a few years ago were considered only as waste.

KEYWORDS

- **agro-industrial waste**
- **coffee**
- **circular economy**
- **revaluation**
- **benefits**

REFERENCES

Afolabi, O. O. D.; Sohail, M.; Cheng, Y. Optimisation and Characterisation of Hydrochar Production from Spent Coffee Grounds by Hydrothermal Carbonisation. *Renew. Energy* **2020,** *147,* 1–23. https://doi.org/10.1016/j.renene.2019.09.098

Araújo Teles, T. N. de A.; Tavares Cavalcanti Liberato, M. da C., Eds.. *Produção Acadêmica do Laboratório de Bioquímica e Biotecnologia da Universidade Estadual do Ceará—Volume 1;* 2021. Editora Poisson. https://doi.org/10.36229/978-65-5866-043-9

Arias Lamos, D.; Montaño Díaz, L. N.; Velasco Sánchez, M. A.; Martínez Girón, J. Alimentos Funcionales: Avances de Aplicación en Agroindustria. *Tecnura* **2018,** *22* (57), 55–68. https://doi.org/10.14483/22487638.12178

Atabani, A. E.; Al-Muhtaseb, A. H.; Kumar, G.; Saratale, G. D.; Aslam, M.; Khan, H. A.; Said, Z.; Mahmoud, E. Valorization of Spent Coffee Grounds into Biofuels and Value-Added Products: Pathway Towards Integrated Bio-Refinery. *Fuel* **2019,** *254,* 115640. https://doi.org/10.1016/j.fuel.2019.115640

Atila, F. *A Useful Way to Dispose of Phenolic-rich Agro- industrial Wastes: Mushroom Cultivation;* 2019, Vol 3 (2).

Barbosa, G. C.; Ferrão, A. M. de A. Refuncionalização de fazendas de café a partir da atividade turística em Campinas (SP). *Terrae Didatica* **2020,** *16,* e020023–e020023. https://doi.org/10.20396/td.v16i0.8657347

Buck, N.; Wohlt, D.; Winter, A. R.; Ortner, E. Aroma-Active Compounds in Robusta Coffee Pulp Puree—Evaluation of Physicochemical and Sensory Properties. *Molecules* **2021,** *26* (13), Article 13. https://doi.org/10.3390/molecules26133925

Cabrera Rodríguez, E.; León Fernández, V.; Montano Pérez, A. de la C.; Dopico Ramírez, D. Caracterización de Residuos Agroindustriales Con vistas a su Aprovechamiento. *Centro Azúcar* **2016,** *43* (4). http://scielo.sld.cu/scielo.php?script=sci_arttext&pid=S2223-48612016000400003

Cárdenas, C.; Quesada, A. R.; Medina, M. Á. Chapter 70—Kahweol, a Coffee Diterpene with Anti-inflammatory Properties. *Coffe Healt Disease Preven.* **2015,** 627–633. https://doi.org/10.1016/B978-0-12-409517-5.00070-X.

Chadnurn, H.; Davis, A. P. IUCN Red List of Threatened Species: Coffea Canephora. *IUCN Red List of Threatened Species* [Online] 2017. https://dx.doi.org/10.2305/IUCN.UK.2017-3.RLTS.T18290186A18539466.en

Chavarro, C. F. G.; Gutiérrez, M. C.; Blanco, V. C. P.; Ruiz, X. M. C. Amaryllidaceae: Fuente Potencial de alcaloides. Actividades biológicas y farmacológicas. *Cienc. Agric.* **2020,** *17* (3), 3. https://doi.org/10.19053/01228420.v17.n3.2020.11379

Chaves-Ulate, E. C.; Esquivel-Rodíguez, P. Ácidos Clorogénicos Presentes en el Café: Capacidad Antimicrobiana y Antioxidante. *Agron. Mesoam.* **2019,** 299–311. https://doi.org/10.15517/am.v30i1.32974

Cid Carda, M. C.; Concheiro Nine, ??ngel. *El caf?? Como Fuente de Antioxidantes y Compuestos Bioactivos: Discurso de Ingreso Como Acad??Mico Correspondiente*; NINO, 2013.

Clark de Lara, B.; Speckman Guerra, E. *La república de las Letras. Asomos a la Cultura Escrita del México Decimonónico* (III) [Online], 2005, Universidad Autónoma de México. http://www.libros.unam.mx/la-republica-de-las-letras-asomos-a-la-cultura-escrita-del-mexico-decimononico-9789703210886-libro.html

Corredor Vargas, Y. A.; Pérez Pérez, L. I. Aprovechamiento de Residuos Agroindustriales en el Mejoramiento de la Calidad del Ambiente. *Revista Facultad Ciencias Básicas* **2018,** 59–72. https://doi.org/10.18359/rfcb.3108

Corvellec, H.; Stowell, A. F.; Johansson, N. Critiques of the Circular Economy. *J. Ind. Ecol.* **2022,** *26* (2), 421–432. https://doi.org/10.1111/jiec.13187

Cury, R. K.; Aguas, M. Y.; Martinez, M. A.; Olivero, V. R.; Chams, C. L. Residuos Agroindustriales su Impacto, Manejo y Aprovechamiento. *Rev. Colombiana Cienc. Anim. Recia* **2017,** *9* (S1), 122–132. https://doi.org/10.24188/recia.v9.nS.2017.530

Echeverria, M. C.; Nuti, M. Valorization of the Residues of Coffee Agro-industry: Perspectives and Limitations. *Open Waste Manag. J.* **2017,** *10,* 13–22. https://doi.org/10.2174/1876400201710010013

Figueroa Hernández, E.; Pérez Soto, F.; Godinez Montoya, L. *La Producción y el Consumo Del Café*. ECOFARN; 2015. https://www.ecorfan.org/spain/libros/LIBRO_CAFE.pdf

Freitas, L. C.; Barbosa, J. R.; da Costa, A. L. C.; Bezerra, F. W. F.; Pinto, R. H. H.; Carvalho Junior, R. N. de. From Waste to Sustainable Industry: How can Agro-Industrial Wastes Help in the Development of New Products? *Resour. Conserv. Recy.* **2021,** *169,* 105466. https://doi.org/10.1016/j.resconrec.2021.105466

Ganesh, D. S.; Rahul, B.; Shuta, S.; J Rajesh, B.; Arivalagan, A.; Eyas, M.; Ranjna, S.; Shashi, K. B.; A E, A.; Vicenzo, M.; Jeong-Jun, Y.; Han, S. S.; Gopalakrishnan, K. A review on Valorization of Spent Coffee Grounds (SCG) Towards Biopolymers and Biocatalysts Production. *Bioresour. Technol.* **2020,** *314,* 123800. https://doi.org/10.1016/j.biortech.2020.123800

Godoy Padilla, D. J.; Daza La Plata, R.; Fernández Curi, L. M.; Layza Mendiola, A. E.; Alcarraz, R. E.; Hidalgo Lozano, V.; Gamarra Carrillo, S. G.; Gómez Bravo, C. A. Caracterización Del Valor Nutricional de los Residuos Agroindustriales para la Alimentación de Ganado Vacuno en la región de San Martín, Perú. *Ciencia y Tecnología Agropecuaria* **2020,** *21* (2), 2. https://doi.org/10.21930/rcta.vol21_num2_art:1374

Grafström, J.; Aasma, S. Breaking Circular Economy Barriers. *J. Clean. Prod.* **2021,** *292,* 126002. https://doi.org/10.1016/j.jclepro.2021.126002

Hellström, J. K.; Törrönen, A. R.; Mattila, P. H. Proanthocyanidins in Common Food Products of Plant Origin. *J. Agric. Food Chem.* **2009,** *57* (17), 7899–7906. https://doi.org/10.1021/jf901434d

Hendroko Setyobudi, R.; Krido Wahono, S.; Gamawati Adinurani, P.; Ahmad Wahyudi, Widodo, W.; Mel, M.; Nugroho, Y. A.; Prabowo, B.; Liwang, T. In *Characterisation of Arabica Coffee Pulp—Hay from Kintamani—Bali as Prospective Biogas Feedstocks*, MATEC Web of Conferences, 2018, Vol. 164, 01039. https://doi.org/10.1051/matecconf/201816401039

Hou, K.; Wang, Z. Application of Nanotechnology to Enhance Adsorption and Bioavailability of Procyanidins: A Review. *Food Rev. Int.* **2022,** *38* (sup1), 738–752. https://doi.org/10.1080/87559129.2021.1888970

Imene Ousaadi, M.; Merouane, F.; Berkani, M.; Almomani, F.; Vasseghian, Y.; Kitouni, M. Valorization and Optimization of Agro-Industrial Orange Waste for the Production of Enzyme by Halophilic Streptomyces sp. *Environ. Res.* **2021,** *201,* 111494. https://doi.org/10.1016/j.envres.2021.111494

Ji Young, K.; Dong Hee, K.; Hye Gwang, J. Inhibitory Effect of the Coffee Diterpene Kahweol on Carrageenan-Induced Inflammation in Rats. *Biofactors* **2006,** *26* (1), 17–28.

Karmee, S. K. A spent coffee Grounds Based Biorefinery for the Production of Biofuels, Biopolymers, Antioxidants and Biocomposites. *Waste Manag.* **2018,** *72,* 240–254. https://doi.org/10.1016/j.wasman.2017.10.042

Kc, Y.; Subba, R.; Shiwakoti, L. D.; Dhungana, P. K.; Bajagain, R.; Chaudhary, D. K.; Pant, B. R.; Bajgai, T. R.; Lamichhane, J.; Timilsina, S. Utilizing Coffee Pulp and Mucilage for Producing Alcohol-Based Beverage. *Fermentation* **2021,** *7* (53), 1–13. https://doi.org/10.3390/fermentation7020053

Khochapong, W.; Ketnawa, S.; Ogawa, Y.; Punbusayakul, N. Effect of *In vitro* Digestion on Bioactive Compounds, Antioxidant and Antimicrobial Activities of Coffee (*Coffea arabica* L.) Pulp Aqueous Extract. *Food Chem.* **2021,** *348,* 129094. https://doi.org/10.1016/j.foodchem.2021.129094

Kieu Tran, T. M.; Kirkman, T.; Nguyen, M.; Van Vuong, Q. Effects of Drying on Physical Properties, Phenolic Compounds and Antioxidant Capacity of Robusta wet Coffee Pulp (*Coffea canephora*). *Heliyon* **2020,** *6* (7). https://doi.org/10.1016/j.heliyon.2020.e04498

Kotowski, U.; Heiduschka, G.; Seemann, R.; Eckl-Dorna, J.; Schmid, R.; Kranebitter, V.; Stanisz, I.; Brunner, M.; Lill, C.; Thurnher, D. Effect of the Coffee Ingredient Cafestol on Head and Neck Squamous Cell Carcinoma Cell Lines. *Strahlentherapie Onkologie* **2015,** *191* (6), 511–517. https://doi.org/10.1007/s00066-014-0807-x

Leite, P.; Silva, C.; Salgado, J. M.; Belo, I. Simultaneous Production of Lignocellulolytic Enzymes and Extraction of Antioxidant Compounds by Solid-State Fermentation of Agro-Industrial Wastes. *Ind. Crop. Prod.* **2019,** *137,* 315–322. https://doi.org/10.1016/j.indcrop.2019.04.044

León Carrasco, J. C. USDA: Producción Mundial de café alcanzaría los 175.4 Millones de sacos en la campaña 2020/2021. *USDA: Producción Mundial de café alcanzaría los 175.4 millones de sacos en la campaña 2020/2021.* 2021. https://agraria.pe/noticias/usda-produccion-mundial-de-cafe-alcanzaria-los-175-4-millone-23516

Loukri, A.; Tsitlakidou, P.; Goula, A.; Assimopoulou, A. N.; Kontogiannopoulos, K. N.; Mourtzinos, I. Green Extracts from Coffee Pulp and Their Application in the Development of Innovative Brews. *Appl. Sci.* **2020,** *10* (6982), 1–12. https://doi.org/doi:10.3390/app10196982

Magoni, C.; Bruni, I.; Guzzetti, L.; Dell'Agli, M.; Sangiovanni, E.; Piazza, S.; Regonesi, M. E.; Maldini, M.; Spezzano, R.; Caruso, D.; Labra, M. Valorizing coffee pulp by-Products as Anti-inflammatory Ingredient of Food Supplements Acting on IL-8 release. *Food Res. Int.* **2018,** *112,* 129–135. https://doi.org/10.1016/j.foodres.2018.06.026

Marín Tello, C.; Zelada Castillo, L.; Vázquez Arquero, A.; Vieria, A.; Siche, R. Coffee Pulp: An Industrial By-product with Uses in Agriculture, Nutrition and Biotechnology. *Rev. Agric. Sci.* **2020,** *8,* 323–342. https://doi.org/10.7831/ras.8.0_323

Martínez Alemán, S. R.; Castillo Hernández, F. D.; Aguilar González, C. N.; Rodriguez Herrera, R. Extractos de Pulpa de Café: Una Revisión Sobre Antioxidantes Polifenólicos y su Actividad Antimicrobiana. *Investigación y Ciencia: de la Universidad Autónoma de Aguascalientes* **2019,** *77,* 73–79.

McNutt, J.; He, Q. (Sophia). Spent Coffee Grounds: A Review on Current Utilization. *J. Ind. Eng. Chem.* **2019,** *71,* 78–88. https://doi.org/10.1016/j.jiec.2018.11.054

Méndez dos Santos, É.; Malvezzi de Macedo, L.; Lacalendola Tundisi, L.; Artem Ataide, J.; Camargo, G. A.; Alves, R. C.; P Oliveira, M. B.; Gava Mazzola, P. G. Coffee by-Products in Topical Formulations: A Review. *Trends Food Sci. Technol.* **2021,** *111,* 280–291. https://doi.org/10.1016/j.tifs.2021.02.064

Mies, A.; Gold, S. Mapping the Social Dimension of the Circular Economy. *J. Clean. Prod.* **2021,** *321,* 128960. https://doi.org/10.1016/j.jclepro.2021.128960

Morales, M. R.; Cortés, A. M. R.; Cortés, G. R. *Recuperación de Acido Cafeico a Partir de Residuos de Pulpa de Café*; Participación de la Mujer en la Ciencia, 2016.

Munguía Ameca, G.; Ortega Cerrilla, M. E.; Zetiga Córdoba, P.; Díaz Cruz, A.; Soto Hernández, M.; Herrera Haro, J. Chemical Composition and Antioxidant Capacity of Coffee Pulp. *Ciência e Agrotecnologia* **2018,** *42,* 307–313. https://doi.org/10.1590/1413-70542018423000818

Muzaifa, M.; Rahmi, F.; Syarifudin. Utilization of Coffee By-Products as Profitable Foods—A Mini Review. *IOP Conf. Ser.: Earth Environ. Sci.* **2021,** *672* (1), 012077. https://doi.org/10.1088/1755-1315/672/1/012077

Myo, H.; Nantarat, N.; Khat-Udomkiri, N. Changes in Bioactive Compounds of Coffee Pulp through Fermentation-Based Biotransformation Using *Lactobacillus plantarum* TISTR 543 and Its Antioxidant Activities. *Fermentation* **2021,** *7* (4), 4. https://doi.org/10.3390/fermentation7040292

Pacheco Machicado, T. M.; Almanza Vega, G. R. [tutor. Estudio químico de residuos de café (*Coffea arábica*) como base Científica Para Determinar su Potencial uso en Productos con Valor Agregado. Thesis, 2018. http://repositorio.umsa.bo/xmlui/handle/123456789/19076

Pacheco, T. M.; Torrez Álvarez, S.; Almanza, G. R. Cuantificación de Compuestos Bioactivos en Cáscara de Coffea Arábica en Bolivia. *Rev. Bol. Quím.* **2018,** *35* (5), 117–126.

Peñaranda González, L. V.; Montenegro Gómez, S. P.; Giraldo Abad, P. A. Aprovechamiento de Residuos Agroindustriales en Colombia. *Revista Investigación Agraria Ambiental* **2017,** *8* (2), 2. https://doi.org/10.22490/21456453.2040

Perea Ramírez, L. Estudio Teórico Del Mecanismo de Oxidación de Sistemas Antioxidantes Tipo Bis- Derivados de Ácido Cafeico y ácido Ferúlico. Thesis, Universidad Veracruzana. 2014. https://cdigital.uv.mx/bitstream/handle/123456789/46860/PereaRamirezLuis.pdf?sequence=5&isAllowed=y

Perez Sandoval, N. M. *Valoración del Biocombustible Obtenido Mediante Fermentación de Residuos Agroindustriales* [Universidad Científica del Sur], 2020. https://repositorio.cientifica.edu.pe/handle/20.500.12805/1503

Ravindran, R.; Hassan, S. S.; Williams, G. A.; Jaiswal, A. K. A Review on Bioconversion of Agro-Industrial Wastes to Industrially Important Enzymes. *Bioengineering* **2018,** *5* (4), 4. https://doi.org/10.3390/bioengineering5040093

Ren, Y.; Wang, C.; Xu, J.; Wang, S. Cafestol and Kahweol: A Review on Their Bioactivities and Pharmacological Properties. *Int. J. Mol. Sci.* **2019,** *20* (17), 17. https://doi.org/10.3390/ijms20174238

Riera, M. A.; Maldonado, S.; Palma, R. Residuos Agroindustriales Generados en Ecuador Para la Elaboración de Bioplásticos. *Rev. Ing. Ind.* **2019,** *17* (3), 227–246. https://doi.org/10.22320/S07179103/2018.13

Ríos-pérez, F.; Soto-simental, S.; Quintero-lira, A.; Piloni-martini, J.; Güemes-vera, N. Cocoa Pod Hull Flour: An Option for the use of Agroindustrial Waste. *Boletín de Ciencias Agropecuarias del ICAP* **2020,** *6* (11), 5–7. https://repository.uaeh.edu.mx/revistas/index.php/icap/issue/archive

Rojas-González, A. F.; Flórez-Montes, C.; López-Rodríguez, D. F.; Rojas-González, A. F.; Flórez-Montes, C.; López-Rodríguez, D. F. Prospectivas de Aprovechamiento de Algunos Residuos Agroindustriales. *Rev. Cuba. Quím.* **2019,** *31* (1), 31–52.

Romero, J. M.; Camilo, J. *Manual de Producción Sostenible de café en la República Dominicana/Instituto Interamericano de Cooperación para la Agricultura. – República Dominicana* [Online]. IICA, 2015. https://repositorio.iica.int/bitstream/handle/11324/8726/BVE20037756e.pdf?sequence=1&isAllowed=y

Romualdo, G. R. [UNESP. Cafeína, trigonelina e ácido clorogênico: Modulação da expressão de miRNAs na hepatocarcinogênese associada à fibrose [Online], 2020. https://repositorio.unesp.br/handle/11449/191731

Rosales Delgado, S.; Alzate Arbelaez, A. F.; Rojano, B. Antioxidant Capacity, Bioactive Compounds in Coffee Pulp and Implementation in the Production of Infusions. *Acta Sci. Pol. Technol. Aliment.* **2019,** *18* (3), 235–248. https://doi.org/10.17306/J.AFS.0663

Rue, E. A.; Rush, M. D.; van Breemen, R. B. Procyanidins: A Comprehensive Review Encompassing Structure Elucidation Via Mass Spectrometry. *Phytochem. Rev.* **2018,** *17* (1), 1–16. https://doi.org/10.1007/s11101-017-9507-3

Ruiz-Pompa, K.; González-Hernández, J. C. ¡Lo que debes saber del ácido elágico! *Milenaria, Ciencia y Arte* **2021,** *11* (18), 31–34.

Saberian, M.; Li, J.; Donnoli, A.; Bonderenko, E.; Oliva, P.; Gill, B.; Lockrey, S.; Siddique, R. , Recycling of Spent Coffee Grounds in Construction Materials: A Review. *J. Clean. Prod.* **2021,** *289,* 125837. https://doi.org/10.1016/j.jclepro.2021.125837

SADER. México, Onceavo Productor Mundial De café [Online], 2018, marzo 2. gob.mx. http://www.gob.mx/agricultura/es/articulos/mexico-onceavo-productor-mundial-de-cafe?idiom=es

Sadh, P. K.; Duhan, S.; Duhan, J. S. Agro-industrial Wastes and their Utilization using Solid State Fermentation: A Review. *Bioresour. Bioprocess.* **2018,** *5* (1), 1. https://doi.org/10.1186/s40643-017-0187-z

SAGARPA. Plan de manejo de residuos generados en actividades agrícolas primera etapa: Diagnóstico nacional [Onlie], 2015. https://www.gob.mx/cms/uploads/attachment/file/346978/Manejo_de_residuos_Detallado.pdf

SAGARPA. Café Mexicano Planeación Agrícola Nacional [Online], 2017. https://www.gob.mx/cms/uploads/attachment/file/256426/B_sico-Caf_.pdf

Sangta, J.; Wongkaew, M.; Tangpao, T.; Withee, P.; Haituk, S.; Arjin, C.; Sringarm, K.; Hongsibsong, S.; Sutan, K.; Pusadee, T.; Sommano, S. R.; Cheewangkoon, R. Recovery of Polyphenolic Fraction from Arabica Coffee Pulp and Its Antifungal Applications. *Plants* **2021,** *10* (7), 7. https://doi.org/10.3390/plants10071422

Saval Bohórquez, S. Aprovechamiento de Residuos Agroindustriales: Pasado, *Presente Futuro* **2012,** *16* (2).

Schmidt Rivera, X. C.; Gallego-Schmid, A.; Najdanovic-Visak, V.; Azapagic, A. Life Cycle Environmental Sustainability of Valorisation Routes for Spent Coffee Grounds: From Waste to Resources. *Resour. Conserv. Recycl.* **2020,** *157* (September 2019), 104751. https://doi.org/10.1016/j.resconrec.2020.104751

SEDARH. Se llevó a cabo la 121 Sesión del Consejo Internacional del Café [Online]. SAGARPA, 2018. http://www.gob.mx/agricultura|sanluispotosi/articulos/se-llevo-a-cabo-la-121-sesion-del-consejo-internacional-del-cafe?idiom=es

Sharma, B.; Larroche, C.; Dussap, C.-G. Comprehensive Assessment of 2G Bioethanol Production. *Bioresour. Technol.* **2020,** *313,* 123630. https://doi.org/10.1016/j.biortech.2020.123630

SIAP. Producción de Cítricos en México [Online]. 2022, marzo 1. gob.mx. http://www.gob.mx/publicaciones/articulos/produccion-de-citricos-en-mexico

Siedentopp, U. El Café, Planta Medicinal y de Deleite. *Rev. Int. Acupunt.* **2009,** *3* (3), 143–147. https://doi.org/10.1016/S1887-8369(09)72145-3

Taleb, F.; Ammar, M.; Mosbah, M. B.; Salem, R. B.; Moussaoui, Y. Chemical Modification of Lignin Derived from Spent Coffee Grounds for Methylene blue Adsorption. *Sci. Rep.* **2020,** *10* (1), 1. https://doi.org/10.1038/s41598-020-68047-6

Thaiphanit, S.; Wedprasert, W.; Srabua, A. Conventional and Microwave-Assisted Extraction for Bioactive Compounds from Dried Coffee Cherry Peel by-products and Antioxidant Activity of the Aqueous Extracts. *ScienceAsia* **2020,** *46S* (1), 12. https://doi.org/10.2306/scienceasia1513-1874.2020.S002

Torres Ugalde, Y. C.; Romero Palencia, A.; Román Gutierrez, A. Metilxantinas, Café y Cafeína: Amigos o Enemigos. *Educación y Salud Boletín Científico Instituto de Ciencias de la Salud Universidad Autónoma del Estado de Hidalgo* **2018,** *6* (12), 12. https://doi.org/10.29057/icsa.v6i12.3099

Valencia-avilés, E.; Ignacio-figueroa, I.; Sosa-martínez, E.; Bartolomé-camacho, M. C. Polyphenols: Antioxidant and Toxicological Properties. *Revista de la Facultad de Ciencias Químicas* **2017,** *16,* 15–29.

Valencia-Hernandez, L. J.; Wong-Paz, J. E.; Ascacio-Valdés, J. A.; Chávez-González, M. L.; Contreras-Esquivel, J. C.; Aguilar, C. N. Procyanidins: From Agro-Industrial Waste to Food as Bioactive Molecules. *Foods* **2021,** *10* (12), Article 12. https://doi.org/10.3390/foods10123152

Valladares Juárez, A. G. Recuperación de Ácido Ferúlico de Subproductos y Desechos de la Industria de la Tortilla. Thesis, Universidad Autónoma de Nuevo León, 2006. http://eprints.uanl.mx/21245/1/1020156133.pdf

Vega, A.; De León, J. A.; Reyes, S. M. Determinación del Contenido de Polifenoles Totales, Flavonoides y Actividad Antioxidante de 34 Cafés Comerciales de Panamá. *Inf. Tecnol.* **2017,** *28* (4), 29–38. https://doi.org/10.4067/S0718-07642017000400005

Venkitasamy, C.; Zhao, L.; Zhang, R.; Pan, Z. Chapter 6—Grapes. *En Integrated Processing Technologies for Food and Agricultural By-Products*. Elsevier Inc, 2019; pp 133–163. https://doi.org/10.1016/B978-0-12-814138-0.00006-X

Wen, L.; Zhang, Z.; Rai, D.; Sun, D.-W.; Tiwari, B. K. Ultrasound-assisted Extraction (UAE) of Bioactive Compounds from Coffee silverskin: Impact on Phenolic Content, Antioxidant Activity, and Morphological Characteristics. *J. Food Proc. Eng.* **2019,** *42* (6), e13191. https://doi.org/10.1111/jfpe.13191

Wong-Paz, J. E.; Guyot, S.; Aguilar-Zárate, P.; Muñiz-Márquez, D. B.; Contreras-Esquivel, J. C.; Aguilar, C. N. Structural Characterization of Native and Oxidized Procyanidins (condensed tannins) from Coffee Pulp (*Coffea arabica*) using Phloroglucinolysis and Thioglycolysis-HPLC-ESI-MS. *Food Chem.* **2021,** *340* (May 2020), 127830. https://doi.org/10.1016/j.foodchem.2020.127830

Wuerges, K. L.; Dias, R. C. E.; Viegas, M. C.; de Benassi, M. T. Kahweol and Cafestol in Coffee Brews: Comparison of Preparation Methods. *Rev. Ciênc. Agron.* **2020,** *51*. https://doi.org/10.5935/1806-6690.20200005

Zhang, X.; Zhang, Y.; Ngo, H. H.; Guo, W.; Wen, H.; Zhang, D.; Li, C.; Qi, L. Characterization and Sulfonamide Antibiotics Adsorption Capacity of Spent Coffee Grounds based Biochar and Hydrochar. *Sci. Total Environ.* **2020,** *716,* 137015. https://doi.org/10.1016/j.scitotenv.2020.137015

Zou, Y.; Wu, C.; Ma, C.; He, S.; Brennan, C. S.; Yuan, Y. LWT—Food Science and Technology Interactions of Grape Seed Procyanidins with soy Protein Isolate: Contributing Antioxidant and Stability Properties. *LWT - Food Sci. Technol.* **2019,** *115* (July), 108465. https://doi.org/10.1016/j.lwt.2019.108465

CHAPTER 3

Sustainable Technologies for Resource Recovery from Waste: A Case Study for Better Integrating Environmental and Economic Issues

ALFREDO ÁVILA-GALARZA[1], CLAUDIA YAZMÍN ORTEGA-MONTOYA[2], NORMA LILIA LÓPEZ-DÍAZ[1], and ANY LUCELY SIERRA-NEGRETTE[1*]

[1]*Universidad Autónoma de San Luis Potosí. Programa Multidisciplinario de Posgrado en Ciencias Ambientales, San Luis Potosí, México*

[2]*Universidad Autónoma de Coahuila. Facultad de Ciencias Biológicas Torreón, México*

ABSTRACT

The current linear mode of production and consumption generates high environmental impacts at its final stage. In Mexico, urban waste is managed by local authorities, which is a relevant challenge considering the continuous increase of waste generation, the low efficiency in waste collection, and the high investment needed for a proper final disposal. Organic waste in Mexico is estimated at 56,427 TPD. Anaerobic decomposition of municipal solid waste generates mainly methane, a toxic gas that pollutes the air and has a global warming potential of 22 CO_2 eq. At disposal sites, it is important to manage waste properly to control, manage, exploit, and/or mitigate GHG emissions. The use of biogas for power generation can be an alternative to valorize it, reducing GHG emissions to meet the international GHG reduction commitments, considering the importance of this sector in the national emissions inventory. This

Sustainable Engineering and Agro-Food Processing: Renewable and Nonrenewable Resources.
José Juan Buenrostro-Figueroa, Julio César Tafolla-Arellano, Mónica Chávez-González, Pedro Aguilar-Zárate, A. K. Haghi (Eds.)

chapter presents an overview of the problem of air emissions related to the waste sector in Mexico and the challenges related with biogas management faced by Mexican municipalities throughout the case study of the generation and energy potential of biogas in San Luis Potosí, a Mexican municipality, analyzing social, legal, and technological barriers obstructing progress and proposing proper solutions.

3.1 INTRODUCTION

Cities are the stage of sustainability. According to The World Bank, currently 55% of the world's population lives in cities, with 4.2 billion inhabitants, and by 2050, this percentage is expected to increase to approximately 70% (Urban Development, 2022). According to Flachenecker and Rentschlehler (Flachenecker and Rentschlehler, 2019), there is a need for the public service to make effective use of resources available to coordinate activities geared toward sustaining healthy wellbeing for citizens. In this regard, efforts must be made to replenish renewable resources for the benefit of present and future generations. As the engine for the implementation of all the identified SDGs in an economy, resources at the disposal of the public service must be judiciously utilized to create a balance in the manner in which the earth's resources are accessed (Jackson, 2021).

Waste is one of the main implications of environmental concern for the linear production and consumer model that causes natural resource depletion. Waste is generated in the production, distribution, and consumption stages; in the case of developing countries, it represents a serious local problem in relation to waste disposal sites.

In Latin America and the Caribbean, only 54.4% of municipal solid waste (MSW) is deposited in a sanitary landfill; 18.5% is placed in controlled sites; 23.3% ends up in landfills; 2% is burned in the open air; and 1.8% is disposed directly into the water, causing significant environmental damage (Toro et al., 2016). Proper minimization and confinement, as well as the treatment and valorization of MSW, are uncommon practices (Sáez and Urdaneta, 2014).

Mexico, as well as many other developing countries, is facing solid waste management problems related with urban waste generation, valorization of materials, collection, and waste disposal. According to the Secretariat of Environment and Natural Resources, the domestic waste generation has increased from 0.69 kg/inhabitant/day in 1992 to 0.99 kg/inhabitant/day in 2012 (SEMARNAT 2023). This trend is consistent with the 2.8% yearly growth of natural resource use from 1970 to 2008 (West and Schandl, 2013). According to UNEP, the rise in resource use has been coupled with growth in waste and emissions, contributing to a series of pressure points, including climate change, reduced food security, water scarcity, and air pollution (UNEP 2015).

The loss of materials and the energy embodied in waste are enormous issues for a planet with limited resources and a population that is constantly growing. Valorization of materials at the end of their life cycle is important to avoid the complete loss of valuable materials; however, since products are not designed to be reincorporated at the end of their cycle but programed to be obsolescent after a period, this solution is not the complete answer to the waste problem.

In Mexico, the amount of recyclables for 2012 was 448063 t, with paper and cardboard, PET, and glass being the principal materials recovered (SEMARNAT 2023). In developing countries, there is a social problem related with the recovery of valuable waste because this work is usually carried out by informal recyclers that work on the streets collecting household waste or in dumpsites and controlled sites. This situation highlights the high levels of economic vulnerabilities that exist in developing economies where there are sectors of the society that subsist from the sales of these valuable materials.

In their research, Zolnikov et al. (2021) reported that the most common reported health outcomes of informal recyclers on a systematic review were epidermal, communicable disease, musculoskeletal, respiratory disease, noncommunicable diseases, physiological, gastrointestinal, and waterborne diseases. Another important health problem related with informal recyclers is the recycling of electronic waste, where the presence of halogenated compounds in e-wastes may result in the formation of persistent organic pollutants like polychlorinated biphenyls (PCBs), polychlorinated dibenzo-p-dioxins, and dibenzofurans (PCDD/Fs) during recycling processes (Chakraborty et al., 2018).

3.1.1 PROBLEMS IN THE FINAL DISPOSAL OF MSW IN DEVELOPING COUNTRIES

In Mexico, municipalities are responsible for the cleaning, collection, transfer, treatment, and final disposal of waste and thus are the ones that feel the budgetary and social burden of waste management (SEMARNAT 2019). This is a main issue considering that most of Mexico's 2443 municipalities lack legal, financial, and human resources for the treatment of municipal solid waste (UNEP 2006).

Additionally, inequalities abound in the coverage of collection and disposal services between communities at the national level. For instance, localities with a population of more than 10,000 inhabitants have an average service coverage of 80%, whereas those with less than 10,000 inhabitants only have a service coverage of 23% (UNEP 2006). The collection of domestic waste at the municipal level was carried out in 88.2% for public services, 9.1% for the private sector, and in 2.7% for social services, with 2544 being the number of service providers or operators of the urban solid waste service (INEGI 2021).

Except for some cities, such as Monterrey and Querétaro, municipalities generally do not charge for the collection of MSW from households. Only businesses and large generators pay private (authorized) collectors to take their waste. The municipalities, based on other taxes paid by the population, define and operate a "limited" budget for waste management. This is the reason why they do not have the money necessary to train workers related to waste management nor do they have adequate machinery and instruments to operate a sanitary landfill properly.

Municipalities also lack infrastructure for the recovery of recyclable materials because only 1.4% of the municipalities have a segregation plant for the recovery of valuable materials (INEGI 2021). Another major concern is the lack of infrastructure for the appropriate disposal of waste to protect the environment. In Mexico, the National Solid Urban Waste Atlas accounted for 2225 final sites for the disposal of urban waste, of which 90.5% are dumpsites, 5.6% are controlled sites, and only 4.2% are landfills (Ruiz Suárez et al., 2021).

The environmental impact of dumpsites is the production of heavy metals, leachate that pollutes groundwater, nitrogen oxides (NO_x), furans, and dioxins from multiple uncontrolled combustions that have very harmful effects on human health (Mediavilla Merchán, 2018). In their research, Reyna-Bensusan et al. (2018) state the importance of black carbon generation for uncontrolled burning of waste in developing countries and also suggest that open burning of waste could be a major reason for the apparent poorer respiratory health status of the rural population.

3.1.2 METHANE PRODUCTION FROM DOMESTIC WASTE

Studies have shown that the composition of waste in Mexico varies depending on the region, having an effect on the calorific value, which is higher in central states such as Mexico City and Mexico with a value of 14.9 MJ/kg, as opposed to southern states with values of 12.7 MJ/kg (Fernández et al., 2009).

The generation of methane due to the anaerobic decomposition of organic matter is of great relevance in the context of clean energy sources. In México, according to INEGI, only 1 municipality, Atlacomulco in the State of Mexico, has an anaerobic digestion treatment plan for organic waste with a methane production of 592,512 Nm^3/year (Reyes et al., 2015).

According to Mediavilla Merchán (2018), of the 42.9 Mt of MSW generated each year in Mexico, up to 14 Mt can be used as feedstock for anaerobic digestion reactors, with an estimated energy production in the range of 25–29 PJ. However, four of the five principal barriers identified in his study are related with waste management at the municipal level: the collection of municipal solid waste, the size of municipalities, the sort of collection, and the recovery of organic waste in mechanical biological treatment plants (Weber et al., 2012).

On the other hand, the main results of the consultation on opportunities for biogas in Mexico carried out in the framework of the Mexico-Denmark Energy and Climate Change Programme 2017–2020 (EnergyAnalysis, 2018) were the low-cost opportunities as co-digestion of food waste (waste from markets, slaughterhouses, or dairies) in biodigesters already installed in parts, the use of landfill biogas, and—under favorable and well-managed conditions—biodigester lagoons on pig farms for electricity. Additionally, the consultation concludes that a legal requirement for Clean Thermal Energy Certificates (cetel)—as already exists with Clean Energy Certificates (cel) for electricity—would open up a huge market, enabling the massive entry of biogas into the energy market, large reductions in GHG emissions, and accelerating investments in solid and liquid waste treatment by agro-industrial and agricultural companies. This new policy would be the most effective way to promote waste treatment by biodigestion (Gutierrez, 2018).

3.2 PROBLEMS THAT REDUCE THE GENERATION AND ACCUMULATION OF BIOGAS AT DISPOSAL SITES AND LIMIT ITS ENERGY VALUATION

In the anaerobic digestion of MSW, one of the main products is biogas. This is a flammable mixture of gases containing a proportion of methane (CH_4) ranging from 50 to 70%. The rest is a combination of other gases, mainly carbon dioxide (CO_2). Both are very important precursor gases for global warming.

According to data from the National Institute of Ecology and Climate Change, MSW is responsible for 4% of the country's total GHG emissions. Of this total, final disposal sites contribute 2.9% (equivalent to 21,920.61 t of CO_2e). Of this amount, landfills contribute 17,007.473 t of CO_2e, uncontrolled sites: 2,456.286 t of CO_2e, and open dumps: 2,456.851 t of CO_2e (INECC 2015).

In developing countries, the generation and accumulation of biogas in final disposal sites (for valorization purposes) is greatly reduced due to various situations ranging from the lack of adequate environmental regulations and their correct application to the scarcity of knowledge and the lack of economic resources to build and operate a sanitary landfill correctly with the appropriate personnel, machinery, and technical equipment.

In Mexico, for example, modern environmental regulations date back to the late 1980s, with the publication of the General Law for Ecological Balance and Environmental Protection (Ley General del Equilibrio Ecológico y la Protección Ambiental or LGEEPA in Spanish) in 1988. The General Law for the Prevention and Integral Management of Waste (LGPGIR, for its initials in Spanish) was published in 2003 and its Regulations in 2006. Since then, state environmental laws and municipal regulations on ecology and public sanitation have also been passed, making it mandatory for generators of urban solid waste to deliver it to the collection service separately, at least in its organic and inorganic form. The NOM-083-SEMARNAT-2003 is the legal instrument that indicates the characteristics that a landfill must meet, as well as the environmental studies and engineering calculations that must be carried out, the details of its construction, operation, maintenance, closure, and adequate monitoring.

Unfortunately, waste separation is not enforced, and if some households or other waste generation sites do carry it out, organic waste ends up mixed with inorganic waste since they are all transported in the same collection truck. Municipalities do not have the money to operate differentiated collections. Upon arrival at the final disposal site, the collection trucks unload their waste. The treatment given there depends on whether it is an open dump, a controlled dump, or a sanitary landfill.

In the so-called "open dumps", waste accumulates continuously and remains there without being compacted or covered for several days, emitting its aerobic and anaerobic digestion gases into the air without any control. These sites are also used to dump special waste such as tires, mattresses, electrical appliances, construction waste, etc., which increases the variability of the composition of the materials and the size of the waste. Open dumps also generate other environmental problems, such as leachate contamination of the soil and subsoil, which, when increased by rainfall, sometimes contaminate surface and/or subway water bodies. It is common that, as the landfill approaches its saturation point, the site is set on fire to reduce the accumulated volume of MSW and provide space to continue depositing waste. At these sites, the biogas generated is emitted without any control and is consumed when there is a fire.

On the other hand, in the "controlled dumps", the deposited waste is compacted with a certain regularity and covered with soil, with varying frequencies: once a week or month, before the rainy season, or when there is a complaint from society to the municipal authorities. In these sites, biogas formation and accumulation are poor, as it is difficult to generate anaerobic conditions because the waste cover is porous and irregular, and there are many "air pockets" between the waste. If the cover soil is coarse-grained, air enters the air pockets at the site and leaks biogas.

Finally, in the case of landfills (MSW disposal sites that do comply with NOM-083-SEMARNAT-2003), the problems that determine the variation in biogas generation and accumulation for use are as follows: 1) the diversity of waste deposited; 2) deficiencies in compaction and a daily cover layer due to the use of inadequate stone materials, work techniques, and compaction equipment.

Other aspects to consider are the design and operation of the landfill. Generally, the biogas collection system is designed only to release it to the atmosphere and not with the purpose of capturing and valorizing it. In most cases, biogas is flared to convert CH_4 to CO_2 and reduce the polluting effect of CH_4 as a greenhouse gas. Both the capture and use of biogas are aspects that are already being considered in the proposal to update NOM-083-SEMARNAT-2003.

Once the biogas capture efficiency of the landfill has been estimated, the way in which biogas generation will evolve at the site throughout its useful life and after its closure must be calculated. Depending on the characteristics of the landfill, normally at least two years must pass before the amount of biogas produced can be valued from an energy point of view. The maximum biogas generation will occur shortly after the landfill closes, and from there, the volume of biogas produced will begin to decrease.

3.2.1 ISSUES THAT CONDITION THE USAGE OF BIOGAS GENERATED IN LANDFILLS

To take advantage of the biogas captured in a landfill, it is necessary to subject it to a conditioning process, which consists mainly of removing excess moisture and acidity (sulfur and nitrogen acids), leaving a concentrate of combustible gases, mainly methane (Castells, 2012).

Once the volume of methane available throughout the different stages of biogas generation has been quantified, it is necessary to identify and, if necessary, select the appropriate technology available on the market to convert the biogas into electricity. Modular equipment consisting of engines or microturbines can be chosen. The costs of the generation equipment, its operation, and maintenance, as well as the salaries of the personnel in charge of these actions, must be added to the above costs.

Finally, other legal aspects that must be complied with are as follows:

1. The types of contracts that municipalities establish with the companies that operate the landfills (through a concession) since, as absurd as it may seem, it is common for them to state that biogas and waste should not be valorized; and
2. If the use of biogas is technically and legally possible, the costs for the conduction of electric energy, its use on site, and, if necessary, its storage must be contemplated. The following should be considered: a) compliance with energy regulations (for the use of electricity produced by private individuals) and b) the conditions stipulated in contracts with the electricity distribution company in the respective locality or country; for example, the costs of interconnection with the local electricity system, payment of tariffs per unit of energy uploaded to the grid, etc.

Although the experience of developed countries has shown that the operating costs of landfills are high and unsustainable, these sites continue to be the option with the least environmental impacts for the final disposal of MSW in developing countries (Townsend et al., 2015).

3.2.2 MSW DECOMPOSITION

In landfills, two processes occur during MSW decomposition with the following characteristics (Johannessen, 1999):

In the aerobic digestion process, microorganisms continue the degradation of the matter, taking advantage of the air trapped between the waste. Oxygen and nitrogen are rapidly consumed by nitrification processes, and CO_2 begins to be produced. Chemical reactions raise the temperature of the waste. This phase lasts until the oxygen is depleted (about two weeks).

An acidic phase follows, representing the transition to the anaerobic process. The new phases occur in the absence of oxygen; they produce other gases, such as CH_4, and CO_2 formation decreases. The leachates have a high pH and significant concentrations of ammonia.

TABLE 3.1 MSW Decomposition.

Processes	Requirements	Temperature (°C)	Reactions	Products
Aerobic digestion	Oxygen availability, lower humidity	50–70	Oxidation, nitrification	CO_2, H_2O, and oxidation products
Anaerobic digestion	Absence of oxygen, higher humidity	35–50	Reduction, denitrification	Organic acids, CH_4, CO_2, NH_3, H_2S, and leachate

3.3 MODELING AS A TOOL TO ESTIMATE THE AMOUNT OF BIOGAS THAT CAN BE PRODUCED BY MSW DECOMPOSITION

As a tool to estimate the amount of biogas that can be produced at MSW disposal sites, mathematical models describing the anaerobic digestion of waste have been developed. The first efforts date back to the 1970s (Buhr and Andrews, 1977).

In the late 1990s, zero-order, first-order, multiphase, and modified multiphase models were used in the United States to determine the biogas generation from 18 disposal sites. The results obtained from the application of a first-order model reached values of correlation coefficients (R2) between 0.914 and 0.955 (Vogt and Augenstein, 1977; Khan, 2011; Karanjekar, 2012).

In Mexico, for the study of biogas generation in final disposal sites, it is recommended to use the Mexican Biogas Model since its use is simple, the necessary input data are usually available, it considers specific environmental variables for different areas of the country, and its results are similar to those of more complex models. Currently, the most recognized model in the literature is LandGem (Aguilar-Virgen et al., 2014; Urrego-Martínez and Rodríguez-Miranda, 2016).

In the National Inventory of Renewable Energies (INERE) by SENER, it was reported that, in Mexico, in 2015, there were eight energy generation projects by means of biogas extraction in landfills. The following table shows the details of eight projects in operation using biogas for energy purposes. The one in Salinas Victoria, Nuevo León, which has an authorized and installed capacity of 15.95 MW, stands out. Unfortunately, since 2018, there has been a slowdown in the opening and investment of this type of project, mainly due to the uncertainty of national policies on electric energy and the development of alternative energies (Diaz and Lilia, 2019).

The Mexican Biogas Model (MMB, in Spanish) was developed in 2003 by SCS Engineers and updated in 2009 (version 2.0). The mathematical expression of the model is as follows (Murray et al., 2009):

TABLE 3.2 Existing Projects in Mexico {Data Selected from Diaz and Lilia, 2019}.

Year	Name of the landfill	Municipality and State	Daily MSW reception (tons)	Power generation company	Authorized capacity (MW)	Biogas (m^3/year)	Operator of the landfill
2003	Salinas Victoria	Monterrey, Nuevo León	5000	Bionergía de Nuevo Léon (Benlesa)	16.95	83,157	Particular
2010	Ciudad Juárez	Ciudad Juárez, Chihuahua	1200	Transformadora de Energía Eléctrica de Juárez	6.4	13,811	Particular
2008	Durango	Durango, Durango	400	Ener-G	1.6	6000	City Hall
2009	Puerto de Chivos	Atizapán, Edo. De México	500	Energreen	0.6	3040	City Hall
2010	San Nicolás	Aguascalientes, Aguascalientes	914	Ener-G	3.20	10,556	City Hall
2012	Querétaro	Mompaní, Querétaro	750	TQM Generación Energía Renovable	2.746	12,614	Particular
2013	La Perseverancia	Cuautla, Morelos	800	Energía Renovable de Cuautla	1.065	2000	Particular
2013	Saltillo	Saltillo, Coahuila	600	Lorean Energy Group	2.122	7992	City Hall

$$Q = \sum_{t=1}^{n} 0 \sum_{j=0.1}^{1} 2kL_{\circ} \frac{Mi}{10} e^{-ktij} (MCF)(F) \tag{3.1}$$

where:

Q:	Maximum expected biogas flow (m^3/year).
i:	Increase in time of 1 year.
n:	(year of calculation) — (year in which the waste deposit started).
j:	Time increment of 0.1 years.
k:	Methane generation rate (1/year).
L0:	Potential methane generation (m^3/Mg).
Mi:	Mass of waste deposited in year i (Mg).
tij:	Age of section j of the waste mass Mi, disposed in year i (decimal years).
MCF:	Methane correction factor
F:	Fire adjustment factor.

The MMB was developed in an Excel spreadsheet and is based on a 1 (first order) equation (Ludwig et al., 2009). The model needs certain input data such as the year the site was opened, the year of closure, the annual disposal rates, and the site location. Questions such as the previous and current physical conditions of the site must also be answered. Based on this information, the model automatically defines the values of the methane generation rate (k) and potential methane generation (L0) and estimates the biogas capture efficiency. The values used by the model were developed based on specific data on climate, waste characterization, and waste disposal practices in Mexico; the effects of these conditions on the amounts and rates of biogas generation were also estimated (Murray et al., 2009).

The MMB estimates annual biogas generation based on the amount of deposited waste accumulated during a year. Forecasts for multiple years are made by varying the annual projection and then iterating the equation. The resulting total biogas generation is twice the calculated methane generation. The MMB expresses biogas generation and recovery in m^3/h or ft^3/min.

3.4 CASE STUDY IN SAN LUIS POTOSI, MEXICO

The following is a case study of the modeling of biogas generation and its possible use in the degradation of urban solid waste from a landfill located in a state in central Mexico.

The landfill is operated by a private company and receives about 800 t/day of waste; therefore, according to NOM-083-SEMARNAT-2003, it is type A, subclassification A1. In order to obtain environmental impact and construction permits, the following studies were required: a) regional geological and geohydrological; b) hydrological; c) topographical; d) MSW generation and composition; e) biogas generation; and f) leachate generation.

The site where the landfill is located has an arid climate with a hot summer temperature of BS0 (kw). The average annual temperature ranges from 12 to 18°C. The rainiest season is summer; in winter, it is only 5%–10.2% of the annual total.

The landfill began operations in 2015 and is estimated to have a useful life of 25 years. At the site, MSW management begins with weighing the amount of incoming MSW. The MSW is then taken to the shooting area, where it is spread and compacted. Finally, they are covered with a layer of tepetate measuring approximately 40 cm. The biogas is concentrated and released into the atmosphere through chimneys. The leachate is recirculated weekly. A topographic survey is done monthly, and every six months measurements and analysis of the biogas and leachate composition, as well as the quality of the aquifer water (upstream and downstream of the landfill), are done (Red Ambiental 2018).

To estimate the amount of biogas that can be produced by the decomposition of MSW accumulated in landfills (the objective of this study), the Mexican Biogas Model 2.0 (MMB 2.0) was used.

In the case study, the input data to the model were as follows:

TABLE 3.3 Input Data to the Mexican Biogas Model 2.0 to Calculate Biogas Generation.

1. Name of the site	San Juanico el Grande
2. City	San Luis Potosí
3. State	San Luis Potosí
4. Region	5 (Northwest-Interior North)
5. Year of opening of the site	2015
6. Annual disposition of the most recent year	305,112 Mg
7. Projected year of closure	2041
8. Estimated annual disposition increase	1.5%
9. Average landfill depth	12 m
10. Percentage of area with waste covered on a daily basis.	90%
11. Percentage of the area with waste without daily cover	10%

The model automatically calculated values for the methane generation index (k) and potential methane generation (L0) for the state of San Luis

Potosi and estimated the capture efficiency based on the answers provided (MMB Manual).

Given the indications of the MMB user's manual, the methane correction factor (MCF) was assigned a value of 1.0. This was determined considering the characteristics of the landfill such as the depth of the MSW confinement cells and the site's operation methodology. Finally, since no fires have been reported at the landfill that consume the biogas produced, the fire adjustment factor was given a value of "0".

3.4.1 MODELING RESULTS

The following table shows the amount of MSW accumulated in the landfill (in tons), starting in 2015 and until its possible closure in 2042. It also shows the amount of biogas expected to be generated, expressed in m^3/year.

TABLE 3.4 Modeling of Biogas Generation at the Study Site from MMB 2.0.

Year	Accumulated MSW (t)	Biogas (m^3/year)	Year	Accumulated MSW (t)	Biogas (m3/year)
2015	35,265	0	2029	4,521,594	22,398,951
2016	263,207	268,304	2030	4,886,394	23,754,000
2017	542,682	1,984,451	2031	5,256,694	25,076,439
2018	847,794	3,977,298	2032	5,632,594	26,369,811
2019	1,157,494	6,033,014	2033	6,014,094	27,637,362
2020	1,471,794	7,989,260	2034	6,401,294	28,881,302
2021	1,790,794	9,854,839	2035	6,794,294	30,104,401
2022	2,114,594	11,638,421	2036	7,193,194	31,309,204
2023	2,443,294	13,347,902	2037	7,598,094	32,498,040
2024	2,776,894	14,990,478	2038	8,009,094	33,673,052
2025	3,115,494	16,571,945	2039	8,426,294	34,836,206
2026	3,459,194	18,098,336	2040	8,849,794	35,989,310
2027	3,808,094	19,575,152	2041	9,279,694	37,134,028
2028	4,162,194	21,007,415	2042	9,279,694	38,271,892

Figure 3.1 represents the variations: a) of the MSW accumulation projected for the landfill under study and b) of the amount of biogas that will be generated, as calculated by MMB 2.0.

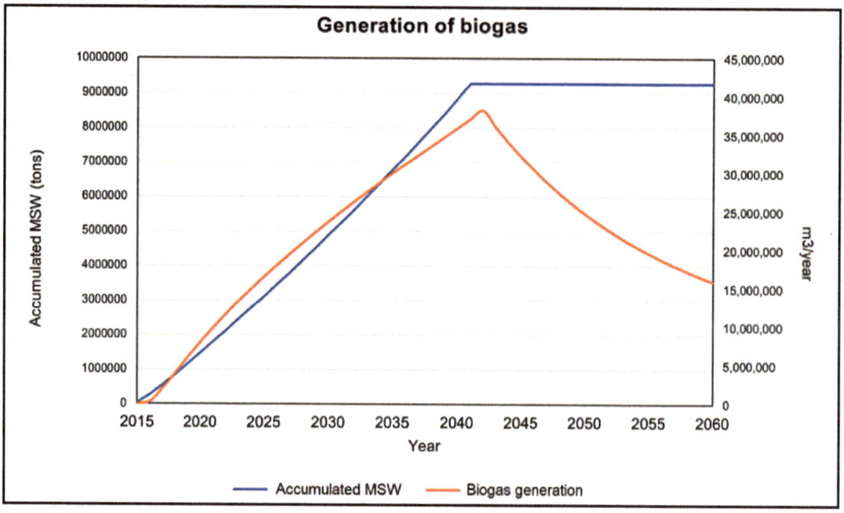

FIGURE 3.1 Representation of the variation of MSW accumulation and biogas generation (according to MMB 2.0) in the studied landfill.

The model indicates that the highest biogas generation will occur in the year 2042, that is, one year after the landfill closes. This may be the result of the maximum amount of waste accumulated in 2041. In addition, the MMB 2.0 model assumes that the decomposition of the organic fraction of waste occurs approximately one year after landfilling.

3.4.2 BIOGAS CAPTURE EFFICIENCY AT THE LANDFILL

According to the United States Environmental Protection Agency (USEPA), in a landfill, the biogas capture efficiency is a function of the proportion of the landfill surface that is covered by the capture system and the efficiency in the operation and maintenance of the system. Based on its studies, the USEPA also points out that, under the best conditions, landfills in Mexico can reach maximum capture efficiencies of 90%, whereas in unmanaged disposal sites, even with a good capture system installed, they do not reach 50% biogas capture (Ludwig et al., 2009).

Given the characteristics of the landfill under study, three possible scenarios of biogas capture efficiency were proposed as follows: 50%, 60%, and 70%. Considering these scenarios, in the year 2042 (the year of highest biogas generation), biogas capture could be achieved in the following quantities: 19,135,946 m^3/year (at 50%), 22,963,135 m^3/year (at 60%), and 26,790,325 m^3/year (at 70%) (Fig. 3.2).

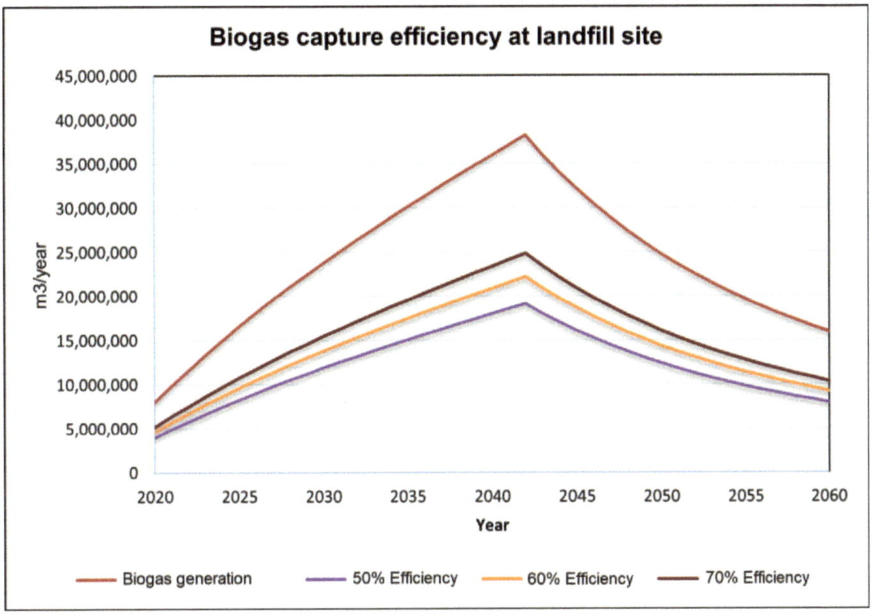

FIGURE 3.2 Modeling of biogas generation and capture at the study site for biogas capture efficiencies of 50%, 60%, and 70%.

3.5 ELECTRICAL MEASUREMENT

The most commonly used equipment in landfill biogas power generation projects are internal combustion engines, turbines, and microturbines. The advantages of the engines are their low cost and scalability. Their disadvantages are their high maintenance costs and the fact that they generate more emissions into the air. Turbines, on the other hand, have greater resistance to corrosion, and their operation and maintenance are less expensive than those of internal combustion engines (Mutz et al., 2017; Blanco et al., 2017).

In the case study to determine the electrical sizing, aspects such as: a) the Lower Calorific Power (LCP) of biogas of 16.9 MJ/m^3 were considered; b) the biogas capture efficiency was assumed to be 50%. According to the equipment available on the market, a Jenbacher-J320 internal combustion engine was proposed, which runs on biogas and has an electrical power of 1062 kW, an electrical efficiency of 39.4%, and a biogas consumption of 575 m^3/h.

The following table contains the most representative results of the electrical sizing calculation:

TABLE 3.5 Electrical Measurement of the Studied Landfill.

Year	Accumulated MSW	Biogas m^3/h	Biogas capture (m^3/h)	Total available power (kW)	Electrical power (kWe)
2016	263,207	31	15	71.9	28
2017	542,682	227	113	532	207
2018	847,794	454	227	1066	416
2019	1,157,494	689	344	1617	630
2020	1,471,794	912	456	2141	835
2021	1,790,794	1125	563	3119	1030
2022	2,114,594	1329	664	3577	1216
2023	2,443,294	1524	762	4017	1395
2024	2,776,894	1711	856	4440	1566
2025	3,115,494	1892	946	4849	1732
2026	3,459,194	2066	1033	5245	1891
2027	3,808,094	2235	1117	5629	2046
2028	4,162,194	2398	1199	6002	2195
2029	4,521,594	2557	1279	6365	2341
2030	4,886,394	2712	1356	6719	2482
2031	5,256,694	2863	1431	7066	2620
2032	5,632,594	3010	1505	7405	2756
2033	6,014,094	3155	1578	7739	2888
2034	6,401,294	3297	1649	8066	3018
2035	6,794,294	3437	1718	8389	3146
2036	7,193,194	3574	1787	8708	3272
2037	7,598,094	3710	1855	9023	3396
2038	8,009,094	3844	1922	9334	3519
2039	8,426,294	3977	1988	9643	3640
2040	8,849,794	4108	2054	9950	3761
2041	9,279,694	4239	2120	10225	3880
2042	9,279,694	4369	2185	9669	3999
2043	9,279,694	4119	2060	9128	3771

According to what is shown in the table, in 2022, given that the electrical power of the Jenbacher-J320 ICM is 1,062 kW and the availability of biogas to operate is 664 m^3/h, the study site could start using biogas with one unit. Subsequently, in 2028, another 1 MW module could be added and another one in 2036, to reach an installed capacity of 3,186 MWh. From 2043 onward, biogas generation will start to decrease and therefore the electricity potential.

3.6 CONCLUSIONS

At the end of this chapter, the following conclusions can be drawn:

- Despite the existing regulatory framework, the integrated management of MSW in Mexico is deficient. In the sources of generation, the separation of organic and inorganic fractions has not yet been established.
- Municipal authorities do not charge the population for providing MSW collection services, so they generally do not have the economic resources to comply with their part of MSW management: segregated collection, transportation, reuse or recycling, and final disposal.
- The stage of integrated MSW management that presents the most environmental problems is final disposal.
- In Mexico, the poor final disposal of MSW favors the release of polluting gases into the air and reduces the possibility of capturing biogas to be used for energy purposes.
- The proper design, construction, operation, and maintenance of sanitary landfills are basic elements that determine the amount of biogas that can be produced, captured, and valorized for energy purposes.
- The Mexican Biogas Model (MMB v2.0) is a very useful tool that allows knowing, with an acceptable approximation, the amount of biogas that can be generated in a landfill, as well as the electrical power that can be generated from the use of biogas.
- The use of MMB v2.0, applied to the case study, made it possible to identify that the use of biogas for energy purposes is possible after the first 7 years of landfill operation and during the following 40 years, achieving an installed capacity of 3,186 MWh.
- Mexico has important commitments to reduce its greenhouse gas emissions into the air, so the proper management of its waste in landfills and the valorization of the biogas produced there is an opportunity that should be taken advantage of.
- The application of environmental regulations and the obtaining of economic resources for the adequate management of waste are fundamental elements to ensure the adequate integrated management of MSW.

KEYWORDS

- **biogas**
- **organic waste**
- **landfill**
- **Mexican biogas model**

REFERENCES

Aguilar-Virgen, Q.; Taboada-González, P.; Ojeda-Benítez, S.; Cruz-Sotelo, S.; Power Generation with Biogas from Municipal Solid Waste: Prediction of Gas Generation with *In Situ* Parameters. *Renew. Sustain. Energy Rev.* **2014,** *30,* 412–419, doi: https://doi.org/10.1016/j.rser.2013.10.014.

Arvizu Fernández, J. L.; Huacuz Villamar, J. M.; Saldaña Méndez, J. L. Evaluación del Potencial Energético de los Rellenos Sanitarios. *Rev. AIDIS Ing. Ciencias Ambient. Investig. Desarro. Práctica* **2009,** *1* (1) SE-Residuos Sólidos, , [Online]. https://www.revistas.unam.mx/index.php/aidis/article/view/14446.

Blanco, G.; Santalla, E.; Córdoba, V.; Levy, A. Generación de Electricidad a Partir de Biogás Capturado de Residuos Sólidos Urbanos: Un Análisis Teórico-Práctico. Washington, DC [Online] 2017. https://publications.iadb.org/es/publicacion/13967/generacion-de-electricidad-partir-de-biogas-capturado-de-residuos-solidos-urbanos.

Buhr, H. O.; Andrews, J. F.; The Thermophilic Anaerobic Digestion Process. *Water Res.* **1977,** *11* (2), 129–143, doi: https://doi.org/10.1016/0043-1354(77)90118-X.

Castells, X. *Valorización e Residuos Procedentes De Grandes Industrias: Reciclaje De Residuos Industriales* [Online]. Diaz de Santos: Madrid, 2012.

Chakraborty, P.; Selvaraj, S.; Nakamura, M.; Prithiviraj, B.; Cincinelli, A.; Bang, J. J. PCBs and PCDD/Fs in Soil from Informal E-waste Recycling Sites and Open Dumpsites in India: Levels, Congener Profiles and Health Risk Assessment. *Sci. Total Environ.* **2018,** *621,* 930–938, doi. https://doi.org/10.1016/j.scitotenv.2017.11.083.

Clemente Reyes, A.; Estradas Romero, R. Planta de Biodigestión Atlacomulco, Estado de México. *Foro Internacional 2015. Valorización Energética de Resiudos Urbanos. Experiencias y Estratégias Globales* [Online] 2015. https://www.gob.mx/cms/uploads/attachment/file/197648/20._Experiencia_Nacional_1er_Biodigestor_de_RSU_en_M_xico.pdf (accessed Apr 05, 2023).

EA EnergyAnalysis. Biogas in México. Lessons Learned from Partnership Program in 2018-2019. Mexico, 2018. [Online]. https://ens.dk/sites/ens.dk/files/Globalcooperation/Publications_reports_papers/biogas_in_mexico_lessons_learned_eng.pdf.

Flachenecker, F.; Rentschlehler, J. From Barriers to Opportunities Enabling Investments in Resource Efficiency for Sustainable Development. *Public Sect. Econ.* **2019,** *43* (4), 345–373, doi: https://doi.org/10.3326/pse.43.4.2.

Gutierrez, J. P. Situación Actual y Escenarios Para el Desarrollo del biogás en México hacia 2024 y 2030. Mexico 2018. [Online]. https://rembio.org.mx/wp-content/uploads/2020/11/Situacion-actual-y-escenarios-para-el-desarrollo-del-biogas-en-Mexico.pdf.
INECC. Inventario Nacional de Emisiones de Gases y Compuestos de Gases de Efceto Invernadero. 2015.
INEGI. Censo Nacional de Gobiernos Municipales y Demarcaciones Territoriales de la Ciudad de México 2021. *Programas* 2021. https://www.inegi.org.mx/programas/cngmd/2021/ (accessed Apr 10, 2023).
Jackson, E. A. Importance of the Public Service in Achieving the UN SDGS BT—Decent Work and Economic Growth. Leal Filho, W., Azul, A. M., Brandli, L., Lange Salvia, A., Wall, T., Eds.; Springer International Publishing: Cham, 2021; pp 551–561.
Johannessen, L. M. Guidance Note on Recuperation of Landfills Gas from Municipal Solid Waste Landfills. [Online]. Washington, DC, 1999. https://documents1.worldbank.org/curated/en/892171468766767158/pdf/multi-page.pdf.
Karanjekar, R. An Improved Model for Predicting Methane Emissions from Landfills based on Rainfall, Ambient Temperature and Waste Composition. Texas 2012.
Khan, V. R. F. Z. S. E. *From Landfill Gas to Energy Technologies and Challenges*; 1st ed.; CRC Press, 2011.
Ludwig, V.; Stege, G. A.; Engineers, S. C. S.; Dávila, J. L. Manual del Usuario Modelo Mexicano de Biogás [Online] 2009. https://www.globalmethane.org/documents/models/pdfs/manual_del_usuario_modelo_mexicano_de_biogas_v2_2009.pdf.
Mediavilla Merchán, V. Waste-to-Energy in Mexico. Technical Potential for Biogas Production and Greenhouse Gas Mitigation from the Anaerobic Digestion of Municipal Solid Waste. 2018. [Online]. https://www.giz.de/en/downloads/giz2017-en-waste-to-energy.pdf.
Murray, D. L.; L. V. P.E., Stege, G. A.; Dávila, J. L. Manual del Usuario Modelo Mexicano de Biogás. Versión 2.0. Phoenix, Arizona. [Online] 2009. https://www.globalmethane.org/documents/models/pdfs/manual_del_usuario_modelo_mexicano_de_biogas_v2_2009.pdf.
Mutz, D.; Hengevoss, D.; Hugi, C.; Gross, T. Opciones Para el Aprovechamiento Energético de Residuos en la Gestión de Residuos Sólidos Urbanos. Bonn[Online] 2017. https://www.giz.de/en/downloads/Guia GIZ 2017 WasteToEnergy - SP.pdf.
Norma Lilia, L. D.; Evaluación del potencial de generación de biogás en sitios de disposición final del estado de San Luis Potosí, y sus opciones de valoración. [Online] 2019. https://ninive.uaslp.mx/xmlui/bitstream/handle/i/5722/MCA1LDN201901.pdf?sequence=1&isAllowed=y.
Red Ambiental. *Informe del sitio de Disposición Final de San Juanico el Grande*; San Luis Potosí, 2018.
Reyna-Bensusan, N.; Wilson, D. C.; Smith, S. R. Uncontrolled Burning of Solid Waste by Households in Mexico is a Significant Contributor to Climate Change in the Country. *Environ. Res.* **2018,** *163,* 280–288, doi: https://doi.org/10.1016/j.envres.2018.01.042.
Ruiz Suárez, J.; Gavilán García, L. G.; Mendoza Cantú, A.; Ramírez Muñóz, A.; Araiza Aguilar, T. Atlas Nacional de Residuos Sólidos Urbanos. 2021.
Sáez, A.; J. A. Urdaneta, G. Manejo de Residuos Sólidos en América Latina y el Caribe. *Omnia* **2014,** *20* (3), 121–135. https://www.redalyc.org/articulo.oa?id=73737091009.
SEMARNAT. Indicadores Básicos del Desempeño Ambiental - Residuos sólidos. *Datos Abiertos* [Online] 2023. https://datos.gob.mx/busca/dataset/indicadores-basicos-del-desempeno-ambiental--residuos-solidos (accessed Apr 13, 2023).

SEMARNAT. Visión Nacional hacia una Gestión Sustentable: cero residuos. México [Online] 2019. https://www.gob.mx/cms/uploads/attachment/file/435917/Vision_Nacional_Cero_Residuos_6_FEB_2019.pdf.

Toro, E. R.; Nare, M.-S.; Pacheco, J. F.; Contreras, E.; Gálvez, A. Guía General Para la Gestión de Residuos Sólidos Domiciliarios [Online]. 2016. https://www.cepal.org/es/publicaciones/40407-guia-general-la-gestion-residuos-solidos-domiciliarios.

Townsend, T. G.; Powell, J.; Jain, P.; Xu, Q.; Tolaymat, T.; Reinhart, D.. *Sustainable Practices for Landfill Design and Operation*;Springer: Nueva York, 2015.

UNEP. Sustainable Consumption and Production. A Handbook for Policymakers[Online] 2015. https://sustainabledevelopment.un.org/content/documents/1951Sustainable Consumption.pdf.

UNEP. Sustainable Innovation and Technology Transfer Industrial Sector Studies: Recycling - From E-Waste to Resources. United Nations Environment Programme: Berlin, 2006; p 90, [Online]. Available: https://wedocs.unep.org/20.500.11822/33112.

Urrego-Martínez, E.; Rodríguez-Miranda, J. P.; Aplicación de las Metodologías EPA, Mexicano e IPCC Para la Estimación de Biogás, Caso de Estudio Relleno Sanitario Doña Juana, Bogotá - Colombia. *Universidad y Salud* **2016,** *18*, scieloco, 338–344.

Vogt, D.;Augenstein, W. G. Comparison of Models for Predicting Landfill Methane Recovery. Final Report. 1977. [Online]. https://doi.org/10.2172/314088. https://www.osti.gov/servlets/purl/314088.

WB. Urban Development. *The World Bank* [Online] 2020. https://www.worldbank.org/en/topic/urbandevelopment/overview#1 (accessed May 28, 2022).

Weber, L.; Rojas Oropeza, B.; Torres Bernal, M.; Pampillón González, M. Producción de biogas en México. Estado actual y perspectivas. México 2012. [Online]. Available: https://sswm.info/sites/default/files/reference_attachments/WEBER et al 2012. Producción de biogás en México.pdf.

West, J.; Schandl, H. Material use and Material Efficiency in Latin America and the Caribbean. *Ecol. Econ.* **2013,** *94,* 19–27, doi: https://doi.org/10.1016/j.ecolecon.2013.06.015.

Zolnikov, T. R.;, Furio, F.; Cruvinel, V.; Richards, J.; A Systematic Review on Informal Waste Picking: Occupational Hazards and Health Outcomes. *Waste Manag.* **2021,** *126,* 291–308, doi: https://doi.org/10.1016/j.wasman.2021.03.006.

PART 2

Sustainable Spray Drying Technology for Preservation of Bioactive Compounds

CHAPTER 4

Spray Drying: An Alternative for Seasonal Fruit Peservation

GISELA RODRÍGUEZ-MONTES[1], DIANA BEATRIZ MUÑIZ-MÁRQUEZ[2*], ABIGAIL REYES-MUNGUÍA[2], and JORGE ENRIQUE WONG-PAZ[2]

[1]*Tecnológico Nacional de México, Campus Instituto Tecnológico de Ciudad Valles, Ciudad Valles, San Luis Potosí, México*

[2]*Universidad Autónoma de San Luis Potosí. Facultad de Estudios Profesionales Zona Huasteca, Ciudad Valles, San Luis Potosí, México*

ABSTRACT

Currently, a large number of fruits produced in the backyard are very attractive for commercial purposes; however, due to their short seasonality and the fact that most of these products have a high-water content when ripe, their shelf life is limited by factors such as degradation by microorganisms and chemical and enzymatic reactions. As a result, these products are highly perishable, making it difficult for them to be marketed or exported as fresh products, generating large postharvest losses each year. Faced with this problem, it is necessary to search for new technologies to transform these fresh fruits into new products such as powdered juices. This idea could extend its storage life under ordinary conditions since drying the juice results in a more stable product that is easier to use and, when rehydrated, has properties similar to those of the original juice. One of the main methods used to achieve this goal is the spray drying, which consists of atomizing a liquid material in a drying chamber through which a stream of hot air circulates to transform it almost instantaneously into a solid product. However, spray drying of fruit juices, which contain a high percentage of sugar and organic acids, presents technical difficulties such as low yields due to product loss

Sustainable Engineering and Agro-Food Processing: Renewable and Nonrenewable Resources.
José Juan Buenrostro-Figueroa, Julio César Tafolla-Arellano, Mónica Chávez-González, Pedro Aguilar-Zárate, A. K. Haghi (Eds.)

and operational problems due to their hygroscopicity and thermoplasticity at high temperatures and humidity. Nevertheless, given the great potential of using fruits rich in sugars and adding value to them by spray drying, it is essential to study and understand the variables that influence this process in order to obtain safe and quality products. As a valuable contribution, this chapter is dedicated to the topic of spray drying of fruit juices, where this drying technique and the main difficulties associated with the nature of this type of product will be analyzed in detail.

4.1 INTRODUCTION

The main problem in humanity since ancient times has been how to keep food in good condition for a longer time. From this arose the need to develop different techniques to solve this problem, transforming or preserving food by means of thermal or chemical treatments. One of the safest methods used to extend the shelf life of food is drying. This process, by eliminating almost all the water present in this type of product, prevents microorganisms that cause spoilage from growing and multiplying. In addition to many of the enzymes that cause chemical changes in food, it is impossible for them to function without the presence of water. For this reason, it is one of the most effective techniques for preserving fruits (Magaña-Barrera et al., 2019; Geankoplis, 1998). The latter represents a solid source of nutrients that can promote good health. However, as they have a high moisture content, they generate fungi or molds, yeasts, and bacteria, which cause the food to gradually deteriorate, so it is necessary to process them to make them suitable for long-term consumption (FAO, 2021; Hernández-Gómez et al., 2017). In view of this, drying not only ensures the microbiological and physicochemical stability of the food but also considerably reduces its weight, thus facilitating its storage, handling, and transportation without causing major alterations to its nutritional value (Hernández-Gómez et al., 2017).

Conventionally, fruits are dried in the sun; however, this process is usually slow as it is mainly influenced by climatic uncertainties, which is detrimental to product quality due to exposure to long drying times. At present, there are different alternative technologies for drying fruits and vegetables, such as osmotic dehydration, vacuum drying, natural convective drying, forced convective drying, microwave drying, freeze-drying, and spray drying (Tontul and Topuz, 2017; Karam et al., 2016). Spray drying is a technique widely used to quickly transform liquid foods into powder, which sounds simple; however, the real problem occurs when trying to dry food products that, by their nature, are complex. A clear example is fruit juices since they contain high concentrations of sugar and organic acids. Due to their high thermosensitivity and low glass transition temperature, it is difficult to properly control the operating conditions of drying, and these, in turn, allow obtaining a product with acceptable quality characteristics. Therefore, this is the main topic of this chapter: to provide readers with the basics of spray drying as a technique for the preservation and use of seasonal fruits and to delve into the complications involved in drying this type of product.

4.2 SPRAY DRYING BASICS

Spray drying is a unitary process in which a liquid formulation is atomized to form a shower of small droplets that are spread in a stream of hot air, usually at temperatures ranging from 150 to 300°C depending on the material to be dried, to be converted directly into a solid product in the particulate form (Gharsallaoui et al., 2007; Siccha and Ugaz, 1995; Wolska, 2021).

Spray-dried foods can be classified into two groups: nonsticky and sticky. Nonsticky materials can be dried by simple dryers, and the products obtained flow freely. On the other hand, sticky materials are difficult to dry under normal conditions because they can adhere to the wall of the drying chamber. These products include mainly foods rich in sugars and organic acids, such as honey and juices derived from fruits and vegetables (Adhikari et al., 2004; Bhandari et al., 1997; Goula and Adamopoulos, 2010).

Tontul and Topuz (2017) point out that five important stages can be identified during the spray drying process. These stages are shown in Figure 4.1.

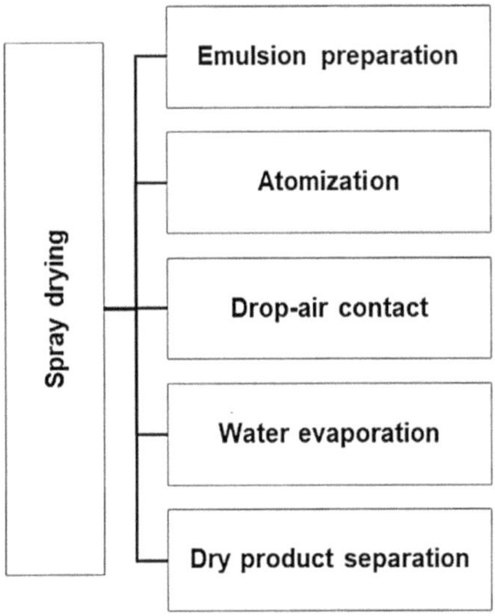

FIGURE 4.1 Important stages in spray drying.

The preparation of the emulsion usually consists of increasing the concentration of the liquid to be atomized and adding carriers or encapsulating materials, such as carbohydrates, gums, lipids, and proteins, among others, in order to increase the product yield and improve the physical properties of the final product (Tontul and Topuz, 2017).

Atomization is achieved by applying energy to the fluid to disperse it into fine particles. The size, distribution, velocity, and final size of the dry particle will depend on the type of atomizer. The objective of this stage is to increase the heat transfer surface area between the dry air and the liquid to achieve evaporation and optimize mass transfer. The finer the droplets, the larger the surface area, the easier the evaporation, and consequently, the higher the thermal efficiency (Caez and Jaraba, 2012; Ángeles, 2009).

The contact of the droplet with the hot air occurs in the drying chamber immediately after atomization. There are three basic forms of contact between the hot air and the droplet: parallel or co-current flow, counter-current flow, and mixed flow. Co-current drying is where the material is atomized and the drying air passes through the dryer in the same direction. In the co-current model, the atomization direction is opposite to the drying air flow, so the product to be dehydrated is exposed to higher temperatures and longer exposure times, so this model is not suitable for thermosensitive products. In the mixed model, both atomization methods are combined, where the fluid is atomized upward and only remains in contact with the hot air for a short period of time to eliminate residual moisture (Tontul and Topuz, 2017; Caez and Jaraba, 2012; Ángeles, 2009; Reineccius, 2004).

Water evaporation occurs immediately after the hot air comes in contact with the liquid, so the hot air raises the temperature of the droplets to a constant value, leading to water evaporation. Finally, when the droplet reaches the critical water content, a dry layer forms on the surface of the droplet, and the drying rate decreases rapidly and becomes dependent on the diffusion of water through the dry layer (Gharsallaoui et al., 2007; Shishir and Chen, 2017). Once the drying process is finished, the separation of the dry product from the humid air is carried out by means of a cyclone, so that the dry particles are dragged by the hot air through this cyclone separator to a collection bottle where they are deposited for later recovery.

4.2.1 PHYSICAL PROPERTIES OF THE DRY PRODUCT

The physical properties of the dry product resulting from the spray drying process depend on the physical and chemical characteristics of the formulation being atomized, as well as on dryer operating variables such as atomization, drying air temperature, and dryer design (Goula and Adamopoulos, 2008). The physical properties of the dry product are shown in Figure 4.2.

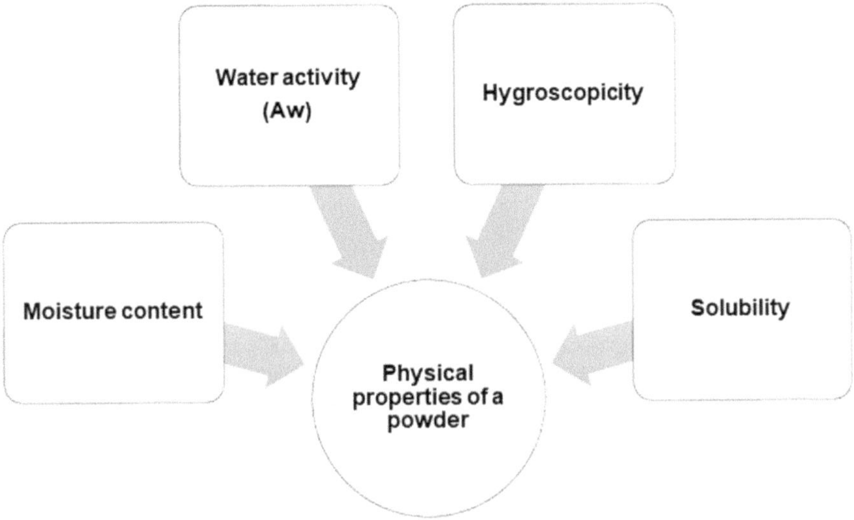

FIGURE 4.2 Physical properties of importance in a powdered product.

4.2.1.1 MOISTURE CONTENT

The stability and shelf life of a powdered product are directly related to the amount of moisture in the product (Bhusari et al., 2014). Generally, powders produced by spray drying have a moisture content below 5% and are considered microbiologically stable since the lower the moisture content, the longer the shelf life, which reduces costs, facilitates transportation, and reduces problems of caking and agglomeration during storage (Tontul and Topuz, 2017; Ramakrishnan et al., 2018). This parameter is mainly influenced by the temperature of the inlet air in the drying chamber. Several researchers affirm that with the use of high temperatures during the drying process, the moisture content of the final product decreases because there is a greater heat transfer to the particles, which provides a greater driving force to evaporate the water, obtaining powders with lower humidity (Quek et al., 2007; Daza et al., 2016).

4.2.1.2 *WATER ACTIVITY (AW)*

Water activity is a parameter that has considerable effects on the shelf life of foods since high water activity values indicate a greater amount of free water available for spoilage reactions to occur, which consequently shortens the shelf life of the food (Tontul and Topuz, 2017; Quek et al., 2007). Dust with a water activity below 0.6 could be considered microbiologically safe since different microorganisms tend to grow and develop at water activity values above this value (Badui, 2006; Saifullah et al., 2016).

4.2.1.3 *HYGROSCOPICITY*

Hygroscopicity can be defined as the ability of a material to absorb moisture in a high relative humidity environment. Santhalakshmy et al. (2015) mention that the hygroscopicity values increase with an increase in moisture content, so the lower the moisture content, the less hygroscopic the powders turn out to be. This parameter is also influenced by the carrier agent added to the sample before atomization since high concentrations of an agent such as maltodextrin result in powders with low hygroscopicity values, which could be related to the less hygroscopic nature of maltodextrin (Braga et al., 2020; Igual et al., 2014; Mishra et al., 2014).

4.2.1.4 *SOLUBILITY*

Solubility is considered an important property for evaluating the quality of powder products as it can directly influence the reconstitution behavior of atomized powders (Shishir and Chen, 2017). The measurement of this property serves to determine the rate and degree to which the components of powder particles dissolve in an aqueous medium (Serna-Cock et al., 2015). Some authors mention that the solubility of a powder is mainly influenced by the moisture content since as the moisture content decreases, the product becomes drier, so less time is required for dissolution, which means that the powder presents a better solubility in water. However, this property is also affected by other variables such as the characteristics of the product to be processed, the type of carrier material to be used to facilitate and improve the drying process, and the physical properties of the powdered product obtained (Arebo et al., 2023).

4.2.2 SPRAY DRYING APPLICATION IN THE INDUSTRY

Spray drying is a simple, fast, and economical technology that has become a widely used alternative for obtaining powdered products not only in the food industry but also in the pharmaceutical and chemical industries due to the good physical properties of the products obtained with it, the short drying time, and the economic potential it represents (Wolska, 2021; Shishir and Chen, 2017).

This method is directly or indirectly linked to a wide variety of everyday products. In the food industry, it is used to produce various products such as instant coffee, milk powder, cheese powder, soups, baby food, and sweeteners (Nath and Satpathy, 1998).

Recently, this drying method has been well-practiced to form microcapsules containing bioactive and nutraceutical food ingredients such as vitamins, minerals, phenolic compounds, carotenoids, essential fatty acids, and essential oils to supplement food products by adding one or more of these essential ingredients that can provide human health benefits (Jafari et al., 2021; Cardona et al., 2021). Compared to other conventional microencapsulation techniques, spray drying offers the attractive advantage of obtaining microcapsules in a relatively simple, continuous operation (Ré, 1998).

4.3 PRESERVATION OF SEASONAL FRUITS BY SPRAY DRYING

Fruits and their derivatives represent an excellent source of nutritional attributes since they contain vitamins, minerals, carbohydrates, and phytochemicals such as polyphenols, flavonoids, carotenoids, and anthocyanins, among others, components which, besides being essential for the correct functioning of the human body, help in the prevention of various cardiovascular, coronary, metabolic, and degenerative diseases (Karam et al., 2016). Unfortunately, the fact that fruits are seasonal raw materials and foods with a very short shelf life proves to be the main drawback for their production and marketing since postharvest deterioration causes rapid degradation of quality, resulting in a large amount of waste. Therefore, it is necessary to transform them into new, less perishable products such as dehydrated fruits, vacuum-packed fruit pulps, bottled juices, frozen fruit, preserves, liqueurs, or juice powders (Adnan et al., 2018).

In this case, the production of fruit juice powders represents a promising approach due to its multiple benefits and economic potential since it not only allows for reduced packaging, storage, and transportation costs but also reduces postharvest losses. In addition, the powders obtained find use in many food and pharmaceutical products. Particularly, in the food industry, fruit powders are used as functional additives, flavoring agents, flavorings, and natural colorings because they have greater stability by physically eliminating an important part of the moisture content present in the fresh fruit, being able to be preserved from months to years according to their storage conditions, compared to the liquid juice or fresh product, which normally remains in good condition for shorter periods of time that can vary from weeks to months (Goula and Adamopoulos, 2010; Shishir and Chen, 2017).

Several drying methods have evolved from the need to produce dehydrated food products and ingredients that are convenient and of better quality (Sabarez, 2021). Among these techniques, spray drying is an established and widely used technology for processing a wide range of liquid food products, such as fruit juices. However, the processing of fruit by this method presents certain complications due to the presence in its composition of a high proportion of low molecular weight sugars, such as fructose, glucose, and sucrose, as well as organic acids, such as citric acid and malic acid, among others, which together consequently cause the problem of stickiness (Igual et al., 2014; Oberoi and Sogi, 2015). The use of high temperatures during drying can cause this phenomenon, causing the particles to adhere to each other, forming thick deposits on the walls of the drying chamber and the cyclone of the equipment, probably because they exceed the glass transition temperature (Tg) of the sugars and acids present; for example, the Tg of fructose, glucose, and sucrose is 16, 31, and 62°C, respectively (White and Cakebread, 1996; Fang and Bhandari, 2012). This change of state can contribute to the adherence of powders in the equipment, which leads to loss of material, reducing the yield. This is one of the main concerns of the spray drying process as it is closely related to the cost and efficiency of production (Braga et al., 2020; Muzaffar and Kumar, 2015). Therefore, it becomes important to optimize the drying process parameters to obtain good-quality powders and a high process yield.

4.3.1 APPLICATION OF SPRAY DRYING TO FRUIT JUICES: A REVIEW

4.3.1.1 BLACKBERRY JUICE

Bermúdez (Bermúdez, 2012) evaluated the effect of drying air temperature and maltodextrin concentration on process yield, anthocyanin microencapsulation efficiency, as well as the whiteness index (IB) and water activity (Aw) of microcapsules obtained from blackberry juice. Maltodextrin was used as an encapsulating agent in concentrations of 30%, 40, and 50. A Mini Spray Dryer B-290 spray dryer was used for the drying process using inlet air temperatures of 120°C, 130°C, and 140°C with a feed flow of 10 mL/min. The encapsulated blackberry juices were characterized, and it was found that maltodextrin concentration and drying air temperature did have a significant effect on anthocyanin microencapsulation efficiency and process yield. It was concluded that it is possible to obtain blackberry powders by spray drying using low drying temperatures and maltodextrin concentrations lower than 30%. It was also determined that it is possible to maximize microencapsulation efficiency and process yield by lowering the drying air temperature.

4.3.1.2 LULO JUICE

Igual et al. (2014) investigated spray drying conditions for lulo pulp (*Solanum quitoense* L.) in which it was possible to favor the yield of the process and produce a stable powder with high nutritional and functional values. As process variables, they considered the inlet air temperature and the concentration of maltodextrin and gum arabic. The concentration of solutes was determined according to the response surface methodology (RSM) experimental design. The response variables considered were outlet temperature, product yield, drying ratio, productivity, water content, hygroscopicity, ascorbic acid content, vitamin C content, phenolic content, flavonoid content, and antioxidant capacity. During drying, they found that when drying lulo juice samples without adding high-molecular-weight solutes, large deposits formed on the walls of the drying chamber and the cyclone of the equipment, which led to low yields for these samples. The addition of maltodextrin and gum arabic not only improved the yield of the product but also reduced the hygroscopicity and water content of the powders obtained and contributed to the retention and increase of the nutritional and functional properties of the fruit.

4.3.1.3 TAMARIND JUICE

Muzaffar and Kumar (2015) studied the potential of soy protein isolate (SPI) in the production of tamarind pulp powder to optimize process conditions and obtain a quality powder with desirable characteristics. The responses evaluated to decide the optimum conditions were process yield, moisture content, hygroscopicity, and solubility. Drying was carried out in a laboratoryscale spray dryer of the high-flow downstream type. SPI concentrations ranged from 15 to 25%, inlet air temperature ranged from 150 to 170°C, and feed flow rate ranged from 400 to 600 mL/h. The results showed that soy protein isolate proved to be an efficient carrier agent since increasing the concentration resulted in a high process yield by decreasing the stickiness in the drying chamber. They also found that the moisture content of the powders was influenced by the concentration of SPI, the inlet air temperature, and the feed flow rate since increasing these three variables reduced the moisture content of the product. Regarding the hygroscopicity of the powders, these authors point out that the low hygroscopicity of SPI contributed to the low values obtained for this parameter when its concentration was increased. As for solubility, SPI concentration and inlet air temperature showed a positive effect on the water solubility of the powder samples since the highest solubility values were obtained at a higher SPI concentration. Finally, they concluded that the optimum spray drying conditions for obtaining tamarind pulp powders with good properties were at a SPI concentration of 25%, an inlet temperature of 170°C, and a feed flow rate of 400 mL/h.

4.3.1.4 WATERMELON JUICE

Oberoi and Sogi (2015) conducted an investigation in which they studied the effect of two drying methods, spray drying and freeze drying, as well as different concentrations of maltodextrin (3, 5, 7, and 10%) on the physicochemical properties and lycopene content of watermelon juice powder. Based on the results, they found that powders obtained by spray drying presented low moisture content, low aqueous activity, high dissolution, and lower concentrations of reducing sugars compared to powders obtained by freeze-drying. They also pointed out that the freeze-dried powder retained more pigment than the spray-dried powder; however, they presented high water activity, a short shelf life, low flowability, and a highly hygroscopic nature. Finally, they noted that the addition of maltodextrin as a drying aid was effective in reducing the stickiness of the powders and improving their physicochemical properties; however, they pointed out that adding maltodextrin in concentrations higher than 5% caused the loss of the characteristic redness of the watermelon.

Quek et al. (2007) obtained watermelon powders by spray drying using maltodextrin as an encapsulating material. They evaluated two different concentrations of maltodextrin, 3 and 5%, and four air inlet temperatures, 145, 155, 165, and 175°C. A Büchi model B-191 spray dryer was used for the drying process. Some parameters, such as moisture content, water activity, dissolution, color, lycopene, and β-carotene analysis, were determined on the atomized powders. They found that, when drying only the watermelon juice solution without drying adjuvants, the particles became sticky and adhered to the walls of the drying chamber and the cyclone of the equipment, so it was necessary to add maltodextrin, obtaining a better yield with 5% of this additive. On the other hand, the results also showed that the inlet temperature has a great influence on the physicochemical properties of the atomized powders since as the inlet temperature increased, the moisture content decreased while the reconstitution time increased. They also noted that the lightness and shade of the powders decreased when using high inlet temperatures, which could be related to the loss of lycopene and β-carotene. They concluded that the optimum drying temperature was 155°C since the powders obtained showed better colorimetric results, low moisture content, low aqueous activity, and good retention of lycopene and carotene.

4.3.1.5 GUAVA JUICE

Osorio et al. (2011) obtained microencapsulates from pink-fleshed guava fruit by spray drying. Maltodextrin, gum arabic, and a combination of the two were used as encapsulating agents. The spray drying process was carried out in a LabPlant SD-06 labscale spray dryer. The guava powders obtained were subjected to sensory analysis to select the best powders. Subsequently, the selected powders underwent physical, structural, and thermal characterization. The results showed that spherical microencapsulates were successfully obtained. They also found a strong influence of relative humidity on the stability of the powders obtained and on the release of aromas. They concluded that by obtaining an innovative and natural product, it is possible to add it to different food products due to its sensory properties.

4.3.1.6 PINEAPPLE JUICE

Abadio et al. (2004) investigated the influence of the effects of maltodextrin concentration and atomization rate on the physical properties of pineapple

(*Ananas comosus*) juice powder. They added 10 DE maltodextrin to the pineapple juice and dried it in a pilotscale spray dryer. The atomization was then sprayed in a rotary atomizer with parallel air flow. Drying was carried out at an air inlet temperature of 190°C and an outlet temperature of 90°C, with a fan speed of 25,000 rpm and a feed rate of 0.18 kg/min. The resulting powders were characterized for moisture content, color, bulk density, true density, and solubility. The results showed that by increasing the maltodextrin concentration, the moisture content decreased. They considered that this may be due to the increase of solids in the feed and a lower amount of free water for evaporation. They also point out that as the amount of maltodextrin increased, the actual density decreased, probably due to the lower moisture content in the product. As for the effect of atomization speed on this same parameter, they mention that by using a higher atomization speed, the real density was lower. Regarding the solubility of the pineapple powders, they point out that, in general, the solubility improved when the maltodextrin concentration was reduced. They emphasized that color parameters did not vary significantly with the levels tested. Finally, they recommended the use of a low atomization rate and the addition of 10% maltodextrin to obtain a product with good solubility.

Hashib et al. (2015) also pulverized pineapple juice, where they studied the effect of inlet temperature and suspension concentrations on the glass transition temperature of pineapple powder. The drying process was carried out in a laboratoryscale spray dryer, where different inlet temperatures of 130, 140, and 150°C were evaluated. Maltodextrin was added at concentrations of 15, 20, and 25%. A Labplant Spray Dryer SD-Basic model was used for the drying process. The glass transition temperature, moisture content, pigmentation, pH value, and particle morphology of the powders obtained were analyzed. The results showed that by using high inlet temperatures and increasing the maltodextrin content, the moisture content of the pineapple powder was reduced. They also noted that the drying air inlet temperature and maltodextrin concentration had effects on the glass temperature of the pineapple powders since the addition of maltodextrin to the pineapple juice caused an increase in the glass transition temperature of the pineapple juice. They concluded that increasing the maltodextrin concentration in the juice increased the yield of the process, the glass transition temperature decreased, the moisture content decreased, and the powder had a lighter color.

These investigations allow us to prove that the incorporation of the correct proportion of drying agents such as maltodextrin and gum arabic to fruit juices before the spray drying process, besides improving the process yield by reducing stickiness problems, also considerably improves the physical properties of the products and allows the preservation of the different nutritional and bioactive compounds present in the fruit juices.

4.3.2 INFLUENCE OF THE GLASS TRANSITION TEMPERATURE OF SUGARS DURING DRYING

Until now, glass transition temperature has been widely recognized as an important indicator of stickiness and agglomeration problems in fruit powders. In general, stickiness, which is considered the onset of the agglomeration phenomenon, occurs when liquid bridges are formed between adjacent powder particles by the amorphous composition of sugar, fat, or dissolved material, or by adhesion between the particles and another surface, such as the walls of the drying chamber and in the conveying system, and this stickiness occurs when the powder is exposed to temperatures above its glass transition temperature (Tg) (Foster et al., 2006). The glass transition can be defined as the passage from a disordered amorphous state to a more stable glassy state. It occurs at different temperatures, sometimes depending on each material, and to use a single value, the average value between the beginning and the end is usually used (Mosquera et al., 2014; Carolina et al., 2007; Nijhuis et al., 1998). Far below the glass transition, the molecules are locked in position, and their molecular motions are restricted to rotation and vibration, resulting in a material with stable properties (Roos, 2010). In simpler terms, the glass transition is a second-order phase transition in which a material transitions from a solid state to a rubbery state where molecules suddenly increase in mobility (Roos, 1993; Robaina et al., 2019; Linnenkugel et al., 2021). Thus, as the temperature increases above Tg, there is an amorphous solid in the gummy state; in this state, many of the physicochemical properties of the dry material change drastically, leading to collapse, sticking, caking, agglomeration, crystallization, loss of volatile compounds, darkening, and oxidation (Bhandari and Howes, 1999).

As a consequence of the viscosity, the deposition of particles on the walls of the drying sprayers becomes undesirable since if the thickness of these deposits is several millimeters, they can ignite in the presence of very high temperatures; moreover, by oxidizing, darkening, or burning, they can cause the final product to be of poor quality since these deposits can separate and mix with the product. In addition, deposits on the walls lead to lower product yield, longer cleaning times, and difficulties in powder handling (Ozmen and Langrish, 2002).

On the other hand, the glass transition temperature of an atomized powder can be used as an indicator of long-term stability. It is well-known that a small increase in the moisture content of encapsulated powders will reduce the glass transition temperature below room temperature, causing the product to become sticky; therefore, products obtained by spray drying should be kept below the glass transition temperature to improve their stability (Karrar et al., 2018).

Since the glass transition temperature is a determining factor for obtaining quality fruit powders with good physical characteristics by spray drying, the determination of the Tg of the juice to be processed before drying becomes a key point for the success of the process. Nowadays, there are several ways to determine the glass transition temperature of fruit juices, such as differential scanning colorimetry, modulated differential scanning colorimetry, differential thermal analysis, and the analytical thermogravimetric method (Hashib et al., 2015). It is useful to have a rough estimate of the glass transition temperature so that drying conditions can be established with greater certainty to reduce stickiness problems.

4.4 METHODS FOR PULVERIZING SUGAR-RICH FOODS

Fruit juices are not dried with fruit pulp solution alone due to their sticky behavior. The low glass transition temperature (Tg) of the sugars requires spray drying to be carried out at low temperatures, but this alternative is far from economical for the process, making it difficult to dry them under normal conditions (Oberoi and Sogi, 2015; Ho et al., 2017). Other common approaches are the use of a specially designed dryer with a refrigerated double-walled drying chamber and wall sweeper, cooling the powder to temperatures below which it becomes sticky, the use of dehumidified air, the proper selection of drying parameters, and, the most common today, the addition of some auxiliary drying agent to prevent or reduce powder sticking (Robaina et al., 2019; Papadakis et al., 2006).

The latter solution has become a way to achieve satisfactory drying of sugar-rich foods by increasing the glass transition temperature of the pulverized product, and in addition, the drying aid must be able to maintain the quality and solubility of the resulting powder (Bhandari et al., 1997; Linnenkugel et al., 2021). The addition of high molecular weight additives such as proteins (gelatine, whey, casein, zein, etc.), carbohydrates (starches, maltodextrins, corn syrups), and gums (gum arabic, mesquite gum, etc.) to the product prior to atomization is a widely used alternative because they have a high glass transition temperature, which consequently reduces particle stickiness and deposition on the drying walls (Santhalakshmy et al., 2015; Muzaffar and Kumar, 2015; Keshani et al., 2015).

Currently, maltodextrin and gum arabic are widely used in the encapsulation of fruit juices because they are very hydrophilic and have lower viscosity, which are key factors in spray drying, and because these two materials are so biocompatible with each other that there are even several reports on the combination of these two substances in the encapsulation of food ingredients by this drying technique (Wang et al., 2021; Ribeiro et al., 2020; Sarabandi et al., 2019). As for maltodextrins $(C_6H_{12}O_5)_n$, due to their physicochemical and functional properties, these polysaccharides are one of the most widely used encapsulating agents in the food industry due to their availability, effectiveness, and low cost (Kha et al., 2010). This drying aid is very effective because it is highly water-soluble, has low viscosity at high solid concentrations, has a low sugar content, is colorless, has a mild flavor, retards crystallization, improves drying characteristics, is available in various molecular weights, and decreases the stickiness and hygroscopicity of dried materials (Kosasih et al., 2023; Ross and Karel, 1991; Madene et al., 2006).

On the other hand, gum arabic (GA), one of the oldest and best-known natural gums, is a highly branched polymer composed of galactose, rhamnose, arabinose, and glucuronic acid units. It has excellent emulsifying, stabilizing, binding, and thickening properties and shows high solubility and low viscosity in aqueous systems, making it extremely beneficial in encapsulation by spray drying. It has an extensive use within the food industry; however, in many cases, this use is limited because gum arabic is more expensive than maltodextrin (Madene et al., 2006; Anandharamakrishnan and Ishwarya, 2015; Tiamiyu et al., 2023).

It has been shown that the addition of auxiliary agents to raise the glass transition temperature of the final powdered product and also to properly manipulate the operating variables of the spray dryer, such as air inlet temperature to the drying chamber, outlet temperature, atomization flow rate, and pressure, allow the successful production of high-quality fruit juice powders.

4.5 CONCLUDING REMARKS AND FUTURE PERSPECTIVES

Spray drying is a key technology in fruit juice processing that undoubtedly represents a viable alternative to add high value to seasonal fruits and extend their shelf life; however, there are still notable challenges in the production of powders by this technique, mainly the changes in the physical state from vitreous to gummy of the thermodynamically out-of-equilibrium components in the powders that consequently lead to their agglomeration and stickiness, factors that are the main causes of the low yields of the process and the many technical difficulties that arise. In view of this, it is important to continue to be at the forefront of the research and search for innovations that will help to overcome the limitations that continually arise when using this drying technique, which at the same time contribute to the improvement of current products and to obtaining new top-quality products that also have a high content of flavors and nutrients.

KEYWORDS

- **spray drying**
- **fruit juices**
- **preservation**
- **sugars**
- **glass transition temperature**

REFERENCES

Abadio, F. D. B.; Domingues, A. M.; Borges, S. V.; Oliveira V. M. Physical Properties of Powdered Pineapple (*Ananas comosus*) juice—effect of Malt Dextrin Concentration and Atomization Speed. *J. Food Eng.* **2004,** *64,* 285–287.

Adhikari, B.; Howes, T.; Bhandari, B. R.; Troung, V. Effect of Addition of Maltodextrin on Drying Kinetics and Stickiness of Sugar and Acid-rich Foods During Convective Drying: Experiments and Modelling. *J. Food Eng.* **2004,** *62,* 53–68.

Adnan, A.; Mushtaq, M.; Ul Islam, T. Chapter 12—Fruit Juice Concentrates. In *Fruit Juices*, Academic Press, 2018; pp 217–240.

Anandharamakrishnan, S.; Ishwarya, S. P. Chapter 6—Encapsulation of Flavors and Specialty Oils. In *Spray Drying Techniques for Food Ingredient Encapsulation*; John Wiley & Sons, Ltd, 2015; pp 127–155.

Ángeles, L. M. *Dimensionamiento y Simulación de un Secador por Aspersión de Nivel Piloto*; Tesis Maestría. Instituto Politécnico Nacional. México, D.F., 2009.

Arebo, M. A.; Feyisa, J. D.; Tafa, K. D.; Satheesh, N. Optimization of Spray-Drying Parameter for Production of Better Quality Orange Fleshed Sweet Potato (*Ipomoea batatas L.*) Powder: Selected Physiochemical, Morphological, and Structural Properties. *Heliyon,* **2023,** *9,* 1–14.

Badui, D. S. Chapter 1—Agua. In *Química de los Alimentos*; Pearson Educación: México, 2006; pp 1–28.

Bermúdez, I. D. H. Microencapsulación de jugo de Zarzamora (*Rubus fruticosus*) Mediante Secado Por Aspersión. Tesis Ingeniería Bioquímica. Instituto Tecnológico de Tuxtla Gutiérrez, Tuxtla Gutiérrez, Chiapas, 2012.

Bhandari, B. R.; Howes, T. Implication of Glass Transition for the Drying and Stability of Dried Foods. *J. Food Eng.* **1999,** *40,* 71–79.

Bhandari, B. R.; Datta, N.; Howes, T. Problems Associated with Spray Drying of Sugar-rice Foods. *Drying Technol.: An Int. J.* **1997,** *15,* 671–684.

Bhusari, S. N.; Muzaffar, K.; Kumar, P. Effect of Carrier Agents on Physical and Microstructural Properties of Spray Dried Tamarind Pulp Powder. *Powder Technol.* 2014, doi: http://dx.doi.org/10.1016/j.powtec.2014.06.038.

Braga, V.; Guidi, L. R.; De Santana, R. C.; Zotarelli, M. F. Production and Characterization of Pineapple-Mint Juice by Spray Drying. *Powder Technol.* **2020,** *375,* 409–419.

Caez, K. M. F.; Jaraba, N. C. Microencapsulación del Jugo de Mango (*Mangifera Indica L.*) Para la Obtención de un Concentrado con la Utilización de Maltodextrina Como Material Pared. Tesis Licenciatura. Universidad de Cartagena: Cartagena de Indias, 2012.

Cardona, D. P.; Patiño Arias, L. P.; Ormaza-Zapata, A. M. Aspectos Tecnológicos de la Microencapsulación de Compuestos Bioactivos en Alimentos Mediante Secado por Aspersión. *Ciencia Tecnología Agropecuaria,* **2021,** *22,* 1–22.

Carolina, B. C.; Carolina, S.; Zamora, M. C.; Jorge, C. Glass Transition Temperatures and Some Physical and Sensory Changes in Stored Spray-Dried Encapsulated Flavors. *LWT* **2007,** *40,* 1792–1797.

Daza, L. D.; Fujita, A.; Fávaro-Trindade, C. S.; Rodrigues-Ract, J. N.; Granato, D.; Genovese, M. I. Effect of Spray Drying Conditions on the Physical Properties of Cagaita (*Eugenia dysenterica DC.*) Fruit Extracts. *Food Bioprod. Process.* **2016,** *97,* 20–29.

Fang, Z.; Bhandari, B. Comparing the Efficiency of Protein and Maltodextrin on Spray Drying of Bayberry Juice. *Food Res. Int.* **2012,** *48,* 478–483.

FAO. 2020. Frutas y Verduras – Esenciales en tu Dieta. *Año Internacional de las Frutas y Verduras*; Documento de antecedentes: Roma, 2021. https://doi.org/10.4060/cb2395es

Foster, K. D.; Bronlund, J. E.; Paterson, A. H. J. Glass Transition Related Cohesion of Amorphous Sugar Powders. *J. Food Eng.* **2006,** *77,* 997–1006.

Geankoplis, C. J. Chapter 9. Drying Process Materials. In *Conveying Processes and Unit Operations*, 3 ed.; Continental S.A. de C.V.: México, 1998; pp 579–634.

Gharsallaoui, A.; Roudaut, G.; Chambin, O.; Voilley, A.; Saurel, R. (2007). Applications of Spray Drying in Microencapsulation of Food Ingredients: An Overview. *Food Res. Int.* **2007,** *40,* 1107–1121.

Goula, A. M.; Adamopoulos, K. G. A New Technique for Spray Drying Orange Juice Concentrate. *Innov. Food Sci. Emerg. Technol.* **2010,** *11,* 342–351.

Goula, A. M.; Adamopoulos, K. G. Effect of Maltodextrin Addition during Spray Drying of Tomato Pulp in Dehumidified Air: II. Powder Properties. *Drying Technol.* **2008,** *26,* 726–737.

Hashib, S. A.; Rahman, N. A.; Suzihaque, M. U. H.; Ibrahim, U. K.; Hanif, N. E. Effect of Slurry Concentration and Inlet Temperature Towards Glass Temperature of Spray Dried Pineapple Powder. *Procedia - Soc. Behav. Sci.* **2015,** *195,* 2660–2667.

Hernández-Gómez, V.; Olvera-García, O.; Guzmán-Tinajero, P.; Morillon-Gálvez, D. Secado de Frutas y Verduras Con Energía Solar. *Rev. Sist. Exp.* **2017,** *4–11,* 22–33.

Ho, T. M.; Truong, T.; Bhandari, B. Chapter 5 - Spray-Drying and Non-Equilibrium States/Glass Transition. In *Non-Equilibrium States and Glass Transition,* . Woodhead Publishing Series in Food Science, Technology and Nutrition, 2017; pp 111–136.

Igual, M.; Ramires, S.; Mosquera, L. H.; Martínez-Navarrete, N. Optimization of Spray Drying Conditions for Lulo (*Solanum quitoense L.*) Pulp. *Powder Technol.* **2014,** *256,* 233–238.

Jafari, S. M.; Arpagaus, C.; Cerqueira, M. A.; Samborska, K. Nano Spray Drying of Food Ingredients; Materials, Processing and Applications. *Trends Food Sci. Technol.* **2021,** *109,* 632–646.

Karam, M. C.; Petit, J.; Zimmer, D.; Djantou, E. B.; Scher, J. Effects of Drying and Grinding in Production of Fruit and Vegetable Powders: A Review. *J. Food Eng.* **2016,** *188,* 32–49.

Karrar, E.; Mahdi, A. A.; Sheth, S.; Ahmed, I. A. M.; Manzoor, M. F.; Wei, W.; Wang, X. Effect of Maltodextrin Combination with Gum Arabic and Whey Protein Isolate on the Microencapsulation of Gurum Seed Oil Using a Spray-Drying Method. *Int. J. Biol. Macromol.* 2018, https://doi.org/10.1016/j.ijbiomac.2020.12.045.

Keshani, S.; Daud, W. R. W.; Nourouzi, M. M.; Namvar, F.; Ghasemi, M. Spray Drying: An Overview on Wall Deposition, Process and Modeling. *J. Food Eng.* **2015,** *146,* 152–162.

Kha, T. C.; Nguyen, M. H.; Roach, P. D. Effects of Spray Drying Conditions on the Physicochemical and Antioxidant Properties of the Gac (*Momordica cochinchinensis*) fruit aril powder. *J. Food Eng.* **2010,** *98,* 385–392.

Kosasih, E. A.; Dzaky, M. I.; Zikri, A.; Rachmanudiputra, A.; Abizar, F.; Fauzi, M. B.; Suharyadi, Y. S. Microencapsulation of Maltodextrin and Gelatin Using Spray Drying with Double-Condenser Compression Refrigeration Systems. *Case Stud. Therm. Eng.* **2023,** *45,* 1–11.

Linnenkugel, S.; Paterson, A. H. J.; Huffman, L. M.; Bronlund, J. E. Prediction of the Glass Transition Temperature of Low Molecular Weight Components and Polysaccharide Mixtures. *J. Food Eng.* **2021,** *292,* 1–8.

Madene, A.; Jacquot, M.; Scher, J.; Desobry, S. Flavour Encapsulation and Controlled Release - A Review. *Int. J. Food Sci. Technol.* **2006,** *41,* 1–21.

Magaña-Barrera, S. M. J.; López-Salazar, B.; Palma-López, U.; Hidalgo-Leal, H. D. Use of Seasonal Fruits and Vegetables from the Tabasco Region, by Dehydrating the Product, Using a Solar Stove. r. *Revista de Energías Renovables.* **2019,** 3–9, 35–40.

Mishra, P.; Mishra, S.; Mahanta, C. L. Effect of Maltodextrin Concentration and Inlet Temperature During Spray Drying on Physicochemical and Antioxidant Properties of amla (*Emblica officinalis*) Juice Powder. *Food Bioprod. Process.* **2014,** *92,* 252–258.

Mosquera, L. F. G.; Giraldo, S. A.; Jiménez, D. G.; Velazco, D. M. L.; Alturo, A. O. Transición vítrea en alimentos: sistemas binarios agua-carbohidratos. *Vector*, **2014,** *9,* 21–28.

Muzaffar, K.; Kumar, P. Parameter Optimization for Spray Drying of Tamarind Pulp Using Response Surface Methodology. *Powder Technol.* **2015,** *279,* 179–184.

Nath, S.; Satpathy, G. R. A systematic Approach for Investigation of Spray Drying Processes. *Drying Technol.: An Int. J.* **1998,** *16,* 1173–1193.

Nijhuis, H. H.; Torringa, H. M.; Muresan, S.; Yuksel, D.; Leguijt, C.; Kloek, W. Approaches to Improving the Quality of Dried Fruit and Vegetables. *Trends Food Sci. Technol.* **1998,** *9,* 13–20.

Oberoi, D. P. S.; Sogi, D. S. Effect of Drying Methods and Maltodextrin Concentration on Pigment Content of Watermelon Juice Powder. *J. Food Eng.* **2015,** *165,* 172–178.

Osorio, C.; Forero, D. P.; Carriazo, J. G. Characterisation and Performance Assessment of Guava (*Psidium guajava L.*) Microencapsulates Obtained by Spray-Drying. *Food Res. Int.* **2011,** *44,* 1174–1181.

Ozmen, L.; Langrish, T. A. G. Comparison of Glass Transition Temperature and Sticky Point Temperature for Skim Milk Powder. *Drying Technol.: An Int. J.* **2002,** *20,* 1177–1192.

Papadakis, S. E.; Gardeli, C.; Tzia, C. Spray Drying of Raisin Juice Concentrate. *Drying Technol.: An Int. J.* **2006,** *24,* 173-180.

Quek, S. Y. Chok, N. K.; Swedlund, P. The Physicochemical Properties of Spray-Dried Watermelon Powders. *Chem. Eng. Process.* **2007,** *46,* 386–92.

Ramakrishnan, Y.; Adzahan N. M.; Yusof, Y. A.; Muhammad, K. Effect of Wall Materials on the Spray Drying Efficiency, Powder Properties and Stability of Bioactive Compounds in Tamarillo Juice Microencapsulation. *Powder Technol.* **2018,** *328,* 406–414.

Ré, M. I. Microencapsulation by Spray Drying. *Drying Technology*, **1998,** *16,* 1195–1236.

Reineccius, G. A. The Spray Drying of Food Flavors. *Drying Technol.: An Int. J.* **2004,** *22,* 1289–1324.

Ribeiro, A. M.; Shahgol, M.; Estevinho, B. N.; Rocha, F. Microencapsulation of Vitamin A by Spray-Drying, Using Binary and Ternary Blends of Gum Arabic, Starch and Maltodextrin. *Food Hydrocoll.* 2020.

Robaina, L. M.; Aragüez-Forte, Y.; Pino, J. A. La transición Vítrea en Relación con la Pegajosidad en el Secado Por Aspersión. *Ciencia y Tecnología de Alimentos*, **2019,** *3,* 62–69.

Roos, Y. H. Glass Transition Temperature and Its Relevance in Food Processing. *Annu. Rev. Food Sci. Technol.* **2010,** *1,* 469–496.

Roos, Y. Melting and Glass Transitions of Low Molecular Weight Carbohydrates. *Carbohydr. Res.* **1993,** *238,* 39–48.

Ross, Y.; Karel, M. Phase Transitions of Mixtures of Amorphous Polysaccharides and Sugars. *Biotechnol. Prog.* **1991,** *7,* 49–53.

Sabarez, H. Chapter 3.04 - Advanced Drying Technologies of Relevance in the Food Industry. In *Innovative Food Processing Technologies* **2021,** 64–61.

Saifullah, M.; Yusof, Y. A.; Chin, N. L.; Aziz, M. G. Physicochemical and Flow Properties of Fruit Powder and Their Effect on the Dissolution of Fast Dissolving Fruit Powder Tablets. *Powder Technol.* **2016,** *301,* 396–404.

Santhalakshmy, S.; Don Bosco, S. J.; Mallela Sabeena, S. F. Effect of inlet temperature on physicochemical properties of spray-dried jamun fruit juice powder. *Powder Technol.* **2015,** *274,* 37–43.

Sarabandi, K.; Jafari, S. M.; Mahoonak, A. S.; Mohammadi, A. Application of gum Arabic and Maltodextrin for Encapsulation of Eggplant Peel Extract as a Natural Antioxidant and Color source. *Int. J. Biol. Macromol.* **2019,** *140,* 59–68.

Serna-Cock, L.; Torres-Leon, C.; Ayala-Aponte, A. Evaluación de Polvos Alimentarios obtenidos de Cáscaras de Mango (*Mangifera indica*) como fuente de Ingredientes Funcionales. *Inform. Tecnol.* **2015,** *26,* 41–50.

Shishir, M. R. I.; Chen W. Trends of Spray Drying: A Critical Review on Drying of Fruit and Vegetable Juices. *Trends Food Sci. Technol.* **2017,** *65,* 49–67.

Siccha, M. A.; Ugaz, O. L. Secado por Atomización (Spray Dryer). *Revista de Química*, **1995,** *1,* 39–48.

Tiamiyu, Q. O.; Adebayo, S. E.; Yusuf, A. A. Gum Arabic Edible Coating and Its Application in Preservation of Fresh Fruits and Vegetables: A Review. *Food Chem. Adv.* **2023,** *2,* 1–7.

Tontul, I.; Topuz, A. Spray-Drying of Fruit and Vegetable Juices: Effect of Drying Conditions on the Product Yield and Physical Properties. *Trends Food Sci. Technol.* **2017,** *63,* 91–102.
Wang, L.; Clardy, A.; Hui, D.; Wu, Y. Physiochemical Properties of Encapsulated Bitter Melon Juice Using Spray Drying. *Bioact. Carbohydr. Diet. Fibre* **2021,** *26,* 1–7.
White, G. W.; Cakebread, S. H. The Glassy State in Certain Sugar-Containing Food Products. *J. Food Technol.* **1996,** *1,* 73–82.
Wolska, E. Fine Powder of Lipid Microparticles – Spray Drying Process Development and Optimization. *J. Drug Deliv. Sci. Technol.* **2021,** *64,* 1–12.

CHAPTER 5

Optimization of Spray-Drying Microencapsulation of Curcumin as a Strategy for Enhancing Water Solubility and Evaluation of Antioxidant Activities

MAYRA AGUILAR-ZÁRATE[1], ALMA YOLANDA VÁZQUEZ-SÁNCHEZ[2], DULCE MARÍA HERNÁNDEZ-MEDINA[3], ROMEO ROJAS[4], GUILLERMO CRISTIAN GUADALUPE MARTÍNEZ-ÁVILA[4], RICARDO GÓMEZ-GARCÍA[5], MARIELA R. MICHEL-MICHEL[3], and PEDRO AGUILAR-ZÁRATE[3*]

[1]*Faculty of Chemistry Sciences-CIEP, Universidad Autónoma de San Luis Potosí, San Luis Potosí, México*

[2]*Área Agroindustrial Alimentaria, Universidad Tecnológica de Xicotepec de Juárez, Xicotepec de Juárez, Puebla, México*

[3]*Engineering Department, Tecnológico Nacional de México/ Instituto Tecnológico de Ciudad Valles, Ciudad Valles, San Luis Potosí, México*

[4]*School of Agronomy, General Escobedo, Universidad Autónoma de Nuevo León, Monterrey, México*

[5]*Universidade Católica Portuguesa, CBQF—Centro de Biotecnologia e Química Fina—Laboratório Associado, Escola Superior de Biotecnologia, Porto, Portugal*

ABSTRACT

Curcumin is a natural pigment obtained from turmeric (*Curcuma longa* L.) rhizome, which has shown different health benefits and potential uses in

Sustainable Engineering and Agro-Food Processing: Renewable and Nonrenewable Resources.
José Juan Buenrostro-Figueroa, Julio César Tafolla-Arellano, Mónica Chávez-González, Pedro Aguilar-Zárate, A. K. Haghi (Eds.)

food and pharmaceutical products, but its application is limited by its insolubility and instability. The present work focuses on microencapsulation of curcumin by spray dryer, in order to improve its solubility and bioactivity. The process includes the use of different biopolymers such as pectin, whey protein and mannitol as wall materials. Thus, a series of curcumin microcapsules were prepared, and the whey protein presented the highest encapsulation capacity (15.77 μg of curcumin). Subsequently, the Taguchi experimental design was used to optimize the operating conditions of the spray dryer, considering the orthogonal array L9 exploring three factors with three levels: feed flow rate (3, 4.5, 6 mL/min), inlet air temperature (130, 150, 170°C), and drying air flow (30, 35, 40 m^3/h). As a result, the optimal operating conditions were a 3 mL/min of feed flow rate, 150°C of inlet air temperature, and 40 m^3/h of drying air flow and in these conditions, the encapsulation efficiency was 20.71 ± 2.83 μg/mL and the solubility 18.92 ± 2.47 μg/mL. Finally, it was evaluated the antioxidant activity of the microcapsules prepared with four different concentrations of curcumin by DPPH• radical scavenging activity and peroxidation inhibition assay. The result of the DPPH method showed a 98% ± 0.015 value which has a similar profile in all treatments, whereas a lower (20.75% ± 1.15) value was obtained of the lipoperoxidation test. This study provides practical information on the selection of wall materials and encapsulation methods for microencapsulating curcumin for various applications, particularly it would be helpful to the industrial food application.

5.1 INTRODUCTION

Curcuma longa L. is a plant which is a member of the ginger family and is widely cultivated throughout the tropical and subtropical regions of the world, mainly in Southern Asia. Particularly, curcumin is a bright yellow-orange polyphenolic pigment present in the *Curcuma longa* L. rhizomes, which exhibits a wide range of therapeutic properties, including antioxidant, anti-inflammatory, antimicrobial, antiviral, antifungal, immunoregulatory, and antitumoral activities (Andrade-Damián et al., 2023; Hewlings and Kalman, 2017; Wilken et al., 2011; Xu et al., 2018; Zorofchian Moghadamtousi et al., 2014).

Regardless of curcumin is an important permitted natural colorant used in food, it presents poor stability and low aqueous solubility. In fact, curcumin is practically insoluble in aqueous solutions, susceptible to heat, light, and extreme acidity and undergoes oxidative degradation which invalidates its use in pharmaceuticals and limits its use in food-processing industries (Nelson et al., 2017; Rafiee et al., 2019; Wang et al., 2009). Thus, curcumin microencapsulation may overcome these drawbacks by enhancing its stability and improving its solubility in aqueous media, facilitating its use as a natural pigment. Microencapsulation is a widely used technology to improve the stability and bioavailability of bioactive labile compounds (Gharsallaoui et al., 2007). Particularly, in the food-processing field, microencapsulation techniques have been widely used to protect food ingredients such as flavors, essential oils, lipids, oleoresins, and colorants against deterioration, volatile losses, or interaction with other ingredients. Besides, microencapsulation enables production of food ingredients in the form of powdered products with new properties (Desai and Park, 2005; I Ré, 1998).

Microcapsules are the product of microencapsulation and contain the bioactives in the core surrounded by a coating of wall materials, where a micro/nano environment is created, developing a physical barrier that prevents molecular diffusion and chemical reactions improving the stability of the encapsulated compounds (Taguchi et al., 2016). Various microencapsulation techniques have been development such as spray-drying, freeze-drying, spray-chilling/cooling, coacervation, fluidized bed coating, liposome entrapment, molecular inclusion extrusion, and centrifugal extrusion among others (Fang and Bhandari, 2010; Gouin, 2004). In this sense, spray-drying is the most commonly used technique owing to its many advantages such as high cost-efficiency, low cost, flexibility, reproducible, allows easy scale-up, and produces microparticles of good quality (Gharsallaoui et al., 2007; Jafari et al., 2008). Furthermore, the spray-drying has the following advantages: it does not need the incorporation of organic solvents to the process, the microparticles formed in the end of the process are in the form of a dried powder, and the temperature increases in the matrix remains limited throughout the drying process. Thus, spray-drying is especially useful for encapsulating various heat-sensitive food ingredients such as polyphenols, flavonoids, anthocyanins, prebiotics, phytosterols, and polysaccharides (Fang and Bhandari, 2010; Murugesan and Orsat, 2012; Ray et al., 2016).

Optimal spray-drying conditions must also be considered. Among the main factors that must be optimized are feed temperature, feed flow rate, and drying air flow (Augustin et al., 2001). As well, the selection of an adequate wall material is important. In this sense, encapsulating agent can be selected from a wide variety of natural and synthetic polymers. Particularly, carbohydrates, such as maltodextrins, starches, corn syrup solids, and gum Arabic, have all been widely used as encapsulating agents (Gharsallaoui et al., 2007; Jafari et al., 2008).

The microencapsulation of polyphenols, and namely of curcumin by a spray-drying process is still limited; a few studies have employed spray-drying techniques to encapsulate curcumin (Cano-Higuita et al., 2015; Coronel-Delgado et al., 2017; Sousdaleff et al., 2013), and only one research optimized microencapsulation using a Taguchi design (O'Toole et al., 2012). Based on these considerations, the aim of this work was to improve curcumin solubility and bioactivity by its microencapsulation using the spray-drying process examining the effect of different wall materials such as pectin, whey protein, and mannitol. Additionally, this study also examined the applicability of the Taguchi experimental design in order to optimize the operating conditions of the spray dryer, considering the orthogonal arrangement L9, where three factors with three levels: feed flow rate (3, 4.5, 6 mL/min), inlet air temperature (130, 150, 170°C), and drying air flow (30, 35, 40 m^3/h) were probed. Finally, we evaluated the antioxidant activity of the microcapsules prepared with four different concentrations of curcumin by DPPH• radical scavenging activity and peroxidation inhibition assay.

5.2 MATERIALS AND METHODS

5.2.1 MATERIALS

Mannitol, pectin from citrus peel, curcumin standard, 2, 2-diphenyl-1-picrylhydrazyl (DPPH), and linoleic acid were purchased from Sigma-Aldrich (St. Louis, MO, USA). Whey protein isolate (ISOPURE) was obtained from General Nutrition Centers (Pittsburg, PA, USA). Ethanol and methanol were for analytical grade (Jalmek, Mexico). The encapsulation matrixes were dried in the Mini Spray Dryer B-290 (BÜCHI, Switzerland).

5.2.2 PREPARATION OF MICROCAPSULES

To prepare the microcapsules for spray-drying, 3 matrixes (pectin, mannitol, and whey protein) were used at 1% (w/v) in 100 mL of ultrapure water. The solutions rested at (4°C) overnight to ensure complete hydration of the polymers and then were used to prepare microparticles. Two mL of a solution composed of curcumin (0.1% (w/v) in ethanol (99%)) were prepared and dispersed in each polymer solution. All solutions were prepared at room temperature and under stirring conditions. The curcumin-loaded microparticles were designed with a theoretical drug load of 0.02% (w/w) in relation to the encapsulating matrix concentration.

5.2.3 MATRIX SELECTION

To select the matrix, we performed the encapsulation efficiency and solubility studies of the different curcumin samples after microencapsulation by spray-drying under the following parameters: 10% of feed flow rate, 150°C of inlet air temperature, and a drying air flow of 40 m^3/h. Then the matrix which presented the best results in both studies was the one selected for the subsequent procedures. The amount of curcumin was quantified by spectrophotometric measurement at 425 nm using a spectrophotometer (Thermo Scientific™ GENESYS 10S). The calibration curve was established using a series of curcumin solutions in 80% ethanol (w/w) ranging from 0 to 250 ppm, which were developed to evaluate the concentration of curcumin released to the solution.

5.2.4 TAGUCHI EXPERIMENTAL DESIGN

Spray-drying parameters were optimized using a Taguchi-based statistical method, which is designed for optimizing different parameters using the fewest number of trials. Parameters to be optimized were feed flow rate, inlet air temperature, and drying air flow. Each parameter was evaluated at three different levels (Table 5.1).

TABLE 5.1 Summary of the Spray-drying Conditions for Optimization Process.

Factor	Level 1	Level 2	Level 3
Feed flow rate (mL/min)	3	4.5	6
Inlet temperature (°C)	130	150	170
Drying flow rate (m^3/h)	30	35	40

An L9 orthogonal array (three parameters, three levels, nine experiments) was used for experimental testing of the method (Table 5.2).

TABLE 5.2 Orthogonal Array for the Optimization of Curcumin Microencapsulation.

Run	Feed flow rate	Inlet air temperature	Drying flow rate
1	1	1	1
2	1	2	2
3	1	3	3
4	2	1	2
5	2	2	3
6	2	3	1
7	3	1	3
8	3	2	1
9	3	3	2

The function higher-the-better was selected as the quality characteristic in the Taguchi methodology. It is expressed as the loss function $L(y) = k * (1/y^2)$ and is represented as follows:

$$E[L(y)] = -10 * log10[(1/n) * \Sigma(1/y_i^2)] \tag{5.1}$$

The factor -10 ensures that this ratio measures the inverse of bad quality, and n is the number of samples.

For significant factors, their percentage of contribution was estimated using the next equation:

$$\text{P} = \left(\frac{SS_i}{SS_T}\right) * 100 = \left(\frac{SS_i - MS_i * df_i}{SS_T}\right) * 100 \tag{5.2}$$

where SS_i (square of sum of i factor), SS_T (total square sum), MS_i (mean square of i factor), and df_i (degrees of freedom of i factor) were obtained from ANOVA. The statistical analysis was carried out using the software Statistica 11 (Statsoft, Tulsa, OK). After analysis, the predicted optimal conditions were obtained, and the experimental validation was carried out.

5.2.5 THE SOLUBILITY OF CURCUMIN

The curcumin solubility in water was determined based on the work by Donsi et al. (2010). Ten mg of microcapsules was dissolved in 1 mL of distillated water. The mixture was heated at 25°C and stirred for 20 h. Later, the mix was centrifuged (Waverly C200) at 5000 rpm for 10 min at room temperature. The absorbance of the supernatant was measured with a UV-visible spectrophotometer (Thermo Scientific^{TM} GENESYS 10S) and the curcumin concentration (μg/mL) was calculated from a previously prepared calibration curve of curcumin.

5.2.6 *ENCAPSULATION EFFICIENCY*

Encapsulation efficiency was estimated according to Gómez-Estaca et al. (2015) with minor modifications. The curcumin loaded in the microparticles was measured by dissolving 10 mg of encapsulated curcumin in 1 mL of ethanol (w/w). The samples were vortexed and centrifuged at 8000 rpm for 5 min at room temperature and subsequently measuring the absorbance of the released curcumin with a UV-visible spectrophotometer at 425 nm and the curcumin concentration was calculated based on a previously prepared calibration curve.

5.2.7 *WHEY PROTEIN SATURATION LIMIT WITH CURCUMIN AND ANTIOXIDANT ACTIVITIES*

In order to know the saturation limit of 1 g of the encapsulating matrix, we evaluated the solubility of 4 treatments loaded with different amounts of curcumin (Table 5.3) microencapsulated by spray-drying under optimal conditions. It was also evaluated the antioxidant activities of each one of the treatments and a control (free curcumin 0.2%) as is depicted next.

TABLE 5.3 Treatments used to Evaluate the Solubility of Curcumin at Different Concentrations in the Same Encapsulating Matrix.

Treatment	Matrix, whey protein (g/100 mL)	Curcumin (mg/100 mL)
1	1	1
2	1	2
3	1	4
4	1	8

5.2.8 *DPPH• RADICAL SCAVENGING ACTIVITY*

The antioxidant activity was evaluated as the DPPH• free radical-scavenging activity according to the methodology proposed by Castro-Lopez et al. (2019) with minor modifications. Briefly, 25 μL of each extract was added to a 775 μL DPPH• methanol solution (200 μM). The reaction mixture was vortexed thoroughly and left in the dark at room temperature for 30 min. The absorbance of the mixture was measured spectrophotometrically at 517 nm. The ability to scavenge the radical was calculated by Equation (5.3):

$$DPPH\ inhibition(\%) = \left[\frac{(A\,control - A\,sample)}{A\,control}\right] * 100 \tag{5.3}$$

where *A control* is the absorbance of the control reaction (containing all reagents except the test compound) and *A sample* is the absorbance with the test compound.

5.2.9 LIPID OXIDATION INHIBITION (LOI) ASSAY

The assay was determined as described by Castro-López et al. (2019). First, the linoleic acid solution was prepared by mixing 0.6 g of linoleic acid and 1.5 g of Tween 20 in 8 mL of ethanol. Then, 50 μL of the curcumin extract was mixed with 100 μL of linoleic acid solution and 1500 μL of acetate buffer (0.02 M, pH 4). The samples were stirred in vortex and incubated (37°C for 1 min). Later, to induce the lipid oxidation, 750 μL of 50 M $FeCl_2$ solution were added and incubated (37°C for 24 h). Two aliquots of 250 μL were withdrawn during this period, at 0 and 24 h, respectively. Each aliquot was processed in the moment as follows: to stop the oxidation process, the aliquot was added to 1 mL of NaOH solution (0.1 M, in ethanol at 10%, v/v). Later, 2.5 mL of ethanol (10%, v/v) was placed to dilute the sample. Then, the absorbance was measured at 232 nm. Ethanol (10%, v/v) was used as blank. Percent inhibition of linoleic acid oxidation was determined with Equation (5.4):

$$LOI(\%) = \left[\frac{(A - B)}{A}\right] * 100 \tag{5.4}$$

where *A* is the difference between the absorbance of distilled water (as control) after 24 h and 0 h of incubation, and *B* is the difference between the absorbance of each extract sample after 24 h and 0 h of incubation.

5.2.10 SCANNING ELECTRON MICROGRAPHY

Morphological characteristics of the microcapsules containing curcumin were identified by scanning electron microscopy (SEM) using a JEOL JSM-6610LV Microscope (JEOL Inc., USA) with an acceleration voltage of 10 kV. Before imaging, each sample was sputter-coated with gold into a JEOL JFC-1100 sputter for 3 min. All the samples were processed and visualized at room temperature (~20°C). Digital images were captured using Quartz PCI imaging software v8 (Quartz Imaging Corp., Vancouver, British Columbia, Canada).

5.3 RESULTS AND DISCUSSION

5.3.1 MICROENCAPSULATION AND MATRIX SELECTION

The encapsulation efficiency with each of the polymers probed, in ethanol 80% and formic acid 5% v/v, and the solubility of the samples in water are summarized in Figure 5.1. In general, mannitol had an encapsulation efficiency of 12.18 ± 0.35 μg/mL in treatment with ethanol, which it increased to 13.18 ± 0.09 μg/mL using formic acid. With respect to the solubility, it showed the lowest value with only 1.45 ± 0.03 μg/mL. Pectin showed the highest values in encapsulation efficiency, 17.27 ± 0.15 μg/mL in ethanol and 15.71 ± 0.09 μg/mL in formic acid, but it showed a solubility value of 5.37 ± 0.37 μg/mL. Finally, the encapsulation efficiency of whey protein with formic acid resulted in 15.77 ± 0.26 μg/mL and by using ethanol an efficiency of 5.11 ± 0.06 μg/mL was obtained. However, the highest value of curcumin solubility (14.14 ± 0.49 μg/mL) was found by using this matrix.

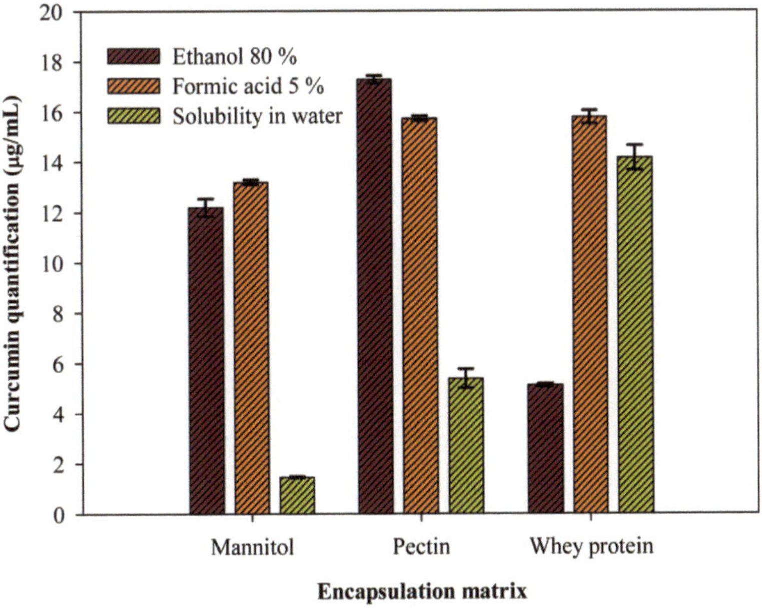

FIGURE 5.1 Matrix evaluation for the microencapsulation of curcumin based on encapsulation efficiency by releasing curcumin using ethanol 80% and formic acid 5%, and curcumin solubility in water.

Therefore, in the present work, whey protein was the better choice for curcumin microencapsulation, as was suggested in other studies. Particularly, Liu et al. (2016) proposed the enhancement of curcumin aqueous solubility when it was encapsulated with whey protein, and Mohammadian et al. (2019) presented the possible utilization of whey protein microgels in improving the applications of curcumin to produce functional food products and drugs. Another study investigating encapsulating carbohydrates indicated a better performance with arabic gum than whey protein, with a difference of 6% between them. This difference may be due to the encapsulation technique, since they encapsulated curcumin by the pH change method (Peng et al., 2020).

5.3.2 TAGUCHI EXPERIMENTAL DESIGN

Three spray-drying parameters were chosen for optimization (feed flow rate, inlet temperature, drying flow rate). The results of the nine trials for both solubility and encapsulation efficiency are shown in Table 5.4. All measurements were taken as the average of three experimental replicates. The impact of each parameter was ranked according to the degree of variance of the experimental result obtained from the analysis in the Statistica 11 software in order to obtain the combination of factors/level for the optimal conditions.

TABLE 5.4 Taguchi Experimental Design Results.

Run	Encapsulation efficiency (μg/mL)	Curcumin solubility in water (μg/mL)
1	15.55 ± 0.22	7.36 ± 0.39
2	15.50 ± 1.18	7.61 ± 0.20
3	14.54 ± 0.20	7.49 ± 0.04
4	14.83 ± 0.35	5.01 ± 0.44
5	14.93 ± 0.82	5.53 ± 0.03
6	14.77 ± 0.19	5.40 ± 0.43
7	15.22 ± 5.20	5.31 ± 0.18
8	15.00 ± 0.21	5.46 ± 1.33
9	15.31 ± 5.57	4.56 ± 0.17

5.3.3 THE SOLUBILITY OF CURCUMIN

According to the ANOVA, the feed flow rate was the factor which impacted most the solubility of curcumin, with a 94.140% contribution, while the inlet temperature and the air flow rate had values of 2.269% and 2.470%, respectively (Table 5.5).

TABLE 5.5 ANOVA with Three Factors (Feed Flow, Temperature, and Air Flow) for the Solubility of Microencapsulated Curcumin by Spray Dryer.

Factor	SS	Df	MS	F	P	Contribution (%)
Feed flow rate	10.39	2	5.19	83.94	0.01	94.14
Inlet air temperature	0.25	2	0.12	2.02	0.33	2.26
Drying flow rate	0.27	2	0.13	2.20	0.31	2.47
Error	0.12	2	0.06			1.12
Total	11.04					100

The effects of the factors at the different levels for the solubility in water of the sample are displayed in Figure 5.2A. Note that in the first factor (feed flow rate) the level 1 has the highest value compared to levels 2 and 3, considering as the optimal level of use. While in the other two factors, the 3 levels are close to the average, which indicates that there is not a significant difference among them.

According to the statistical analysis of our results, we are proposing an ideal set of conditions for encapsulating curcumin via spray-drying to maximize solubility (Table 5.6). The results obtained from the experimental tests (18.91 ± 2.47 µg/mL) were higher than the expected result (7.58 µg/mL). Additionally, this result represents the 94.55% of solubility of curcumin encapsulated, which is higher compared with the 60% of solubility obtained encapsulating it through a fibrillation process (Mohammadian et al., 2019).

5.3.4 ENCAPSULATION EFFICIENCY

The results of the optimization trials showed no differences in the effect produced by all the factors probed (Table 5.7, $p > 0.05$). In this context, we can work with any factor without affecting the expected results (Figure 5.2B), so we decided to use levels 1, 2, and 3 for the spray-drying process to perform a single experiment that allowed evaluating solubility and encapsulation efficiency. As is shown in Table 5.8, the experimental validation values obtained (20.71 ± 2.83 µg/mL) were higher than the optimal conditions expected (15.46 µg/mL). Taguchi methodology allowed finding the experimental trial that enhanced the curcumin encapsulation efficiency. It was demonstrated by encapsulating the total amount of curcumin used in the experiments.

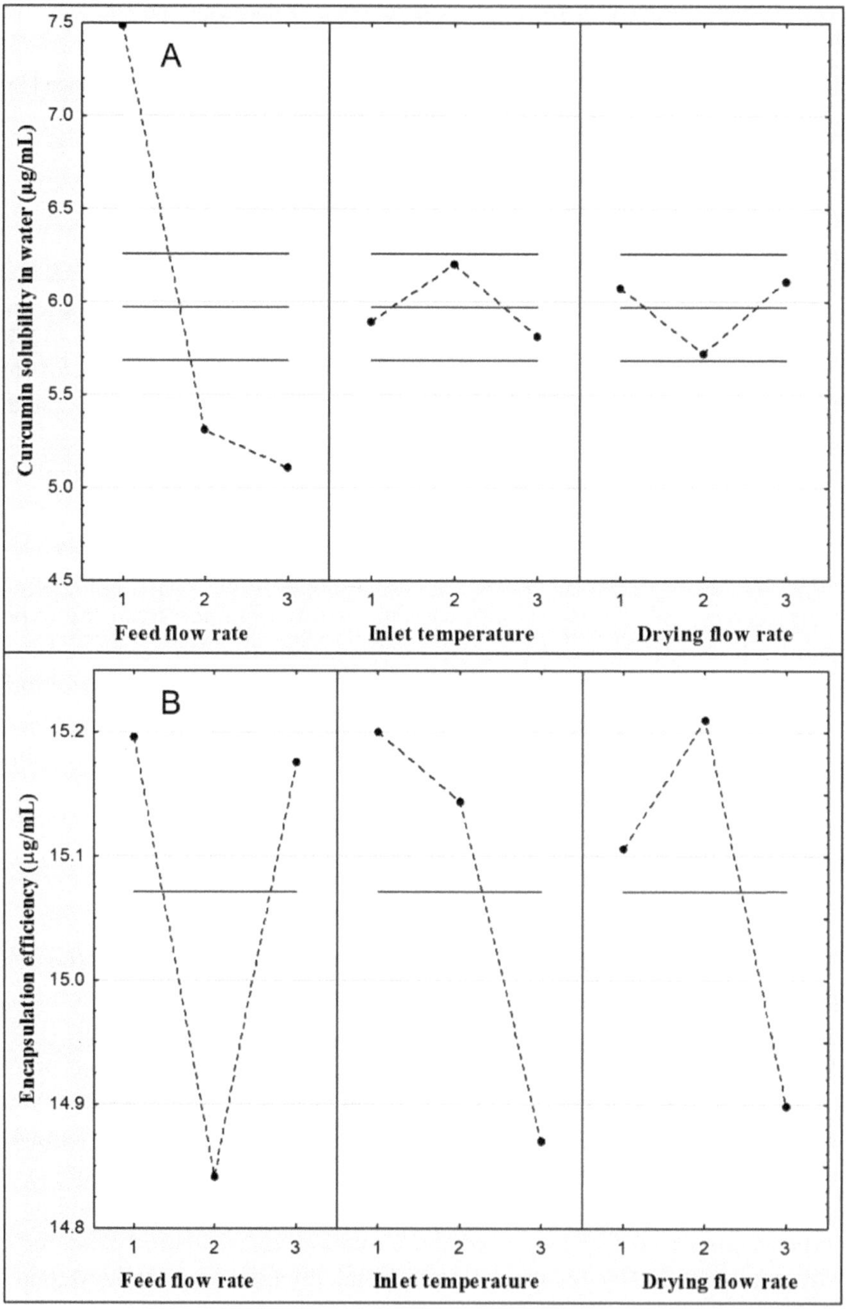

FIGURE 5.2 Performance of individual factors at different levels regarding (A) the solubility of encapsulated curcumin and (B) the encapsulation efficiency.

TABLE 5.6 Spray-drying Optimal Statistical Conditions to Improve the Curcumin Solubility.

Factor	Level	Effect size	Standard error
Feed flow rate	1	1.51	0.14
Inlet air temperature	2	0.23	0.14
Drying flow rate	3	0.14	0.14
Expected value		7.85 μg/mL	
Experimental validation		18.92 ± 2.47 μg/mL	

TABLE 5.7 ANOVA with three Factors (Feed Flow, Temperature, and Air Flow) for the Encapsulation Efficiency of Microencapsulated Curcumin by Spray Dryer.

Factor	SS	Df	MS	F	P	Contribution (%)
Feed flow rate	0.23	2	0.11	0.63	0.61	25.08
Inlet air temperature	0.18	2	0.09	0.50	0.66	19.73
Drying flow rate	0.15	2	0.07	0.40	0.71	15.84
Error	0.37	2	0.18			39.33
Total	0.94					100

TABLE 5.8 Spray-Drying Optimal Conditions in Order to Improve the Encapsulation Efficiency.

Factor	Level	Effect size	Standard error
Feed flow rate	1	0.12	0.24
Inlet air temperature	2	0.07	0.24
Drying flow rate	3	–0.17	0.24
Expected value		15.09 μg/mL	
Experimental validation		20.71 ± 2.83 μg/mL	

5.3.5 INFLUENCE OF LOADED CURCUMIN ON SOLUBILITY AND ANTIOXIDANT ACTIVITIES

The results of the water solubilization of the curcumin microencapsulated are shown in Figure 5.3, where it can see that the highest solubility was found by using 0.8% (w/w) curcumin. It is worth mentioning that theoretically, the 10 mg of microcapsules analyzed should contain 80 μg of curcumin; however, they were only quantified 31.05 ± 4.36 μg/mL of soluble curcumin. Similarly when were used 0.4% (w/w) of curcumin, from theoretically 40 μg loaded in 10 mg of microcapsules only was quantified 24.34 ± 3.55 μg/mL of soluble curcumin. The highest curcumin solubility in water was obtained by encapsulating 0.1 and 0.2% (w/w) of curcumin in 1 g of whey protein. Those treatments reached the solubilization of 8.92 ± 0.89 and 19.09 ± 1.77 μg/mL from the theoretical loading of 10 and 20 μg of curcumin in 10 mg of microcapsules, respectively. It represents the 89.20 and 95.45% of the loaded curcumin. It means that the optimal ratio of curcumin in whey protein is 0.2% w/w.

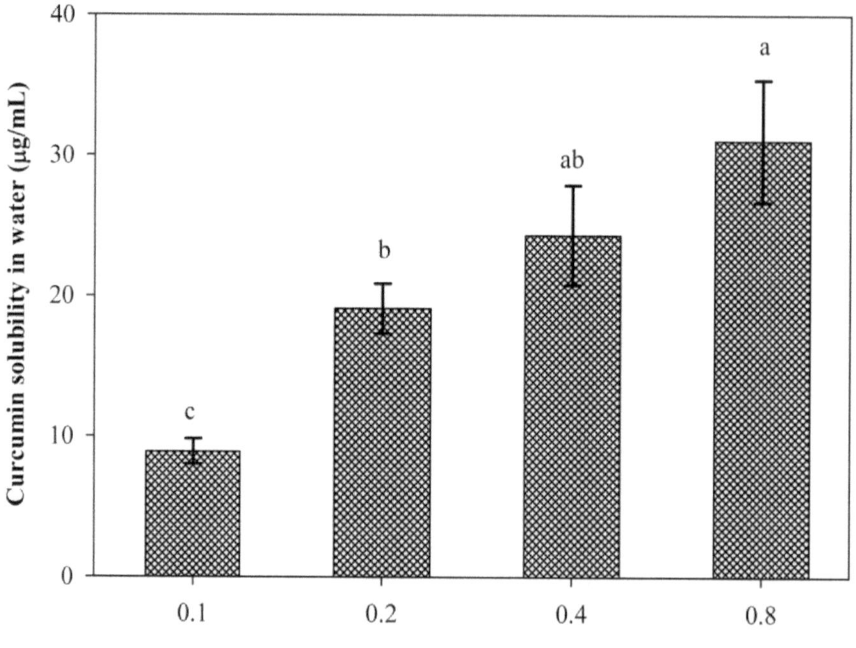

FIGURE 5.3 Solubility of curcumin microcapsules loaded with different concentrations of curcumin in 1 g of whey protein. Similar letters show no differences (Tukey test $p \leq 0.05$).

The results for the inhibition of DPPH free radical by the different concentrations (0.1%, 0.2%, 0.4%, and 0.8%) of encapsulated curcumin are shown in Figure 5.4. The antioxidant activity observed in the four encapsulated curcumin treatments showed up to 98% of DPPH free radical inhibition with no significant differences (Tukey test $p \leq 0.05$).

Notwithstanding, the free curcumin inhibited 93.11 ± 0.47% of DPPH radical. The results obtained in the present work are superior to the 38.18% of inhibition of the DPPH radical in curcumin microencapsulated with

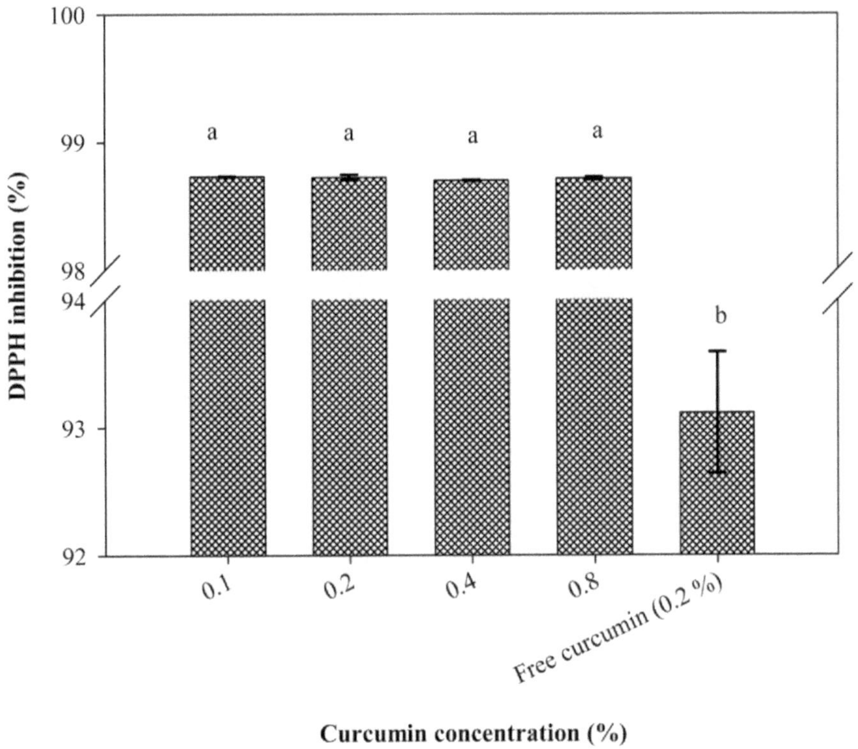

FIGURE 5.4 DPPH free radical inhibition by different concentrations of curcumin encapsulated in whey protein and free curcumin as control. Similar letters show no difference among treatments (Tukey test $p \leq 0.05$).

starch-based wall material reported by Huang et al. (2020). The results obtained with these microcapsules are also compared with the results obtained by a recent study of Hu et al. (2020) based on curcumin encapsulation which reported that the curcumin complexes loaded in whey protein showed 58% inhibition. In this case, our result of the percentage of inhibition increases (40%) significantly.

The antioxidant activity of microparticles also was evaluated by LOI. In terms of the lipid oxidation inhibition capacity in the different concentrations (0.1%, 0.2%, 0.4%, and 0.8%), the maximum average percent (22.02% ± 4.88%) was registered with the treatment 1 (0.1%), followed by treatment 4 (0.8%) with 21.41% (±3.40%) and finally treatments 2 and 3, with 19.58% (±1.29%) and 19.92% (±7.93%), respectively. However, no significant difference between them ($p > 0.05$) was obtained (Fig. 5.5). These results suggested that the capacity of encapsulated curcumin to inhibit lipids decreases compared to free curcumin (69.82 ± 0.99% of inhibition). This was expected since curcumin is high lipid soluble. Notwithstanding, the main goal of the present work was to encapsulate curcumin for its solubilization in water.

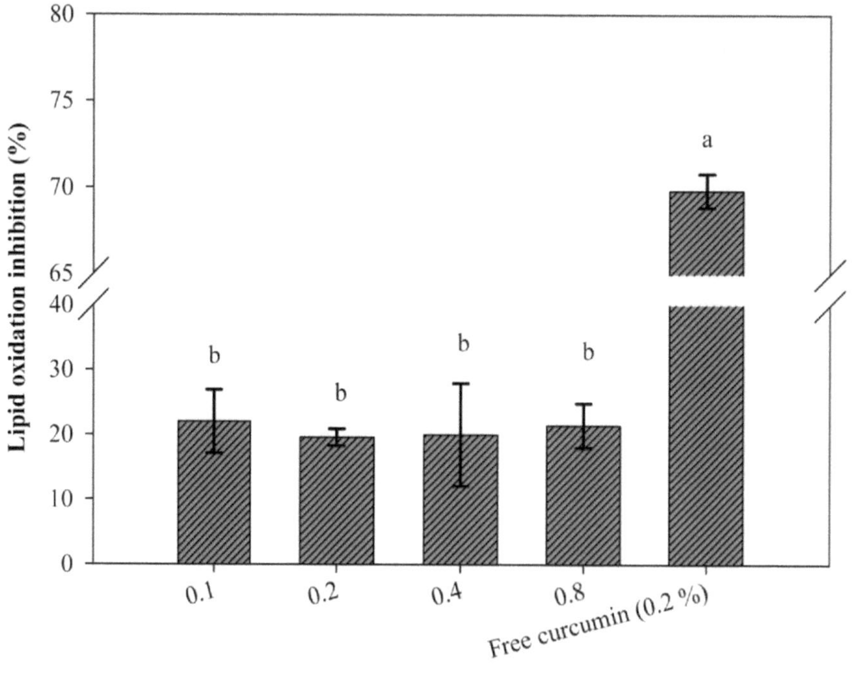

FIGURE 5.5 Lipid oxidation inhibition of curcumin microcapsules loaded with different concentrations of drug and free curcumin tested as control. Similar letters show no difference among treatments (Tukey test $p \leq 0.05$).

On the other hand, Jaguezeski et al. (2019) identified that lipid peroxidation decreased in curcumin-loaded nanocapsules, developed with Eudragit L-100 polymer. This study showed that curcumin nanocapsules potentiated the antioxidant actions (higher total antioxidant capacity and lower lipid peroxidation) of milk used for feeding sheep. In this sense, Ak and Gülçin (2008) determined a high antioxidant activity of curcumin by employing various *in vitro* antioxidant assays such as DPPH, ABTS, and DMPD radical scavenging activity.

5.3.6 MORPHOLOGICAL CHARACTERIZATION OF MICROCAPSULES

The micrographs in Figure 5.6 show the microscopic morphology of the curcumin-loaded microcapsules. A microparticulate dispersion was observed in a circular but not perfect spherical shape (Fig. 5.6a). These microcapsules seem to be porous or composed by agglomerates without fissures and sizes between 1 and 6 µm (Fig. 5.6b). As has been reported in other works, curcumin powder observed in SEM appeared as regular particles which is characteristic of its highly crystalline microstructure (da Rocha et al., 2020; Liu et al., 2016; Wu et al., 2018), and in the micrographs of Figure 5.6 this type of crystals is not seen, so we can deduce that curcumin was efficiently encapsulated into the matrix of whey protein. It is well known that encapsulation efficiency is mostly dependent on interactions between encapsulant and encapsulated compounds and on the encapsulation technique used. In addition, it has been reported that it could be improved by reducing the amount of curcumin for curcumin-saturated process (Liu et al., 2016).

FIGURE 5.6 Micrographs of curcumin-loaded microcapsules prepared with 0.2% w/w. Amplification: (A) 2500× and (B) 5000×.

5.4 CONCLUSION

The present work demonstrates that whey protein was the best wall material for curcumin encapsulated in order to improve curcumin solubility in water compared to mannitol and pectin, achieving a solubility 2.63 times greater than pectin and 9.75 times greater than mannitol. Thus, we used this polymer for the encapsulation of curcumin in subsequent experiments. The optimal conditions for curcumin encapsulation by spray-drying according to the Taguchi experimental design were determined as follows: feed flow rate of 3 mL/min, a temperature of 150°C, and an air flow rate of 40 m^3/h. Moreover, microencapsulated curcumin showed enhanced solubility (18.92 ± 2.47 μg/mL) and encapsulation efficiency (20.71 ± 2.83 μg/mL), high inhibition of the oxidation of the DPPH radical (98 and a low lipid (20.75%) inhibition than free curcumin. This study contributes to the selection of wall materials and processes for future studies on the microencapsulation of curcumin, promoting the application of curcumin in the industry, particularly, in food industry.

KEYWORDS

- **DPPH**
- **curcumin**
- **microencapsulation**
- **microencapsulation efficiency**
- **curcumin solubility**

REFERENCES

Ak, T.; Gülçin, I. Antioxidant and Radical Scavenging Properties of Curcumin. *Chem.-Biol. Interact.* **2008,** *174* (1), 27–37.

Andrade-Damián, F.; Michel, M. R.; Álvarez-Díaz, N. G.; Muñiz-Márquez, D. B.; Wong-Paz, J. E.; Veana-Hernández, F.; Aguilar-Zárate, P. Strategies for the Recovery of Phytochemicals from Turmeric (*Curcuma longa* L.) Cultivated in the Huasteca Potosina. *Biocon. Syst. Plant Physiol. Mod. Agric.* **2023,** 183–199.

Augustin, M. A.; Sanguansri, L.; Margetts, C.; Young, B. Microencapsulating Food Ingredients. *Food Australia* **2001,** *53* (6), 220–223.

Cano-Higuita, D. M.; Malacrida, C. R.; Telis, V. R. N. Stability of Curcumin Microencapsulated by Spray and Freeze Drying in Binary and Ternary Matrices of Maltodextrin, Gum Arabic and Modified Starch. *J. Food Process. Preserv.* **2015,** *39* (6), 2049–2060.

Castro-López, C.; Bautista-Hernández, I.; González-Hernández, M. D.; Martínez-Ávila, G. C. G.; Rojas, R.; Gutiérrez-Díez, A.; Medina-Herrera, N.; Aguirre-Arzola, V. E. Polyphenolic Profile and Antioxidant Activity of Leaf Purified Hydroalcoholic Extracts from Seven Mexican *Persea americana* Cultivars. *Molecules* **2019,** *24* (1), 173. https://doi.org/10.3390/molecules24010173

Coronel-Delgado, A. Y.; Ciro-Velásquez, H. J.; Restrepo-Molina, D. A. Spray Drying of Liquid Extracts of Curcumin: Process Performance and Product Quality Properties. *Ingeniería Competitividad* **2017,** *19* (1), 229–239.

da Rocha, B. A.; Francisco, C. R. L.; de Almeida, M.; Ames, F. Q.; Bona, E.; Leimann, F. V.; Goncalves, O. H.; Bersani-Amado, C. A. Antiinflammatory Activity of Carnauba Wax Microparticles Containing Curcumin. *J. Drug Deliv. Sci. Technol.* **2020,** *59,* 101918.

Desai, K. G. H.; Park, H. J. Recent Developments in Microencapsulation of Food Ingredients. In *Drying Technology,* 2005; Vol. *23* (7), pp 1361–1394. https://doi.org/10.1081/DRT-200063478

Fang, Z.; Bhandari, B. Encapsulation of Polyphenols–A Review. *Trends Food Sci. Technol.* **2010,** *21* (10), 510–523.

Gharsallaoui, A.; Roudaut, G.; Chambin, O.; Voilley, A.; Saurel, R. Applications of Spray-Drying in Microencapsulation of Food Ingredients: An Overview. *Food Res. Int.* **2007,** *40* (9), 1107–1121. https://doi.org/10.1016/j.foodres.2007.07.004

Gómez-Estaca, J.; Gavara, R.; Hernández-Muñoz, P. Encapsulation of Curcumin in Electrosprayed Gelatin Microspheres Enhances Its Bioaccessibility and Widens its uses in Food Applications. *Innov. Food Sci. Emerg. Technol.* **2015,** *29,* 302–307. https://doi.org/10.1016/j.ifset.2015.03.004

Gouin, S. Microencapsulation: Industrial Appraisal of Existing Technologies and Trends. *Trends Food Sci. Technol.* **2004,** *15* (7–8), 330–347.

Hewlings, S. J.; Kalman, D. S. Curcumin: A Review of Its Effects on Human Health. *Foods* **2017,** *6* (10), 92.

Hu, Y.; He, C.; Jiang, C.; Liao, Y.; Xiong, H.; Zhao, Q. Complexation with Whey Protein Fibrils and Chitosan: A Potential Vehicle for Curcumin with Improved Aqueous Dispersion Stability and Enhanced Antioxidant Activity. *Food Hydrocoll.* **2020,** *104,* 105729.

Huang, C.; Wang, S.; Yang, H. Evaluation of Thermal Effects on the Bioactivity of Curcumin Microencapsulated with Porous Starch-Based Wall Material Using Spray Drying. *Processes* **2020,** *8* (2), 172.

I Ré, M. Microencapsulation by Spray Drying. *Dry. Technol.* **1998,** *16* (6), 1195–1236.

Jafari, S. M.; Assadpoor, E.; He, Y.; Bhandari, B. Encapsulation Efficiency of Food Flavours and Oils During Spray Drying. *Dry. Technol.* **2008,** *26* (7), 816–835.

Jaguezeski, A. M.; Gündel, S. S.; Favarin, F. R.; Gündel, A.; Souza, C. F.; Baldissera, M. D.; Cazarotto, C. C.; Volpato, A.; Fortuoso, B. F.; Ourique, A. F. Low-Dose Curcumin-Loaded Eudragit L-100-Nanocapsules in the Diet of Dairy Sheep Increases Antioxidant Levels and Reduces Lipid Peroxidation in Milk. *J. Food Biochem.* **2019,** *43* (8), e12942.

Liu, W.; Chen, X. D.; Cheng, Z.; Selomulya, C. On Enhancing the Solubility of Curcumin by Microencapsulation in Whey Protein Isolate via Spray Drying. *J. Food Eng.* **2016,** *169,* 189–195. https://doi.org/10.1016/j.jfoodeng.2015.08.034

Mohammadian, M.; Salami, M.; Momen, S.; Alavi, F.; Emam-Djomeh, Z.; Moosavi-Movahedi, A. A. Enhancing the Aqueous Solubility of Curcumin at Acidic Condition Through the Complexation with whey Protein Nanofibrils. *Food Hydrocoll.* **2019,** *87,* 902–914. https://doi.org/10.1016/j.foodhyd.2018.09.001

Murugesan, R.; Orsat, V. Spray Drying for the Production of Nutraceutical Ingredients—A Review. *Food Bioprocess Technol.* **2012,** *5,* 3–14.

Nelson, K. M.; Dahlin, J. L.; Bisson, J.; Graham, J.; Pauli, G. F.; Walters, M. A. The Essential Medicinal Chemistry of Curcumin: Miniperspective. *J. Med. Chem.* **2017,** *60* (5), 1620–1637.

O'Toole, M. G.; Henderson, R. M.; Soucy, P. A.; Fasciotto, B. H.; Hoblitzell, P. J.; Keynton, R. S.; Ehringer, W. D.; Gobin, A. S. Curcumin Encapsulation in Submicrometer Spray-Dried Chitosan/Tween 20 Particles. *Biomacromolecules* **2012,** *13* (8), 2309–2314.

Peng, S.; Zhou, L.; Cai, Q.; Zou, L.; Liu, C.; Liu, W.; McClements, D. J. Utilization of Biopolymers to Stabilize Curcumin Nanoparticles Prepared by the pH-Shift Method: Caseinate, Whey Protein, Soy Protein and Gum Arabic. *Food Hydrocoll.* **2020,** *107,* 105963.

Rafiee, Z.; Nejatian, M.; Daeihamed, M.; Jafari, S. M. Application of Curcumin-Loaded Nanocarriers for Food, Drug and Cosmetic Purposes. *Trends Food Sci. Technol.* **2019,** *88,* 445–458. https://doi.org/10.1016/j.tifs.2019.04.017

Ray, S.; Raychaudhuri, U.; Chakraborty, R. An overview of Encapsulation of Active Compounds used in Food Products by Drying Technology. *Food Biosci.* **2016,** *13,* 76–83.

Sousdaleff, M.; Baesso, M. L.; Neto, A. M.; Nogueira, A. C.; Marcolino, V. A.; Matioli, G. Microencapsulation by Freeze-Drying of Potassium Norbixinate and Curcumin with Maltodextrin: Stability, Solubility, and Food Application. *J. Agric. Food Chem.* **2013,** *61* (4), 955–965.

Taguchi, Y.; Saito, N.; Uchida, A.; Tanaka, M. Preparation of Thermosensitive Microcapsules Containing Water Soluble Powder by Melting Dispersion Cooling Method. *J. Encapsul. Adsorp. Sci.* **2016,** *6* (03), 57.

Wang, Y.; Lu, Z.; Lv, F.; Bie, X. Study on Microencapsulation of Curcumin Pigments by Spray Drying. *Eur. Food Res. Technol.* **2009,** *229* (3), 391–396. https://doi.org/10.1007/S00217-009-1064-6

Wilken, R.; Veena, M. S.; Wang, M. B.; Srivatsan, E. S. Curcumin: A Review of Anti-Cancer Properties and Therapeutic Activity in Head and Neck Squamous Cell Carcinoma. *Mol. Cancer* **2011,** *10* (1), 1–19.

Wu, J.; Wang, J.; Zhang, J.; Zheng, Z.; Kaplan, D. L.; Li, G.; Wang, X. Oral Delivery of Curcumin Using Silk Nano-And Microparticles. *ACS Biomater. Sci. Eng.* **2018,** *4* (11), 3885–3894.

Xu, X. Y.; Meng, X.; Li, S.; Gan, R. Y.; Li, Y.; Li, H. Bin. Bioactivity, Health Benefits, and Related Molecular Mechanisms of Curcumin: Current progress, challenges, and perspectives. In *Nutrients*, MDPI AG, 2018; *10* (10), p. 1553. https://doi.org/10.3390/nu10101553

Zorofchian Moghadamtousi, S.; Abdul Kadir, H.; Hassandarvish, P.; Tajik, H.; Abubakar, S.; Zandi, K. A Review on Antibacterial, Antiviral, and Antifungal Activity of Curcumin. *BioMed Res. Int.* **2014,** *2014.*

PART 3

Sustainable Technologies and Engineering

CHAPTER 6

Potentialities from Neglected and Underutilized Species for Sustainable Fermented Beverages Production

FLOR EDITH CALDERÓN-MARTÍNEZ[1],
JORGE LUIS GONZÁLEZ-ESCOBAR[2], RUBÉN F. GONZÁLEZ-LAREDO[1],
ANAHÍ JOBETH BORRÁS-ENRIQUEZ[3], PEDRO AGUILAR-ZÁRATE[2], and
FABIOLA VEANA[2*]

[1]*Tecnológico Nacional de México/Instituto Tecnológico de Durango, Nueva Vizcaya, Durango, México*

[2]*Tecnológico Nacional de México/Instituto Tecnológico de Ciudad Valles, Vista Hermosa, Ciudad Valles, San Luis Potosí, México*

[3]*División de Ingeniería en Industrias Alimentarias, Tecnológico Nacional de México/ TES San Felipe del Progreso, Ejido de SFP, Estado de México, México*

ABSTRACT

Overpopulation, the depletion and exhaustion of natural resources, and food demand have led to the search for sustainability in food systems, not only with a focus on caring for the environment, but also to provide the nutrition that human beings require and, above all, health through functional foods and beverages. For years, fermented beverages have been traditionally produced in Mexico from plant species, among which the following can be found—*Ananas comosus sativus* (tepache and garapiña), *Opuntia streptacantha* (colonche), *Saccharum officinarum* (guarapo), *Acrocomia aculeata* (taberna), *Bromelia*'s species (huapilla

Sustainable Engineering and Agro-Food Processing: Renewable and Nonrenewable Resources.
José Juan Buenrostro-Figueroa, Julio César Tafolla-Arellano, Mónica Chávez-González,
Pedro Aguilar-Zárate, A. K. Haghi (Eds.)

water), *Zea mays* (pozol, tesgüino, sendechó), and *Agave*´s especies, among others. These beverages are considered functional because the presence of probiotic microorganisms and bioactive compounds such as anti-inflammatory, antioxidants, peptides, anti-diabetic, gastro-, neuro-, and cardio-protective, to mention a few, have been evidenced. However, there are plant species such as non-commercial crops that are underutilized or are not appreciated by producers and consumers due to various factors (economic, social, cultural, agronomic, and genetic). These underutilized crops have the potential to be considered in the production of fermented beverages (alcoholics or non-alcoholics) due to the bioactive compounds they present. Among the plant species we can find *Artocarpus*, *Parmentiera aculeata*, *Myrciaria floribunda*, *Solanum betaceum*, *Theobroma bicolor*, *Byrsonima crassifolia*, *Annona purpurea*, and *Cucurbita ficifolia*. More research is needed to revalorize these crops in search of their functional properties to bring healthier products to the consumer.

6.1 INTRODUCTION

The global population projection for the year 2050 is around 10 billion people, and 70% will be living in urban areas (FAO, 2022). These overpopulations will demand high figures of food consumption. However, food waste and loss (FWL) represent the major problems in the world, where 14 and 17%, respectively, are their present rates. In this sense, the United Nations suggests the Sustainable Development Goal (SDG) 12 aims to "ensure sustainable consumption and production patterns" to decrease the FWL and the contribution to other SDGs associated with environmental sustainability and food security (Pasarín and Viinikainen, 2022). Actually, the evaluation of sustainability in food systems is conducted through life cycle assessment (LCA), which is important for quantifying the complex environment interactions in the agri-food system (Cabot et al., 2022). Inherently, circular economy (CE) is associated to chain supply in beverages production for reducing the consumption and waste. A recommended solution is the sustainable beverages, which involve during their production an energy-efficient and low-carb emission (Carley and Yahng, 2018). In addition, sustainable consumption is the major promise for environmental care, because a growth in sustainable behaviors is presented by consumers. Inclusive, they are capable and willing to pay an additional amount for these products (Carley and Yahng, 2018).

However, food systems are limited in vegetable species consumption, since corn, wheat, and rice crops represent around 50% of the caloric intake in diet. In addition, an indifference to the diversity of many nutrient-rich plant species remains in humanity (Hunter et al., 2019). In this sense, it is well known that Mexican biodiversity is wild, semi-domesticated, and domesticated species, where around 2168 have been reported as edible and can be grouped in traditional vegetables, edible flowers, and/or fruits of agave and cacti, and fruits of trees and shrubs (Ulian et al., 2021). The usage of these species have been the development of traditional Mexican fermented beverages (TMFB), such as *Ananas comosus sativus* (tepache and garapiña) (Robledo-Márquez et al., 2021); *Opuntia streptacantha* (colonche) (Cerero-Calvo et al., 2022); *Saccharum officinarum* (guarapo) (Robledo-Márquez et al., 2021); *Acrocomia aculeata* (taberna) (Robledo-Márquez et al., 2021; Romero-Luna et al., 2017); *Bromelia*'s species (huapilla water) (Robledo-Márquez et al., 2021); *Zea mays* (pozol, tesgüino, sendechó) (Robledo-Márquez et al., 2021; Cerero-Calvo et al., 2022); *Agave*´s species (Robledo-Márquez et al., 2021; Cerero-Calvo et al., 2022; Barrios-Gamboa and Pérez-Reyes, 2019), among others. These fermented beverages are considered functional drinks, because principal biological activities, such as anti-inflammatory, antioxidant, anti-diabetic, gastro-, neuro-, and cardio-protective activities, have been evidenced in them (Romero-Luna et al., 2017; Pérez-Armendáriz and Cardoso-Ugarte, 2020). However, other "neglected and underutilized species" (NUS), as well as wild edible plant species and orphan crops, are nutritionally important but have not been appreciated and have the potential for development of fermented beverages, such as *Artocarpus* (Muñiz-Márquez et al., 2019), *Parmentiera aculeate* (Castillo-Ruíz et al., 2022), *Myrciaria floribunda* (Cruz Silva et al., 2020), *Solanum betaceum* (Ramírez and Kallarackal, 2019), *Theobroma bicolor* (Gálvez-Marroquín et al., 2016), *Byrsonima crassifolia* (Agredano-de la Garza et al., 2021), *Annona purpurea* (Vidal-Lezama et al., 2019), *Cucurbita ficifolia* (Moya-Hernández et al., 2020; Barrón-Álvarez et al., 2022), among others. More research is needed to revalorize these crops in search of their functional properties to bring healthier products to the consumer. Specifically, *Artocarpus altilis fosberg* has been used for fermented beverages using *Sacharomyces ellpisoideus* and inuline (Herrera Escobar and Salvador Albores, 2019). Additionally, *Parmentiera aculeata* has been used for distilled alcoholic fermented beverages (Castillo-Ruíz et al., 2022) and has been used as the only carbon source and fermented with acid lactic bacteria as a possible inclusion in the functional beverages thematic (Lara-Cervantes et al., 2022). Other important NUS used for beverage production are *Solanum betaceum*, which has been used for an alcoholic beverage in combination with fruit of *Physalis peruviana* and two yeasts (*Saccharomyces cerevisiae* and *Saccharomyces pastorianus*) (Mejía-Bustamante et al., 2022), and, *Theobroma bicolor*, which are consumed in infusions (Baldera Ocampo et al., 2021). In addition, it should include "tejacate" preparation, a beverage originally from Oaxaca, Mexico, similar to the "pozol" beverage (Pérez-Armendáriz and Cardoso-Ugarte, 2020). Furthermore, the important NUS showing biological activities, such as antimicrobial and antioxidant activities, are *Myrciaria floribunda* (Cruz Silva et al., 2020; de Azevedo et al., 2019), *Annona purpurea* (Anaya-Esparza et al., 2020), and *Byrsonima crassifolia* (Agredano-de la Garza et al., 2021). The latter species also possesses hypoglycemic and anti-tumoral activities (Agredano-de la Garza et al., 2021; Hernández-López et al., 2022). In this sense, this chapter exposes the potentialities of NUS, their biological activities, and their role as alternative substrates for the making of sustainable fermented beverages.

6.2 GENERALITIES OF NUS IN MEXICO

6.2.1 ARTOCARPUS

Artocarpus is a plant genus belonging to the *Moraceae* family. Regarding its origin, it seems unclear because some report being native to Indonesia and others to India. However, most species diversity is mainly in Asia's tropical and subtropical regions, including Bangladesh, Burma, Indonesia, China, and Sri Lanka, as well as parts of Africa and Australia. They are also cultivated in Central and South America like Mexico and Brazil (Hari et al., 2014). The trees are monoecious, that is, male and female inflorescences are found on the same tree (Jagtap and Bapat, 2010). They can measure 21 m at full maturity, although it is more common to be between 12 and 15 m. Some of the best-known species of this fruit are *A. heterophyllus* (jackfruit), *A. altilis* (breadfruit), *A. hirsutus* (wild jack), and *A. lakoocha* (monkey jack) (Baliga et al., 2021).

The jackfruit is typically 30–40 cm in length and 4.5–30 kg in weight (Goswami and Chacrabati, 2015). Inside it is yellow-to-orange with a distinctive sweet and fruity aroma. Its juice is slightly acidic and deeply sweet, with a flavor reminiscent of a mixture of mango and orange and other fruit flavors such as banana, apple, guanabana, papaya, and pineapple (Baliga et al., 2021). Particularly in jackfruit, components such as carotenoids, phenolic compounds (flavonoids, stilbenoids, and arylbenzofurons), tannins, volatile acids sterols, among others, have been determined (Jagtap and Bapat,

2010; Baliga et al., 2021). Through a proximal chemical analysis, it was determined that jackfruit contains carbohydrate (18.9%), protein (1.9%), fat (0.1%), moisture (77%), fiber (1.1%), and minerals (0.8%). It has also reported the presence of other important components such as calcium (20 mg), phosphorus (30 mg), iron (500 mg), vitamin A (540 IU), and thiamin (30 mg) (Jagtap and Bapat, 2010). Regarding volatile compounds, it has propyl isovalerate, 3-methylbutyl acetate, 1-butanol, isopentyl isovalerate, butyl acetate, ethyl isovalerate, butyl isovalerate, and 2-methylbutyl acetate (Jagtap and Bapat, 2010).

FIGURE 6.1 Jackfruit.
Source: Adapted from Morelos-Flores et al. (2022).

6.2.2 *PARMENTIERA ACULEATA*

It is a plant native to Mexico that is distributed from all over the Mexican territory to Honduras. It thrives in savannahs, low deciduous forest, medium forest, and high evergreen forests. It is present, especially, in warm, semi-warm, and temperate climates (Angón-Galván, 2006). The tree can measure up to 15 m in height; however, its average height ranges from 4 to 9 m (Pérez-Morales et al., 2021). This plant has a thick and woody trunk with numerous branches. The leaves are divided into five light green leaflets, the support that joins them to the stem has wings and at the base of each leaf, there are two spines. The flowers are solitary or in groups, greenish cream in color, and grow directly from the trunk or at the ends of the branches (Angón-Galván, 2006; Pérez-Morales et al., 2021).

The *P. aculeata* fruit is mainly used as fodder for livestock; however, its human consumption as a key ingredient in typical dishes of the Yucatan gastronomy has been reported (Pérez-Morales et al., 2021; Dominguez Orta and Herrera Martínez, 2018). It has various common names such as cuajilote, chote, guetoxiga, and others. It is 20–30 cm long and 6.5 cm in diameter and shaped like a long cucumber with prominent ribs, with average weight ranging between 130 and 200 g (Dominguez Orta and Herrera Martínez, 2018; Álvarez-Olivera et al., 2010; Juárez-Martínez et al., 2019). It is yellowish-green outside; when it reaches maturity, it turns brown and reddish. The pulp is fibrous, whitish in color, with numerous black seeds and the juice of the fruit is sweet like sugar cane (Pérez-Morales et al., 2021). Regarding its proximal chemical composition, it has reported a content of moisture (83.8–89.67%), carbohydrate (3.03–11.58%), protein (0.33–8.15%), fat (0.03–6.40%), fiber (3.9–18.36%), and minerals (3.5–4.0 %) (Angón-Galván, 2006; Pérez-Morales et al., 2021; Juárez-Martínez et al., 2019; Villar-Herrera, 2011). Presence of phenolic compounds and fatty acids such as palmitic, oleic, linoleic, and stearic have also been reported (Pérez-Morales et al., 2021). Several strategies have been sought for the valorization of this fruit, such as its incorporation into some products for the preparation of tortillas, production of fermented beverages, and making of sweeting syrups (Castillo-Ruíz et al., 2022; Dominguez Orta and Herrera Martínez, 2018).

FIGURE 6.2 Chote fruit.

***Source*:** Adapted from García Rojas et al.

6.2.3 *MYRCIARIA FLORIBUNDA*

Myrciaria floribunda (H. West ex Willd.) O. Berg belongs to the *Myrtaceae* family. It is found worldwide but develops chiefly in tropical and subtropical regions, such as Central and South America (Cruz Silva et al., 2020). It has been found abundantly in the Atlantic Forest biome in Brazil (Souza and Morim, 2008). It is a tree that grows up to 15 m in height with glabrous, brown branches; the leaves are oblong-elliptical, 6.0–12.0 cm long, 2.5–3.5 cm wide, and white and pink in color (López et al., 2006). It is also known as cabo de chivo, escobillo, guayabillo, mije, mijo, mizto, murta, and guavaberry (Cruz Silva et al., 2020).

Fruits are tiny globular berries 11–13 mm in diameter and have an average weight of 0.76 g. It can of different colors depending on plant or ripened stage, such as yellow, orange, red, or purple (Souza and Morim, 2008; García et al., 2021). The fruits are used for human consumption, either in beverages (liqueurs, juices, soft drinks), jams (jellies and ice creams), or in frozen form. They are commonly known as rumberry, camboim, jabuticabinha, myrtle, duke, goiabarana, araçazeiro, and cambuí (Cruz Silva et al., 2020). The ripe fruit is characterized by being fleshy, succulent, and juicy with a sweet, astringent flavor and an intense citrus aroma. It has been reported that these berries contain a high content of sugar and bioactive

FIGURE 6.3 Guavaberry fruit.

Source: Adapted from Castro et al. (2018) and Oliveira et al. (2021).

compounds such as carotenoids, flavonoids, and phenolic acids (Santos et al., 2017). According to the proximal chemical analysis on a dry basis, the fruit is composed of ashes (1.89%), lipids (2.43%), protein (4.78%), and total carbohydrates (90.89%) (de Oliveira et al., 2018). It has also been reported that the fruit contains important components such as carotenoids, flavonoids (rutin, myricitrin, quercitrin, and catechin), gallic, and ellagic acid de (García et al., 2021; de Oliveira et al., 2018). Regarding the major volatile compounds, β-Z-ocimene (50.8%), 1,8-cineol (3.14%), γ-terpinen (2.51%), and β-caryophyllene (1.16%) have been identified to be present (de Oliveira et al., 2018).

6.2.4 SOLANUM BETACEUM

The tomatillo or tamarillo (*Solanum betaceum* Cav. or *Cyphomandra betacea*) is a small perennial tree species that reaches a height of 2–4 m. It is native mainly to South America (Andean regions of Colombia, Peru, Chile, Ecuador, Bolivia, and Brazil). However, it has been widely spread to other areas worldwide, including New Zealand, Australia, Southeast Asia, Italy, Jamaica, Haiti, Mexico, Puerto Rico, Costa Rica, East Africa, and Southern Europe (Ramírez and Kallarackal, 2019). This tree has important adaptability characteristics so that it can be grown at high altitudes from 2000 to 2800 m and at low altitudes from 100 to 1500 m, with average temperatures between 13 and 24°C (Feicán-Mejía et al., 2016). Tomatillo is also commonly known as the tree tomato. It is considered an economically significant crop because it has increased its consumption in recent years in many countries in America and Europe. Colombia is one of the leading commercial producers; it harvests about 120,000 tons of tree tomato fruits yearly on 6500 ha (Ramírez and Kallarackal, 2019).

The fruits are egg-shaped spheres ranging from 4 to 10 cm long to 3–5 cm in diameter with a weight between 30 and 160 g. It can have different colors (yellow, orange, red, and purple) and its flavor is characterized by acetous, sour, astringent, or sweet. Maturity is reached between 21 and 27 weeks from the flowering stage (Ramírez and Kallarackal, 2019). Regarding chemical composition, the whole fruit is characterized by moisture (86 −92%), carbohydrates (7.0%), proteins (1.4–2.5%), lipids (0.1–0.7%), ash (0.6–0.7%), and fibers (1.1%) (Feicán-Mejía et al., 2016; Wang and Zhu, 2020). Other important components have also been identified, including oleic acid, linoleic acid, squalene, sterol, β-carotene, and phenolic acids and their derivatives (Wang and Zhu, 2020).

FIGURE 6.4 Tomatillo fruit.

Source: Adapted from Feicán-Mejía et al. (2016); Ramírez and Kallarackal (2019); Wang and Zhu (2020).

6.2.5 *THEOBROMA BICOLOR*

This plant belongs to the *Malvaceae* family; it is a close relative to *Theobroma cacao* L which is widely known and traded in the chocolate industry. *Theobroma bicolor* is native to Central and South America; its distribution extends from southern Mexico to the Peruvian and Brazilian Amazon (Lim, 2012). The tree and its fruit are known by different regional names—balamte (jaguar tree), pataxte, white cocoa, malacayo cocoa, wild cocoa, macambo, bacao, and patashte (Kufer and McNeil, 2006). This plant is associated with cacao plantations or in backyards since there are no commercial plantations. One of the most common uses is as a shade tree for developing other species, including *T. cacao*. Regarding flowers, they have five light red lanceolate sepals and five bright red sac-like petals. The ligules connected to the end of each petal have an elongated rounded shape, and they are dark red like the staminodes. The ovaries and the style are light green (Ponce-Sánchez et al., 2021).

The *T. bicolor* fruits are more prominent than *T. cacao*; they measure 25–35 cm long by 12–15 cm wide and can weigh between 0.5 and 3.0 kg. The shell is woody, complex, wrinkled, and with different color degrees from light green to brown. The parts of the fruit are represented by peel (23.8%), pulp (62.5%), and seeds (13.7%). The tree produces approximately 14–40 fruits per harvest (Furlán and Bressani, 1999). The pulp and grains are used locally to make sweets (marzipan, nougat, and marshmallow), chocolate (mixture of *T. bicolor* and *T. cacao* seeds), traditional beverages (atole, pozol, polvillo, and popo), and they are rarely eaten as fresh fruit (Gálvez-Marroquín et al., 2016). In this context, more dissemination and importance have been given to the seeds than to the pulp of the fruit. Regarding proximal chemical composition, the seed presents moisture (20.5%), carbohydrates (34.6%), proteins (15.4%), lipids (24.2%), ash (2.2%), and fibers (3.1%) (Tinajero-Carrizalez et al., 2021). Again, the pulp juice is sweet (12.2°Brix) and is composed of sugars such as glucose (26 mg/g DW), sucrose (77 mg/g DW), and fructose (18 mg/g DW); it has a nearly neutral pH 6.65 and a protein content of 5.1 mg/g DW (Gómez-Ubiaño and Díaz-Díaz, 2018). Some important components have been reported to be present such as fiber (3.29%), ash (6.10%), and phenolic compounds (161.82 (mgGAEa/100 g fw) (González et al., 2016).

FIGURE 6.5 *Theobroma bicolor* fruit.

***Source*:** Adapted from Gálvez-Marroquín et al. (20116) and Ponce-Sánchez et al. (2021).

6.2.6 *BYRSONIMA CRASSIFOLIA*

Byrsonima crassifolia belongs to the *Malpighiaceae* family. It is widely distributed throughout South America, wherein it is known as nanche, nanchi, changunga, nance, murici, etc. (Hernández-López et al., 2022; Maldonado Peralta et al., 2020). It is an evergreen and deciduous shrub tree in dry forests, measuring between 3 and 15 m in height. Its cup is vast, open, and irregular. The tree trunk shows frequently ascending branches from the ground. Their leaves are elongated in sheets, 5–15 cm long by 2–7.5 cm wide (Hernández-López et al., 2022; Vázquez Yanes et al., 1999). They have elliptical shapes with an entire margin and are dark green and greyish-yellowish green. The flowers are actinomorphic, in racemes or narrow terminal panicles 5–15 cm long, and pubescent. They are yellow-reddish in color, 1.5 cm in diameter, with five rounded petals (Hernández-López et al., 2022; Vázquez Yanes et al., 1999).

The fruits are traditionally grown in the wild or in the backyard in several of the tropical countries of The America during May to October (Maldonado Peralta et al., 2020). Its distribution and sale tend to be restricted to the local market. The ripe fruit is traditionally consumed raw, and also in the form of juices, yogurt, jellies, jams, gruels, sweets, desserts, syrups, ice creams, fresh beverages, dehydrated, pickled nanche with sugar water and liquor,

FIGURE 6.6 Nanche fruit.

***Source*:** Adapted from Maldonado et al. (2020).

and mistelas (nanche liquor) (Medina Torres et al., 2017; Hernández-López et al., 2022). These are small drupes from yellow to red in color depending on their maturation state with a strong aroma; the fruit measures 1–2 cm in diameter and weighs an average of 4.65 g; the fresh pulp weight is 4.16 g (Maldonado Peralta et al., 2020). The pulp is soft, pasty, and sweet, with a fruity aroma like rotten cheese. In terms of its proximal composition on dry matter, a content of lipids (14.14%), proteins (6.53%), carbohydrates (68.90%), fiber (7.34%), and ash (2.54%) has been reported (Da Cunha et al., 2020). Regarding its aroma, a study reported that it contains substances such as ethyl butanoate, ethyl hexanoate, 1-octen-3-ol, butyric acid, hexanoic acid, and phenylethyl alcohol. Its pulp is constituted of ethyl, methyl, and phenylethyl esters and by carboxylic acids, terpenoids, delta-lactones, and some sulfur compounds (Rezende and Fraga, 2003). Other constituents such as polyphenols and flavonoids have also been detected (Souza et al., 2018).

6.2.7 ANNONA PURPUREA

The *Annona purpurea* Moc. et Sess. (chincuya, sincuya, guanacona, cabeza de negro) is a species of the Annonaceae family, little known, and native to Mesoamerica. Its distribution is confined from southern Mexico to northern South America (Panama, Colombia, and Venezuela). This plant can be found in the wild or cultivated in home gardens, reaching 6–10 m in height (Vidal-Lezama et al., 2019; Luna-Cazáres and González-Esquinca, 2008). As for its leaves, they are simple, membranous, obovate to elliptic, with a size of 12–30 cm long and 6–14 cm wide. Flowers are extra axillary, solitary, and large. They have three outer petals measuring 5 by 2 cm, are fleshy, ferruginous-sericeous on the outside, and tinged with purple on the inside (Vidal-Lezama et al., 2019; Luna-Cazáres and González-Esquinca, 2008).

Chincuya fruits are oval to spheroidal in shape, with a diameter of 15–20 cm and average weight of 1.23 kg, covered with numerous rigid spike-like protrusions and brown, felt-textured skin. The pulp is orange, fragrant, and fibrous with a combination of multiple flavors such as banana, passion fruit, and papaya. It is used as the main ingredient to make ice creams, sweets, juices, beverages, and concoctions for medicinal purposes (Vidal-Lezama et al., 2019; Luna-Cazáres and González-Esquinca, 2008). The pulp's proximal composition presents lipids (1.03%), proteins (8.0%), carbohydrates (77.43%), fiber (9.17%), and ash (3.83%) (López Martínez et al., 2022). The authors have also reported bioactive compounds such as phenolics, flavonoids, and alkaloids.

FIGURE 6.7 Chincuya fruit.

Source: Adapted from Girón Pérez and González Esquinca (2020).

6.2.8 CUCURBITA FICIFOLIA

Cucurbita ficifolia is a climbing plant of the *Curcubitaceae* family; its distribution extends from the tropical and subtropical regions of Central and South America. It is known as chilacayote, chivere, alcayota, and cidra. The chilacayote belongs to a species group called pumpkins, cultivated for their fruits, flowers, and edible seeds (Lira Saade, 1995; Andres, 1992). This plant is traditionally grown with other species in the milpa (garden). Furthermore, it does not require special care for its development because it grows in different types of soils and is resistant to extreme temperatures and viral diseases (Lira Saade, 1995; Andres, 1992). Chilacayote has angular stems and petiolate leaves. The male flowers are extended with a warm bell-shaped 5–10 mm long, about 5–15 mm wide, linear sepals, and a tubular bell-shaped corolla. The female flowers have strong peduncles; their shape goes from ovoid to elliptical, occasionally foliaceous sepals, and a corolla slightly larger than the male flowers (Lira Saade, 1995; Andres, 1992; Lim, 2012).

The fruits of chilacayote are like watermelons since they measure from 14 to 50 cm long with a diameter of 14–25 cm. They are globose to ovoid-elliptic, light or dark green, with or without white longitudinal stripes or stripes toward the apex, minutely speckled with white or green and white or cream. The shell is rigid, persistent, and without ribs. The pulp is white, sweet, and fibrous, with numerous seeds (Lira Saade, 1995; Andres, 1992. The tender chilacayote is consumed as a fresh vegetable, while in its mature state is used to prepare typical sweets and traditional drinks (Lim, 2012). Also, it is used in elaborating alcoholic beverages such as chicha and craft beers (Mejía-Guadarrama, 2020; Cutler and Cardenas, 1947). Content of carbohydrates (49.76%), protein (13.28%), ash (8.75%), lipids (9.14%), and fiber (19.07%) has been reported (Mejía-Guadarrama, 2020). Other components such as essential fatty acids, phenols, flavonoids, and others have also been identified (Barrón-Álvarez et al., 2002; Lim, 2012).

FIGURE 6.8 Chilacayote fruit.

Source: Adapted from Mejia-Guadarrama (2022) and *Naturalista*.

6.3 POTENTIALITIES FROM NUS IN MEXICO: BIOLOGICAL ACTIVITIES

Humans currently face various diseases related to unhealthy eating habits and sedentary lifestyles. This trend has triggered an interest in finding new food components that promote health, with particular attention to obesity, type 2 diabetes, and cardiovascular and neurodegenerative diseases. Obesity is a positive imbalance between energy intake and expenditure, with excessive weight gain related to fat accumulation in adipose tissue (Association American Diabetes, 2014). On the other hand, diabetes is described as a metabolic disorder in insulin concentrations or resistance to insulin actions, resulting in hyperglycemia. Finally, cardiovascular diseases are related to the accumulation of fat and cholesterol in the walls of blood vessels (Chia and Egan, 2020). As can be seen, these diseases are widely related, as well as others. For that reason, healthy alternatives are currently being sought to deal with these diseases.

For many years, plants, like their leaves, seeds, roots, and fruits, have been widely investigated as sources of new bioactive compounds against emerging diseases (Iqbal et al., 2021). This section summarizes some research on the biological properties of NUS, particularly fruit pulps, which could be used for sustainable fermented beverage production (Table 6.1). To our knowledge, there is little information regarding their biochemical and molecular mechanisms of action. However, identifying several phytochemicals, such as phenolic compounds, flavonoids, and terpenes, would show their potential therapeutic applications. Herein, some studies are presented, identifying antimicrobial, antioxidant, antiobesity, antidiabetic, anti-inflammatory, and anti-urolithiasis activities.

Nonetheless, significant biological activities in other plant species belonging to the same family or genera have been observed. For instance, *T. cacao* belongs to the same genus as *T. bicolor*. In this species, numerous phytochemicals with biological properties have been found. For example, antimicrobial activities against *S. aureus* and *E. coli* have been reported in the peel. Furthermore, in the cocoa bean, bioactive compounds such as proteins, polyphenols, and flavonoids with inhibition activity of pancreatic lipase, α-Amylase, α-Glucosidase, and angiotensin I-converting enzyme (ACE) have been observed. Furthermore, thermogenic activity and lipid metabolism activation have also been found (Oboh et al., 2014; Jang et al., 2018; Jorge Coronado-Cáceres et al., 2020). These findings suggest that *T. bicolor* could have biological properties to control or prevent chronic degenerative diseases such as obesity, diabetes, and hypertension. In this same sense, *A. reticulata* belongs to the same genus as *A. purpurea*. Extract from its seeds exhibited a protective effect in diabetic Wistar rats through its antidiabetic, antihyperlipidemic, antioxidant, and anti-inflammatory properties (Wen et al., 2019). Therefore, NUS resources are a potential source of bioactive compounds, and through biotechnological research, its potential therapeutic applications can be identified.

TABLE 6.1 Biological Properties of Bioactive Compounds Present in Neglected and Underutilized Plant Species (NUS).

Biological activity	Fruit (linked compounds)	Findings	Reference
Antimicrobial	*A. purpurea* (Unknown)	Seed extracts showed antimicrobial activity against *Escherichia coli, Salmonella typhi*, and *Klebsiella pneumoniae*.	Luna-Cazáres et al., 2008
Antioxidant	*P. aculeata* (Polyphenols)	Antioxidant and antimicrobial activity (against *S. typhi)* were observed due to the polyphenols content in the ripe pulp.	Santiago-Ruiz et al., 2021
	B. crassifolia (Polyphenols)	The fruit extract proved efficient for removing reactive O and N species. It also prevents ROO• induced oxidative damage in human erythrocytes, lipid peroxidation, hemoglobin oxidation, lipid peroxidation, glutathione depletion, and oxidized glutathione formation.	Mariutti et al., 2014
	A. heterophyllus (Polyphenols and flavonoids)	Jackfruit pulp shows high antioxidant activity by in-vitro assays, which are related to phenolic compounds content and flavonoid.	Jagtap and Bapat, 2010
Antiobesity	*A. heterophyllus* (Polysaccharides)	Jackfruit consumption in obese rats restricts weight gain and enhances the serum lipid profile. It also causes higher species diversity modulating the microbial composition and reversing the dysbiosis by high-fat diet.	Zeng et al., 2023
Antidiabetic	*C. ficifolia* (Polyphenols and flavonoids)	Natural and fermented chilacayote juice with *L. plantarum* showed antioxidant activity, alpha-glucosidase inhibition, and a hypoglycemic effect in CD-1 mice.	Barrón-Álvarez et al., 2022; Moya et al., 2020
	B. crassifolia (Sesquiterpenes)	Seed extract improved glucose metabolism in mice, reducing insulin resistance and stimulating insulin production. This trend was related to pancreatic β cells' protection.	Pérez Gutiérrez et al., 2016

TABLE 6.1 *(Continued)*

Biological activity	Fruit (linked compounds)	Findings	Reference
	B. crassifolia (Unknown)	In normoglycemic and induced-diabetic rats, consuming Nance fruits and seeds improves hyperlipidemia and hyperinsulinemia. Inhibition of advanced glycation end products and oxidative stress was also observed.	Pérez-Gutierrez et al. 2010
Anti-inflammatory	*A. heterophyllus* (Artocarpesin, norartocarpetin, and oxyresveratrol)	Phenolic compounds isolated and identified from the fruits show inhibitory effects on the production of proinflammatory mediators in lipopolysaccharide (LPS)-activated RAW 264.7 murine macrophage cells.	Fang et al., 2008
	S. betaceum (polysaccharides)	A galactoarabinoglucuronoxylan isolated from tomatillo showed an antinociceptive effect on inflammatory pain models.	do Nascimento, et al., 2013
	S. betaceum or *C. betacea* (Polyphenols and flavonoids)	Cytotoxicity to hepatocellular liver carcinoma (HepG2) and non-hormone-dependent breast carcinoma (MDA-MB-231) were observed in tomatillo fruit extracts.	Mutalib, 2016
Anti-neurodegenerative	*M. floribunda* (Essential oils)	Fruit peel′ essential oil demonstrated the capacity for acetylcholinesterase inhibition, preventing the acetylcholine breakdown, which is directly linked to the progression and development of neurodegenerative diseases	da Silva Barbosa et al., 2020
Anti-urolithiatic	*P. acueleata* (Unknown)	An antiurolytic effect linked to bladder stone fragmenting was observed in rats Wistar with induced urolithiasis administered with *P. aculeata* extracts.	Morales-Sánchez, et al., 2015

6.4 POTENTIAL APPLICATIONS OF NUS IN BEVERAGES

Throughout history and worldwide, ancient ethnic groups usually made fermented beverages from sugar sources available in their local habitats, such as fruit, roots, and other vegetable ingredients. These traditional beverages played a significant role in religious ceremonies, the regional economy, gastronomic delight, and cultural identity (Rubio-Castillo et al., 2021; Tamang et al., 2009). Furthermore, these techniques also helped to preserve and enhance the nutritional value of foods and beverages. Although the technical and cultural relevance, many of these traditional beverages have been lost over time.

Currently, there is a trend for rescuing customs and traditional practices not only to recover the identity of the native people but also to take advantage of ancestral knowledge (Dagne, 2014). On the other hand, the food industry is seeking alternative ingredients derived from unconventional fruits to develop new fermented and non-fermented beverages to meet the constantly growing consumer demand (Liu et al., 2022). In this context, the diversity of mostly unknown underutilized vegetal sources offers a good range of raw materials with bioactive compounds and techno-functional components, which could impact on innovative organoleptic characteristics. In this section, we will focus on discussing the potential applications of undervalued plant resources in beverage design.

Teobroma bicolor, also called Pataxte or white cocoa, is a species close to *T. cacao*. The seeds such as roasted cocoa beans are usually used in traditional beverages such as pozol, tejate, pinol, and tascalate (Soleri et al., 2007). However, little is known about using its pulp as an ingredient in beverages. The *T. bicolor* presents a profile of components and physicochemical characteristics like *T. cacao (*Reisdorff et al., 2004; Pérez-Mora et al., 2018); therefore, both could be used to design beverages. The pulp of *T. cacao* is made up of a group of mucilaginous spongy cells that contain high sugar content (18%) such as sucrose, glucose, fructose, starch, and pectin, as well as various organic acids such as citric acid, lactic acid, malic acid, succinic acid, and acetic acid. These components make the cocoa pulp a viable raw material for being used in the food industry (Nunes et al., 2020). Dias et al. (2007) designed a wine fermenting the *T. cacao* pulp with *Saccharomyces cerevisiae*; this beverage contained concentrations established in the standards of alcohol, methanol, esters, and acetaldehyde. Likewise, the level of acceptance was moderate in the sensory analysis. On the other hand, Nunes et al. (2020) used cocoa pulp (30%) as an adjunct for beer production on a pilot scale. Surprisingly, important volatile compounds that impacted the sensory qualities of the beer were identified since a moderate level of acceptance was observed by sensorial analysis.

Cucurbita ficifolia is better known as chilacayote. Although it is a cosmopolitan and versatile crop, its consumption and commercialization are limited. The ripe pulp contains a high carbohydrate content (49.76%). For that, it was suggested as an ingredient for developing fermented drinks. For example, Chicha is a traditional alcoholic beverage from South America. In Peru, they call it beer; its artisanal elaboration in clay pots can use malted corn kernels and chilacayote pulp before spontaneous fermentation starts (Cutler and Cardenas, 1947). By contrast, in southern Mexico (Chiapas), Chicha is prepared inside the chilacayote by adding piloncillo through a hole. Subsequently, the fruit is closed so that spontaneous fermentation can occur for three days (Tawas Penago, 2019). Chilacayote has also been used as an adjunct for preparing a Porter-type craft beer. Herein, the dehydrated fruit with the malt mixture was used to prepare the must. Then, two yeasts *S. cerevisiae* and *K. marxianus* were used independently and mixed. These changes in craft beer elaboration positively impacted flavor, color, and aromas (Mejía-Guadarrama, 2022). As seen by the research, this raw material has huge potential to be used in the design of fermented beverages.

Jackfruit (*Artocarpus heterophyllus* Lam.) is another fruit with attractive characteristics that could contribute to beverage innovation. It is considered an exotic fruit because of its appearance and tropical flavors, such as pineapple, banana, mango, orange, melon, and papaya. Another important characteristic of the pulp is its high content of polyphenols and sugars (19%). However, the uses of jackfruit pulp are limited. For this reason, strategies have been sought for its use and conservation. One of the most viable approaches is its incorporation into the production of beverages. For example, An et al. (2022) designed a drink using jackfruit pulp fermented with *S. cerevisiae*. An interesting feature was that some important bioactive compounds were retained in the final product, such as the high content of polyphenols, vitamin C, and antioxidant activity. On the other hand, Sharma et al. (2013) produced a jackfruit wine with high alcohol content (18%) and antioxidant activity. Surprisingly, the wine had a soft aroma that impacted the sensory characteristics. As the pulp of the jackfruit, its residues have also been investigated regarding their viability in the design of the beverage. For instance, John and Narasimham (1993) made a carbonated drink from the fruit residue (rind, pod, and unfertilized floral parts) to use the waste. First, they performed an enzymatic treatment of the raw material with pectinases, then made an extraction, clarification, and carbonation. According to the authors, the drink offers similar commercial prospects to pineapple juice. In this same sense, Dam and Nguyen (2013) developed a fermented beverage from fruit rags of jackfruit. The rags are a waste material of latex-like filaments that come from compartments of edible fruit. The residues were treated with pectinases and then fermented with *Saccharomyces oviformis*. The drink showed a low degree of alcohol (5.5%). However, the flavor and aromas impacted the acceptance of the beverage. Finally, these investigations reveal that jackfruit is an exciting resource for developing beverages.

Byrsonima crassifolia is also known locally as Nachi or Murici. It is a delicious fruit with a high carbohydrate content and a strong aroma of rotten cheese (Da Cunha et al., 2020). The most abundant compounds, such as ethanol, ethyl hexanoate, butanoic acid, and hexanoic acid, were identified through a volatile compound analysis (Alves and Franco, 2003). Nachi liqueur (Mistela) is the best-known alcoholic beverage from this fruit, which is prepared by an infusion of whole Nanchi fruit with sugar, water, and liquor for two months (Medina Torres et al., 2017). Notwithstanding, the Nanchi pulp also has been used in fermented beverages. Bizinoto et al. (2015) reported the use of the fruit to produce a semi-dry white table wine using *S. cerevisiae*. This wine displayed a standard alcohol content (11.13%) and mostly complied with Brazilian regulatory standards. In another study, a not fermented beverage fruit pulp-based was standardized. Using a 2^2 factorial experimental design, the conditions were optimized to formulate a beverage with moderate antioxidant activity, attractive sensory qualities, and a high level of consumer acceptance (de Souza et al., 2020). Finally, these reports have shown that this raw material is highly viable for developing new beverages.

There are many plant resources with biotechnological characteristics for the design of new foods. However, collaborative work is required between the local communities, the scientific community, and the consumer to valorize these resources. As seen in this section, there is still a long way to go since research on the design of beverages with undervalued plant resources has been little explored.

6.5 CONSERVATION OF NUS: SOCIAL, ECONOMIC, AND ENVIRONMENTAL IMPACTS

Although NUS, also named orphan or minor crops, are plant species that may be used locally in a partial basis, they are seldom produced in a formal agriculture system. Usually, the causes for this are their lack of value, competitiveness, and consumer appreciation. Broadly, they can be annual or perennials, wild or partly cultivated, including agaves, cactus, cereals, fruits, tubers, etc. (Paludosi et al., 2020). However, some of them can be recognized and promoted as practical raw materials of fermentation processes to produce nutritional attractive beverages, such as those obtained from *Artocarpus altilis fosberg* (Herrera Escobar and Salvador Albores, 2019), *Parmentiera aculeate (*Castillo-Ruíz et al., 2022), and *Solanum betaceum* (Mejía-Bustamante et al., 2022), which have been used for the making of an alcoholic beverage in combination with fruit of *Physalis peruviana* and two yeasts (*Saccharomyces cerevisiae* and *Saccharomyces pastorianus*) and, *Theobroma bicolor*, which has been consumed in infusions (Baldera Ocampo et al., 2021) and is including for "tejacate" (Pérez-Armendáriz and Cardoso-Ugarte, 2020). Furthermore, NUS, such as *Myrciaria floribunda* (Cruz Silva et al., 2020; de Azevedo et al., 2019), *Annona purpurea* (Anaya-Esparza et al., 2020), and *Byrsonima crassifolia* (Agredano-de la Garza et al., 2021; Hernández-López et al., 2021) have important antimicrobial and antioxidant activities. In the end, customers are the key elements to change the NUS standards of the actual food system, allowing steady actions and consistent decision-making. Even though many of the NUS represent high nutrient values, sometimes better than their conventional counterparts, they do not always fulfill the nutritional gaps. However, the fermentation process may enhance the starch digestibility, the availability of nutrients and significant bioactives, and the reduction of anti-nutrients. Therefore, taking these crops with no disdain, they certainly could be considered as foods of the future. Particularly, only a small fraction of the Mexican wild species (about 10% of the whole flora) is currently conserved as seeds (Ulian et al., 2021). Therein, few of the NUS are included and safeguarded and many more waiting to be included in national projects for seed banking. These NUS may be developed into superior phenotypes to improve their nutraceutical value, enhance their plant breeding approaches, and boost their global adaptability (Popoola et al., 2022). Thus, the genetic upgrading of NUS can be accomplished by speed breeding samples, monitored by omics technologies and genome editing methods.

Many of the neglected or underutilized crop species may have nutraceutical and pharmaceutical potential, signifying their likely role as raw material for fermentation foods and phytomedicines. They can conduct multiple objectives, promoting sustainable agricultural strengthening and healthy diets, becoming fundamental sources of food processing systems. Though in some instances, the current wild harvesting and use of several species can be unsustainable, endangering, and facing them to extinction. In other cases, their present production levels are limited to fulfill the demands for industrial production scale. Therefore, cultivation practices typical of sustainable conservation policies are urgently desirable to lessen more population drops of NUS in the wild. Also, research and development is needed on all elements of the identified value chain, from the crops upgrading to their commercialization and consumption. A multi- and transdisciplinary strategy should be required involving a wide range of experts on all aspects of the recognized NUS with the potential utilization and product development in inclusive or new value chains (Mudau et al., 2022). Significant research is demanded, such as the genomic data description of the microorganisms related to the traditional Mexican fermented beverages. The typical artisanal activity would be reinforced with this information in favor of the communities' producers and with the consequent strengthen of the local knowledge sharing. This high-tech information is elemental to succeed in the production of brand quality beverages and diverse biotechnological applications (Robledo-Márquez et al., 2021). The research on fermented beverages from NUS must be in collaborative proposals and in transdisciplinary approaches that allow the drafting of regulatory frameworks to protect the diversity of the fermented beverages, keeping valid the complex interactions of humans with plants and microorganisms. In the Mexican case, ethnic conservation and native restoration can be recognized into the international actions (i.e., UNESCO) for promotion and protection of the Mesoamerican biocultural heritage (Ojeda-Linares et al., 2021).

Traditional Mexican fermented beverages (TMFB) are quite diverse and some inclusive in danger, demanding quick preservation actions. Typically, TMFB are closely related to wild plants that have been adapted and domesticated, as it is appreciated in the importance of fermented beverages based on agave saps, maize, and cacti fruits, among many more wild species that are used in traditional beverages. As the agave saps have demanded most of the attention, the conservation actions on other endangered sources and local know-hows have been limited. Besides, imported technologies, other plant species, tools, and artifacts have been historically combined with traditional procedures, modifying the beverage products, and diversifying their sensory attributes. The TMFB are then vivid examples of how traditional knowledge is represented by a dynamic process, which is regularly changing and evolving. Therefore, any systematic analysis should start with the gathering of information about the cultural background and diversity of the TMFB, recording their geographical distribution, and identifying the major issues and research tendencies required for their recovery and conservation (Ojeda-Linares et al., 2021).

Particularly, the studies related to TMFB have commonly focused on the biotechnological knowledge of such fermented beverages, identifying advances and alternative research opportunities. As most fermented beverage processes worldwide are neither standardized nor completely studied yet. Thus, the emphasis should be set on the isolation and identification of the involved endogenous microorganisms, pointing to the standardization of the fermentation process. In the investigation of the metabolic process and the generation of subproducts. And finally, in the analysis of the nutritional value, and the physical and sensorial attributes of the TMFB (Pérez-Armendáriz and Cardoso-Ugarte, 2020). This way the research scope of TMFB can be expanded, including little-known beverages such as axokot, balché, chorote, sendejo, taberna, and tuba, as well as getting support from the use of Omics technologies and new generation sequencing techniques to optimize DNA determinations.

Many of the NUS of Mexico may be perfect for fermentation having bioactive components. They have the potential to be exploited as unexplored fermentation products with added functional properties. From *Artocarpus* (Muñiz-Márquez et al., 2019), *Parmentiera aculeate* (Castillo-Ruíz et al., 2021), *Myrciaria floribunda* (Cruz Silva et al., 2020), *Solanum betaceum* (Ramírez and Kallarackal, 2019), *Theobroma bicolor* (Gálvez-Marroquín et al., 2016), *Byrsonima crassifolia* (Agredano-de la Garza et al., 2021), *Annona purpurea* (Vidal-Lezama et al., 2019), and *Cucurbita ficifolia* (Moya-Hernández et al., 2020; Barrón-Álvarez et al., 2022), these plants are rich in nutrients and offer a range of delicious flavors and textures. With the growing interest in traditional foods and alternative sources of nutrition, fermented products from these neglected plants are sure to gain popularity in the years to come. However, further research is necessary to explore the full potential of these plants (Cuvas-Limon et al., 2020). Fermentation of these species could lead to the creation of new products, the optimization and design of new food bioprocesses, offering numerous health benefits, creating economic opportunities, and preserving traditional food cultures. Economically, the use of neglected and underutilized plant sources can be a source of income for small-scale farmers, reducing their reliance on single-crop agriculture. Focusing on underutilized crops can also contribute to the development of niche markets and improve trade opportunities both locally and internationally, providing economic opportunities for producers.

Socially, the conservation of NUS sources can improve the livelihoods of rural communities in Mexico. Many of these plant species are part of local traditional diets, and their conservation can contribute to retaining cultural practices and promoting sustainable food systems. The preservation of these plants and their integrated use can also contribute to greater food diversity and better nutrient availability, particularly for marginalized communities.

From an environmental perspective, the conservation of NUS sources can contribute to the protection of biodiversity and promote sustainable use of natural resources. Many of these plants grow naturally in ecological niches with low-intensity agricultural systems, and their incorporation into agroecosystems can enhance biodiversity and promote more sustainable land use practices. Additionally, NUS offer numerous benefits; they have high potential for developing climate-resilient cropping systems, reducing food insecurity and poverty, improving health and nutrition, enhancing biodiversity, and regenerating degraded lands. These species can also contribute significantly to agricultural prosperity and economic competitiveness, thereby fostering agricultural and rural development. However, barriers to the full potential of NUS' contributions to food security, nutrition, and environmental sustainability include a lack of awareness of their potential, scarce knowledge of crop genetics and physiology, limited availability of superior genetic resources, weak seed systems, and inadequate levels of funding in agricultural research, development and adoption.

6.6 CONCLUSION AND FUTURE PROSPECTS

It is important to note that several NUS are suitable for commercialization due to their nutritional, techno-functional, or pharmaceutical properties. Therefore, there is a growing recognition of carrying out studies focused on identifying, characterizing, and commercializing NUS. This information is fundamental for designing strategies that allow sustainable food production, ecosystem conservation, and improving human health and security. However, current research is limited, restricting the development of primary crops, agri-food value chains, and biotechnological potential. In this sense, it is time for academia, agribusiness, policymakers, and development organizations to collaborate to stimulate global research and development of these often-neglected species for significant benefits linked to the United Nations Sustainable Development Goal 12–"Ensure sustainable consumption and production patterns." Therefore, the knowledge far-flung in this field supported by technologies and policies that help unlock the commercial potential of NUS and create inclusive value chains, the latter framework can be beneficial. However, it is outstanding to mention that promoting NUS could lead to food gentrification (when food becomes "cool"), making it inaccessible to the people who depend on it. Therefore, NUS promoters must be careful to ensure the availability of NUS varieties to avoid increasing costs.

KEYWORDS

- **fermented beverages**
- **underutilized vegetable sources**
- **bioactive compounds**
- **functional beverages**
- **sustainability**
- **revalorization**

REFERENCES

Agredano-de la Garza, C. S.; López-Guzmán, G.; Balois-Morales, R.; León-Fernández, A. E.; Bautista-Rosales, P. U.; Palomino-Hermosillo, Y.; Jiménez-Zurita, J.; Jiménez-Ruiz, E.; Juárez-López, P. Compuestos de Interés Funcional Del Nanche (*Byrsonima crassifolia* (L.) H.B.K.). *Acta Agrícola y Pecu.* **2021,** *7,* E0071013. https://doi.org/E0071013 https://doi.org/10.30973/aap/2021.7.0071013.

Álvarez-Olivera, P. A.; Calzada-Almas, E.; Batista-Cruz, C. Etnobotánica y Propagación de Parmentiera Edulis D. C., Árbol de Uso Múltiple En Cuba. *Rev. For. Baracoa* **2010,** *29* (1), 77–86.

Alves, G. L.; Franco, M. R. B. Headspace Gas Chromatography-Mass Spectrometry of Volatile Compounds in Murici (*Byrsonima Crassifolia* L. Rich). *J. Chromatogr. A* **2003,** *985* (1–2), 297–301. https://doi.org/10.1016/S0021-9673(02)01398-5.

An, V. N.; Thinh, P. Van; Do, V. L.; Duy, N. Q.; Thuy, D. T.; Hien, T. T. The Influencing Factors on the Production of Alcoholic Drinking from Jackfruit (*Artocarpus heterophyllus* L.). *Mater. Sci. Forum* **2022,** *1048 MSF* , 476–484. https://doi.org/10.4028/www.scientific.net/MSF.1048.476.

Anaya-Esparza, L. M.; García-Magaña, M. de L.; Abraham Domínguez-Ávila, J.; Yahia, E. M.; Salazar-López, N. J.; González-Aguilar, G. A.; Montalvo-González, E. Annonas: Underutilized Species as a Potential Source of Bioactive Compounds. *Food Res. Int.* **2020,** *138*, 109775. https://doi.org/10.1016/j.foodres.2020.109775.

Andres, T. C. Biosystematics, Theories On The Origin, And Breeding Potential Of *Cucurbita ficifolia*. In *Biology and Utilization of the Cucurbitaceae*; Jeffrey, C., Ed.; Cornell University Press: New York, 1992; pp 102–119. https://doi.org/10.7591/9781501745447-011.

Angón-Galván, P. Caracterización Parcial Del Fruto Parmentiera Edulis, Universidad Tecnológica de la Mixteca, 2006.

Association American Diabetes. Diagnosis and Classification of Diabetes Mellitus. *Diabetes Care* **2014,** *37* (Supplement 1), S81–S90. https://doi.org/10.2337/dc14-S081.

Baldera Ocampo, J. F.; Granda Santos, M. S.; Chavez Quintana, S. G. Capacidad Antioxidante y Polifenoles Totales de Infusión de Cascarilla de Cacao (*Theobroma cacao*) y Macambo (*Theobroma bicolor*). *Rev. Investig. Agroproducción Sustentable* **2021,** *5* (3), 13–19. https://doi.org/10.25127/aps.20213.814.

Baliga, M. S.; Shivashankara, A. R.; Haniadka, R.; Dsouza, J.; Bhat, H. P. Phytochemistry, Nutritional and Pharmacological Properties of *Artocarpus heterophyllus* Lam (Jackfruit): A Review. *Food Res. Int.* **2011,** *44*, 1800–1811. https://doi.org/10.1016/j.foodres.2011.02.035.

Barrios-Gamboa, A. Y.; Pérez-Reyes, Y. G. Elaboración de Bebida Fermentada Tipo Vino Con Mora y Maracuya, Universidad de Ciencias y Artes de Chiapas, 2019.

Barrón-Álvarez, N.; Prado-Barragán, L. A.; Fortis-Barrera, M. de los Á.; Alarcon-Aguilar, F. J. Fermentation of the *Cucurbita ficifolia* Fruit Juice: Its Antioxidant Activity and Effects on the Glycemia. *Beverages* **2022,** *8*, 55. https://doi.org/10.3390/beverages8030055.

Bizinoto, C.; Gonçalves, C.; Tomé, P.; Fragiorge, E. Desenvolvimiento Do Mosto Do Fermentado Alcoólico de Murci (*Byrsonima crassifolia* (L.) (Rich))-Malpighiaceae. In *Anais do Simpósio Latino Americano de Ciências de Alimentos*, 2015.

Cabot, M. I.; Lado, J.; Clemente, G.; Sanjuán, N. Towards Harmonised and Regionalised Life Cycle Assessment of Fruits: A Review on Citrus Fruit. *Sustain. Prod. Consum.* **2022,** *33*, 567–585. https://doi.org/10.1016/J.SPC.2022.07.024.

Carley, S.; Yahng, L. Willingness-to-Pay for Sustainable Beer. *PLoS One* **2018,** *13* (10), e0204917. https://doi.org/10.1371/JOURNAL.PONE.0204917.

Castillo-Ruíz, R.; Castillo-Archila, J. A.; Alemán-Castillo, S. E.; Castillo-Ruiz, O.; Trejo-Díaz, G. N. Alternativas Para El Aprovechamiento Del Cuajilote (*Parmentiera edulis*): Una Especie Agroindustrial Subutilizada. *CienciaUAT* **2022,** *17* (1), 123–138. https://doi.org/10.29059/cienciauat.v17i1.1635.

Castro, J. C.; Maddox, J. D.; Imán, S. A. Camu-Camu-*Myrciaria dubia* (Kunth) McVaugh. In *Exotic Fruits Reference Guide*; Rodrigues, S., de Oliveira Silva, E., Sousa de Brito, E., Eds.; Academic Press: United Kingdom, 2018; pp 97–105. https://doi.org/10.1016/B978-0-12-803138-4.00014-9.

Cerero-Calvo, C.; Sánchez-Medina, M. A.; Pérez-Santiago, A. D.; Matías-Pérez, D.; García-Montalvo, I. A. Probióticos Presentes En Bebidas Fermentadas Mexicanas. *TIP Rev. Espec. en Ciencias Químico-Biológicas* **2022,** *25*, 1–13. https://doi.org/10.22201/fesz.23958723e.2022.436.

Chia, C. W.; Egan, J. M. Incretins in Obesity and Diabetes. *Ann. N. Y. Acad. Sci.* **2020,** *1461* (1 Specail Issue: Tje Year in Diabetes and Obesity), 104–126. https://doi.org/10.1111/nyas.14211.

Cruz Silva, A. V.; Sirqueira Nascimento, A. L.; Neves Muniz, E. Fruiting and Quality Attributes of Cambui (*Myrciaria floribunda* (West Ex Willd.) O. Berg in the Atlantic Forest of Northeast Brazil. *Rev. Agro@mbiente On-line* **2020,** *14*. https://doi.org/10.18227/1982-8470RAGRO.V14I0.5861.

Cutler, H. C.; Cardenas, M. In *Chicha, a Native South American Beer*, 1947; vol 13.

Cuvas-Limon, R. B.; Nobre, C.; Cruz, M.; Rodriguez-Jasso, R. M.; Ruíz, H. A.; Loredo-Treviño, A.; Texeira, J. A.; Belmares, R. Spontaneously Fermented Traditional Beverages as a Source of Bioactive Compounds: An Overview. *Crit. Rev. Food Sci. Nutr.* **2020,** *0* (0), 1–23. https://doi.org/10.1080/10408398.2020.1791050.

Da Cunha, M. C.; Machado, P. da S.; Araújo, A. B. S.; Carvalho, E. E. N.; Vilas Boas, E. V. de B. Impact of Processing and Packaging on the Quality of Murici Jelly [*Byrsonima crassifolia* (L.) Rich] during Storage. *Nutr. Food Sci.* **2020,** *50* (5), 871–883. https://doi.org/10.1108/NFS-09-2019-0293.

Dagne, T. Protecting Traditional Knowledge in International Intellectual Property Law: Imperatives for Protection and Choice of Modalities. *14 Jhon Marshall Rev. Intellect. Prop. Law* **2014,** *25*.

Dam, S. M.; Nguyen, N. T. Production of Fermented Beverage from Fruit Rags of Jackfruit (*Artocarpus heterophyllus*). *Acta Hortic.* **2013,** *989*, 285–292. https://doi.org/10.17660/ActaHortic.2013.989.37.

da Silva Barbosa, D. C.; Holanda, V. N.; de Assis, C. R. D.; de Oliveira Farias de Aguiar, J. C. R.; doNascimento, P. H.; da Silva, W. V.; do Amaral Ferraz Navarro, D. M.; Silva, M. V. da; de Menezes Lima, V. L.; dos Santos Correia, M. T. Chemical Composition and Acetylcholinesterase Inhibitory Potential, *in Silico*, of *Myrciaria floribunda* (H. West Ex Willd.) O. Berg Fruit Peel Essential Oil. *Ind. Crops Prod.* **2020,** *151*, 112372. https://doi.org/10.1016/j.indcrop.2020.112372.

de Azevedo, M. M. L.; Cascaes, M. M.; Guilhon, G. M. S. P.; Andrade, E. H. A.; Zoghbi, M. das G. B.; da Silva, J. K. R.; Santos, L. S.; da Silva, S. H. M. Lupane Triterpenoids, Antioxidant Potential and Antimicrobial Activity of *Myrciaria floribunda* (H. West Ex Willd.) O. Berg. *Nat. Prod. Res.* **2019,** *33* (4), 506–515. https://doi.org/10.1080/14786419.2017.1402311.

de Oliveira, J. S. D.; Pintode Lemos, E. E.; Rezende, L. de P.; Dos Santos, E. F.; Barros Silva, R.; Gallo, C. M. Morphoanatomy, Floral Biology and Reproductive Phenology of Cambuí (*Myrciaria floribunda* (h. West Ex Willd.) o. Berg.). *Rev. Bras. Frutic.* **2021,** *43* (6), 1–10. https://doi.org/10.1590/0100-29452021618.

de Oliveira, L. M.; Porte, A.; de Oliveira Godoy, R. L.; da Costa Souza, M.; Pacheco, S.; de Araujo Santiago, M. C. P.; Gouvêa, A. C. M. S.; da Silva de Mattos do Nascimento, L.; Borguini, R. G. Chemical Characterization of *Myrciaria floribunda* (H. West Ex Willd) Fruit. *Food Chem.* **2018,** *248*, 247–252. https://doi.org/10.1016/j.foodchem.2017.12.053.

de Souza, V. R.; Aniceto, A.; Abreu, J. P.; Montenegro, J.; Boquimpani, B.; de Jesuz, V. A.; Elias Campos, M. de B.; Marcellini, P. S.; Freitas-Silva, O.; Cadena, R.; Teodoro, A. J. Fruit-Based Drink Sensory, Physicochemical, and Antioxidant Properties in the Amazon Region: Murici (*Byrsonima crassifolia* (L.) Kunth and Verbascifolia (L.) DC) and Tapereba (*Spondia mombin*). *Food Sci. Nutr.* **2020,** *8* (5), 2341–2347. https://doi.org/10.1002/FSN3.1520.

Dias, D. R.; Schwan, R. F.; Freire, E. S.; Dos Santos Serôdio, R. Elaboration of a Fruit Wine from Cocoa (*Theobroma cacao* L.) Pulp. *Int. J. Food Sci. Technol.* **2007,** *42* (3), 319–329. https://doi.org/10.1111/j.1365-2621.2006.01226.x.

Dominguez Orta, J. C.; Herrera Martínez, G. Una Delicia En El Árbol: El Pepino Kat. *Desde el Herb. CICY* **2018,** *10*, 156–159.

Do Nascimento, G. E.; Hamm, L. A.; Baggio, C. H.; De Paula Werner, M. F.; Iacomini, M.; Cordeiro, L. M. C. Structure of a Galactoarabinoglucuronoxylan from Tamarillo (*Solanum betaceum*), A Tropical Exotic Fruit, and Its Biological Activity. *Food Chem.* **2013,** *141*, 510–516. https://doi.org/10.1016/J.FOODCHEM.2013.03.023.

Fang, S. C.; Hsu, C. L.; Yen, G. C. Anti-Inflammatory Effects of Phenolic Compounds Isolated from the Fruits of *Artocarpus heterophyllus*. *J. Agric. Food Chem.* **2008,** *56* (12), 4463–4468. https://doi.org/10.1021/jf800444g.

FAO. Voluntary Code of Conduct for Food Loss and Waste Reduction, Rome, 2022.

Feicán-Mejía, C. G.; Encalada-Alvarado, C.; Becerril, A. Descripción Agronómica Del Cultivo de Tomate de Árbol. *Agroproductividad* **2016,** *9* (8), 78–86.

Furlán, A. L.; Bressani, R. Vegetable Resources with Agroindustrial Potential from Guatemala. Chemical Characterization of the Pulp and of the Seeds of *Theobroma bicolor*. *Arch. Latinoam. Nutr.* **1999,** *49* (4), 373–378.

Gálvez-Marroquín, L. A.; Reyes-Reyes, A. L.; Avendaño-Arrazate, C. H.; Hernández-Gómez, E.; Mendoza-López, A.; Díaz-Fuentes, V. . Pataxte (*Theobroma bicolor* Humb. & Bonpl.): Especie Subutilizada En México. *Productividad* **2016,** *9* (1), 41–47.

García, Y. M.; Ramos, A. L. C. C.; de Paula, A. C. C. F. F.; Nascimento, M. H. Do; Augusti, R.; de Araújo, R. L. B.; de Lemos, E. E. P.; Melo, J. O. F. Chemical Physical Characterization and Profile of Fruit Volatile Compounds from Different Accesses of *Myrciaria floribunda* (H. West Ex Wild.) o. Berg through Polyacrylate Fiber. *Molecules* **2021,** *26* (17), 5281. https://doi.org/10.3390/molecules26175281.

García Rojas, C. D.; González Roblero, F. B.; Gorrochotegui Salas, I. A.; López Alegría, A. G.; Vera García, K. L.; Veláquez López, A. A.; Vela Gutiérrez, G. Cuajilote. Alternativa Para Desarrollar Nuevos Productos Alimenticios Por Su Alto Contenido Nutrimental. *Divulg. la Cienc. Junio 2* (19).

Girón Pérez, M.; González Esquinca, A. Gaceta de Divulgación Científica Del Instituto de Ciencias Biológicas de La U. *Gac. Divulg. científica del Inst. Ciencias Biológicas la UNICACH* **2020,** *1* (1), 3.

Gómez-Ubiaño, A.; Díaz-Díaz, M. del C. Microbioma Humano y Salud: Efecto de Probióticos y Prebióticos. In *Salud, alimentación y sexualidad en el ciclo vital.*; Molero-Jurado, M. del M., Gázquez-Linares, J. J., Pérez-Fuentes, M. del C., Simón-Márquez, M. del M., Martos-Martínez, Á., Barragán-Martín, A. B., Eds.; ASUNIVEP: España, 2018; pp 333–340.

González, A. A.; Moncada, J.; Idarraga, A.; Rosenberg, M.; Cardona, C. A. Potential of the Amazonian Exotic Fruit for Biorefineries: The *Theobroma bicolor* (Makambo) Case. *Ind. Crops Prod.* **2016,** *86*, 58–67. https://doi.org/10.1016/j.indcrop.2016.02.015.

Goswami, C.; Chacrabati, R. Jackfruit (*Artocarpus heterophylus*). In *Nutritional Composition of Fruit Cultivars*; Simmonds, M. S., Preedy, V. R., Eds.; Academic Journals, 2015; pp 317–335. https://doi.org/10.1016/B978-0-12-408117-8.00014-3.

Hari, A.; Revikumar, K. G.; Divya, D. Artocarpus: A Review of Iths Phytochemistry and Pharmacology. *J. Pharma Search* **2014,** *9* (1), 7. https://doi.org/10.1126/science.217.4554.7.

Hernández-López, A.; Roldán-Sabino, C.; Parra-Reyes, A.; Ávila-Alejandre, A. X. Revisión Del Estudio Etnobotánico, Fitoquímico y Farmacológico de *Byrsonima crassifolia* (L.) Kunth (Malphigeaceae). *Acta Agrícola y Pecu.* **2022,** *8*, e0081010. https://doi.org/10.30973/aap/2022.8.0081010.

Herrera Escobar, H. D. J.; Salvador Albores, E. R. Bebida Fermentada Funcional a Partir de La Pulpa de Artocarpus A, Universidad de Ciencias y Artes de Chiapas, Tuxtla Gutiérrez, Chiapas, 2019. https://doi.org/10.16/CSS/JQUERY.DATATABLES.MIN.CSS.

Hunter, D.; Borelli, T.; Beltrame, D. M. O.; Oliveira, C. N. S.; Coradin, L.; Wasike, V. W.; Wasilwa, L.; Mwai, J.; Manjella, A.; Samarasinghe, G. W. L.; Madhujith, T.; Nadeeshani, H. V. H.; Tan, A.; Ay, S. T.; Güzelsoy, N.; Lauridsen, N.; Gee, E.; Tartanac, F. The Potential of Neglected and Underutilized Species for Improving Diets and Nutrition. *Planta* **2019,** *250* (3), 709–729. https://doi.org/10.1007/s00425-019-03169-4.

Iqbal, Z.; Iqbal, M. S.; Hashem, A.; Abd_Allah, E. F.; Ansari, M. I. Plant Defense Responses to Biotic Stress and Its Interplay With Fluctuating Dark/Light Conditions. *Front. Plant Sci.* **2021,** *12* (March), 1–22. https://doi.org/10.3389/fpls.2021.631810.

Jagtap, U. B.; Bapat, V. A. Artocarpus: A Review of Its Traditional Uses, Phytochemistry and Pharmacology. *J. Ethnopharmacol.* **2010,** *129*, 142–166. https://doi.org/10.1016/j.jep.2010.03.031.

Jang, M. H.; Kang, N. H.; Mukherjee, S.; Yun, J. W. Theobromine, a Methylxanthine in Cocoa Bean, Stimulates Thermogenesis by Inducing White Fat Browning and Activating Brown Adipocytes. *Biotechnol. Bioprocess Eng.* **2018,** *23* (6), 617–626. https://doi.org/10.1007/s12257-018-0434-y.

John, P. J.; Narasimham, P. Processing and Evaluation of Carbonated Beverage from Jackfruit Waste (*Artocarpus heterophyllus*). *J. Food Process. Preserv.* **1993,** *16* (6), 373–380. https://doi.org/10.1111/j.1745-4549.1993.tb00217.x.

Jorge Coronado-Cáceres, L.; Rabadán-Chávez, G.; Mojica, L.; Hernández-Ledesma, B.; Quevedo-Corona, L.; Lugo Cervantes, E.; Massieu, W.; Manuel Stampa, esq I. Cocoa (*Theobroma cacao* L.) Seed Proteins' Anti-Obesity Potential through Lipase Inhibition Using In Silico, In Vitro and In Vivo Models. *Foods* **2020,** *9*, 1359. https://doi.org/10.3390/foods9101359.

Juárez-Martínez, Y. Y.; Villanueva González, G.; Veana Hernández, F.; Aguilar Zárate, P. Estudio Exploratorio Del Contenido de Nutrimentos de *Parmentiera aculeata. Tectzapic. Rev. Académico-Científica* **2019,** *5* (1), 25–28.

Kufer, J.; McNeil, C. L. The Jaguar Tree (*Theobroma bicolor* Bonpl.). In *Chocolate in Mesoamerica: A Cultural History of Cacao*; Gainesville, F., Ed.; Florida Scholarship Online, 2006; pp 90–104. https://doi.org/10.5744/FLORIDA/9780813029535.003.0004.

Lara-Cervantes, T. de J.; Carrillo-Inungaray, M. L.; Balderas-Hernández, V. E.; Aguilar-Zárate, P.; Veana, F. Valorization of *Parmentiera aculeata* Juice in Growth of Probiotics in Submerged Culture and Their Postbiotic Production: A First Approach to Healthy Foods. *Arch. Microbiol.* **2022,** *204* (11), 1–8. https://doi.org/10.1007/s00203-022-03295-w.

Lim, T. In *Cucurbita Ficifolia. Edible Medicinal And Non-Medicinal Plants*; Springer, Dordrecht: New York, 2012; vol 2.

Lim, T. K. Theobroma Bicolor. In *Edible Medicinal And Non Medicinal Plants*; Springer, Dordrecht: New York, 2012; vol 3, Fruits. https://doi.org/10.1007/978-94-007-2534-8_29.

Lira Saade, R. Estudios Taxonomicos Ecogeograficos de Las Cucurbitaceae Latinoamericanas de Importancia Economica. In *Systematic and Ecogeographic Studies on Crop Genepools.* International Plant Genetic Resource Institute: Rome, Italy, 1995; vol 9, pp 1–294.

Liu, H.; Xu, X.; Cui, H.; Xu, J.; Yuan, Z.; Liu, J.; Li, C.; Li, J.; Zhu, D. Plant-Based Fermented Beverages and Key Emerging Processing Technologies. *Food Rev. Int.* **2022**. https://doi.org/10.1080/87559129.2022.2097256.

López, A.; Bicerra, E.; Díaz, E. Perfil Ecológico de Cuatro Rodales de Camu Camu Árbol *Myciaria floribunda* (H. West. Ex Willd) O. Berg En Ucayali. *Ecol. Apl.* **2006,** *5* (1–2), 45–52.

López Martínez, C. R.; García Mateos, M. del R.; Martínez Damián, M. T.; Sánchez Sánchez, L. Calidad Nutricional y Nutracéutica Del Fruto de Tres Especies de Annonaceae: Guanábana, Chirimoya y Chincuya. *Nov. Sci.* **2022,** *14* (28), 1–15. https://doi.org/10.21640/ns.v14i28.2966.

Luna-Cazáres, L. M.; González-Esquinca, A. R. Actividad Antibacteriana de Extractos de *Annona diversifolia* Safford y *Annona purpurea* Mociño & Sessé Ex Dunal. *Polibotanica* **2008,** *25*, 121–125.

Maldonado Peralta, M. de los Á.; Sánchez Santillán, P.; Rojas García, A. R.; Valenzuela Lagarda, J. L.; Bottini Luzardo, M. B.; Alaniz Gutiérrez, L. Characterization and Evaluation of Fruits of 'Nanche' (*Byrsonima crassifolia* L.). *Rev. Mex. Ciencias Agrícolas* **2020,** *11* (1), 151–160. https://doi.org/10.29312/REMEXCA.V11I1.1950.

Mariutti, L. R. B.; Rodrigues, E.; Chisté, R. C.; Fernandes, E.; Mercadante, A. Z. The Amazonian Fruit Byrsonima Crassifolia Effectively Scavenges Reactive Oxygen and Nitrogen Species and Protects Human Erythrocytes against Oxidative Damage. *Food Res. Int.* **2014,** *64*, 618–625. https://doi.org/10.1016/j.foodres.2014.07.032.

Medina Torres, R.; Ibarra Estrada, M. E.; González Chávez, J. G.; Salazar García, S. Preferencias Gustativas de Cinco Productos Mínimamente Procesados Del Fruto Del Nanche (*Byrsonima Crassifolia* (L.) Kunth). *Rev. Venez. Cienc. y Tecnol. Aliment.* **2017,** *8* (1), 45–056.

Mejía-Bustamante, L.; Vasquez-Cadenillas, Rosi errones-Miranda, M.; Paredes-Goycochea, Milton Salazar-Campos, J. Bebida Alcohólica a Base de Aguaymanto (*Physalis peruviana*) y Tomate de Árbol (*Solanum betaceum*): Caracterización Química y Sensorial. *Agroindustrial Sci.* **2022,** *12* (3), 355–363.

Mejía-Guadarrama, I. Evaluación del impacto del chilacayote (*Cucurbita ficifolia*) y la levadura *Kluyveromyces marxianus* en las propiedades de una cerveza artesanal tipo porter, Tecnológico de Estudios Superiores de San Felipe del Progreso, 2022.

Morales-Sánchez, V.; Osuna-Fernández, H.; Brechú-Franco, A.; Laguna-Hernández, G.; Vargas-Solís, R. Evaluación Del Efecto Antiurolítico Del Fruto de *Parmentiera aculeata* En Rata Wistar. *Bot. Sci.* **2015,** *93* (2), 1–6.

Morelos-Flores, D. A.; Montalvo-González, E.; Chacón-López, M. A.; Santacruz-Varela, A.; Zamora-Gasga, V. M.; Torres-García, G.; García-Magaña, M. de L. Comparative Study of Four Jackfruit Genotypes: Morphology, Physiology and Physicochemical Characterization. *Horticulturae* **2022,** *8*, 1010. https://doi.org/10.3390/horticulturae8111010.

Moya-Hernández, A.; Bosquez-Molina, E.; Verde-Calvo, J. R.; Blancas-Flores, G.; Trejo-Aguilar, G. M. Hypoglycemic Effect and Bioactive Compounds Associated with the Ripening Stages of the *Cucurbita ficifolia* Bouché Fruit. *J. Sci. Food Agric.* **2020,** *100* (14), 5171–5181. https://doi.org/10.1002/jsfa.10566.

Mudau, F. N.; Chimonyo, V. G. P.; Modi, A. T.; Mabhaudhi, T. Neglected and Underutilised Crops: A Systematic Review of Their Potential as Food and Herbal Medicinal Crops in South Africa. *Front. Pharmacol.* **2022,** *12*, 1–31. https://doi.org/10.3389/fphar.2021.809866.

Muñiz-Márquez, D. B.; Teixeira, J. A.; Mussatto, S. I.; Contreras-Esquivel, J. C.; Rodríguez-Herrera, R.; Aguilar, C. N. Fructo-Oligosaccharides (FOS) Production by Fungal Submerged Culture Using Aguamiel as a Low-Cost by-Product. *LWT* **2019,** *102*, 75–79. https://doi.org/ https://doi.org/10.1016/j.lwt.2018.12.020.

Mutalib, M. A. Antioxidant Activities, Polyphenolic Profile and Chemopreventive Properties of *Cyphomandra betacea* L. Ethanolic Extract on Human Breast and Liver Adenocarcinoma Cells, Universiti Putra Malasya, 2016.

Naturalista. *Fotos de Chilacayote (Cucurbita ficifolia)*; Comisión Nacional para el Conocimiento y Uso de la Biodiversidad. https://www.naturalista.mx/taxa/181590-Cucurbita-ficifolia/browse_photos (accessed May 10, 2023).

Nunes, C. S. O.; Da Silva, M. L. C.; Camilloto, G. P.; Machado, B. A. S.; Hodel, K. V. S.; Koblitz, M. G. B.; Carvalho, G. B. M.; Paula, A.; Uetanabaro, T. Potential Applicability of Cocoa Pulp (*Theobroma cacao* L) as an Adjunct for Beer Production. *Sci. World J.* **2020,** 3192585. https://doi.org/10.1155/2020/3192585.

Oboh, G.; Ademosun, A. O.; Ademiluyi, A. O.; Omojokun, O. S.; Nwanna, E. E.; Longe, K. O. Inhibition of α-Mylase, α-Glucosidase, and Angiotensin I-Converting Enzyme by Polyphenol-Rich Extracts from Cocoa (*Theobroma cacao*) Bean. *Patholog. Res. Int.* **2014,** *2014*, 549287. https://doi.org/10.1155/2014/549287.

Ojeda-Linares, C.; Álvarez-Ríos, G. D.; Figueredo-Urbina, C. J.; Islas, L. A.; Lappe-Oliveras, P.; Nabhan, G. P.; Torres-García, I.; Vallejo, M.; Casas, A. Traditional Fermented Beverages of Mexico: A Biocultural Unseen Foodscape. *Foods* **2021,** *10* (10), 1–31. https://doi.org/10.3390/foods10102390.

Paludosi, S.; Meldrum, G.; King, E. I. O.; Hunter, D. NUS: What They Are and Why Need Them More than Ever. In *Orphan Crops for Sustainable Food and Nutrition Security: Promoting Neglected and Underutilized Species*; Paludosi, S., King, E. I. O., Hunter, D., Swanubathan, M., Eds.; Routledge, Taylor & Francis Group: London and New York, 2022; pp 3–18.

Pasarín, V.; Viinikainen, T. Enabling a Legal Environment for the Prevention and Reduction of Food Loss and Waste, Legal Brie.; FAO, Ed.; Rome, 2022.

Popoola, J. O.; Aworunse, O. S.; Ojuederie, O. B.; Adewale, B. D.; Ajani, O. C.; Oyatomi, O. A.; Eruemulor, D. I.; Adegboyega, T. T.; Obembe, O. O. The Exploitation of Orphan Legumes for Food, Income, and Nutrition Security in Sub-Saharan Africa. *Front. Plant Sci.* **2022,** *13*, 782140. https://doi.org/10.3389/fpls.2022.782140.

Pérez-Armendáriz, B.; Cardoso-Ugarte, G. A. Traditional Fermented Beverages in Mexico: Biotechnological, Nutritional, and Functional Approaches. *Food Res. Int.* **2020,** *136*, 109307. https://doi.org/10.1016/j.foodres.2020.109307.

Pérez Gutiérrez, R. M.; Muñiz Ramirez, A. Hypoglycemic Effects of Sesquiterpene Lactones from *Byrsonima crassifolia. Food Sci. Biotechnol. Biotechnol.* **2016,** *25* (4), 1135–1145. https://doi.org/10.1007/s10068-016-0182-8.

Perez-Gutierrez, R. M.; Muñiz-Ramirez, A.; Gomez, Y. G.; Ramírez, E. B. Antihyperglycemic, Antihyperlipidemic and Antiglycation Effects of *Byrsonima crassifolia* Fruit and Seed in Normal and Streptozotocin-Induced Diabetic Rats. *Plant Foods Hum. Nutr.* **2010,** *65* (4), 350–357. https://doi.org/10.1007/s11130-010-0181-5.

Pérez-Mora, W.; Jorrin-Novo, J. V.; Melgarejo, L. M. Substantial Equivalence Analysis in Fruits from Three Theobroma Species through Chemical Composition and Protein Profiling. *Food Chem.* **2018,** *240* , 496–504. https://doi.org/10.1016/j.foodchem.2017.07.128.

Pérez-Morales, S.; Crosby-Galván, M. M.; Ramírez-Mella, M.; Bárcena-Gama, J. R.; Hernández-Mendo, O.; Crosby-Galván, E. M. Cuajilote (*Parmentiera aculeata* (Kunth) Seem.): A Potential Fruit for Ruminant Feed. *Agro Product.* **2021,** 75–79. https://doi.org/10.32854/agrop.v14i10.1985.

Ponce-Sánchez, J.; Zurita-Benavides, M. G.; Peñuela, M. C. Reproductive Ecology of White Cacao (*Theobroma bicolor* Humb. & Bonpl.) in Ecuador, Western Amazonia: Floral Visitors and the Impact of Fungus and Mistletoe on Fruit Production. *Rev. Bras. Bot.* **2021,** *44*, 479–489. https://doi.org/10.1007/s40415-021-00709-9.

Ramírez, F.; Kallarackal, J. Tree Tomato (*Solanum betaceum* Cav.) Reproductive Physiology: A Review. *Sci. Hortic.* **2019,** *248*, 206–215. https://doi.org/10.1016/j.scienta.2019.01.019.

Reisdorff, C.; Rohsius, C.; Das Graças Claret De Souza, A.; Gasparotto, L.; Lieberei, R. Comparative Study on the Proteolytic Activities and Storage Globulins in Seeds of *Theobroma grandiflorum* (Willd Ex Spreng) Schum and *Theobroma bicolor* Humb Bonpl, in Relation to Their Potential to Generate Chocolate-like Aroma. *J. Sci. Food Agric.* **2004,** *84* (7), 693–700. https://doi.org/10.1002/jsfa.1717.

Rezende, C. M.; Fraga, S. R. G. Chemical and Aroma Determination of the Pulp and Seeds of Murici (*Byrsonima crassifolia* L.). *J. Braz. Chem. Soc.* **2003,** *14* (3), 425–428. https://doi.org/10.1590/S0103-50532003000300014.

Robledo-Márquez, K.; Ramírez, V.; González-Córdova, A. F.; Ramírez-Rodríguez, Y.; García-Ortega, L.; Trujillo, J. Research Opportunities: Traditional Fermented Beverages in Mexico. Cultural, Microbiological, Chemical, and Functional Aspects. *Food Res. Int.* **2021,** *147*. https://doi.org/10.1016/j.foodres.2021.110482.

Romero-Luna, H. E.; Hernández-Sánchez, H.; Dávila-Ortiz, G. Traditional Fermented Beverages from Mexico as a Potential Probiotic Source. *Ann. Microbiol.* **2017,** *67* (9), 577–586. https://doi.org/10.1007/s13213-017-1290-2.

Rubio-Castillo, Á. E.; Santiago-López, L.; Vallejo-Cordoba, B.; Hernández-Mendoza, A.; Sáyago-Ayerdi, S. G.; González-Córdova, A. F. Traditional Non-Distilled Fermented Beverages from Mexico to Based on Maize: An Approach to Tejuino Beverage. *Int. J. Gastron. Food Sci.* **2021,** *23*. https://doi.org/10.1016/j.ijgfs.2020.100283.

Santiago Ruiz, C.; Nuricumbo Lievano, V. N.; Chapa Barrios, M. G.; Vela Gutiérrez, G.; Velázquez López, A. A. Antimicrobial Activity, Phenolic and Antioxidant Content of Extracts from Cuajilote (*Parmentiera aculeata* Kunth) Fruits at Different Degrees of Ripening. *J. Mex. Chem. Soc.* **2021,** *65* (2), 161–169. https://doi.org/10.29356/jmcs.v65i2.1270.

Santos, E. F.; Lemos, E. E. P.; Rezende, L. P.; Salvador, T. L.; Araújo, R. R. Caracterização Fisico-Química, Compostos Bioativos e Atividade Antioxidante Total de Frutos de Cambuizeiro. *Rev. Ouricuri* **2017,** *7* (1), 064–079.

Sharma, N.; Palmoo Bhutia, S.; Aradhya, D. Process Optimization for Fermentation of Wine from Jackfruit (*Artocarpus heterophyllus* Lam.). *J. Food Process. Technol.* **2013,** *4* (2), 1000204. https://doi.org/10.4172/2157-7110.1000204.

Soleri, D.; Cleveland, D. A. Tejate: Theobroma Cacao and T. Bicolor in a Traditional Beverage from Oaxaca, Mexico. *Food Foodways* **2007,** *15*, 107–118. https://doi.org/10.1080/07409710701260131.

Souza, M. D. C.; Morim, M. P. Subtribos Eugeniinae O. Berg e Myrtinae O. Berg (Myrtaceae) Na Restinga Da Marambaia, RJ, Brasil. *Acta Bot. Brasilica* **2008,** *22* (3), 652–683. https://doi.org/10.1590/S0102-33062008000300006.

Souza, J. N. S.; Silva, E. M.; Loir, A.; Rees, J. F.; Rogez, H.; Larondelle, Y. Antioxidant Capacity of Four Polyphenol-Rich Amazonian Plant Extracts: A Correlation Study Using Chemical and Biological in Vitro Assays. *Food Chem.* **2008,** *106*, 331–339. https://doi.org/10.1016/j.foodchem.2007.05.011.

Tamang, J. P.; Tamang, B.; Schillinger, U.; Guigas, C.; Holzapfel, W. H. Functional Properties of Lactic Acid Bacteria Isolated from Ethnic Fermented Vegetables of the Himalayas. *Int. J. Food Microbiol.* **2009,** *135* (1), 28–33. https://doi.org/10.1016/j.ijfoodmicro.2009.07.016.

Tawas Penago, M. Comportamiento Cinético de Crecimiento de *Saccharomyces cerevisiae* Meyen Ex E. C. Hansen, Aisladas de La Bebida Tradicional "Chicha de Chilacayote" (*Cucurbita ficifolia* Bouché)). https://www.studocu.com/es-mx/document/

universidad-de-ciencias-y-artes-de-chiapas/biologia-celular/tesis-de-licenciatura/9348124 (accessed May 06, 2023).

Tinajero-Carrizalez, C.; González-Pérez, A.; Rodríguez-Castillejos, G.; Castañón-Nájera, G.; Ruíz-Salazar, R. Comparación Proximal En Cacao (*Theobroma cacao*) y Pataxte (*T. bicolor*) de Tabasco y Chiapas, México. *Polibotánica* **2021,** *52*, 135–149.

Ulian, T.; Pritchard, H. W.; Casas, A.; Mattana, E.; Liu, U.; Castillo-Lorenzo, E.; Way, M.; Aranda, P. D.; Lira, R. Neglected and Underutilized Species of Mexico: Status and Priorities for Their Conservation and Sustainable Use. In *Orphan Crops for Sustainable Food and Nutrition Secururity Promoting Neglected and Underutilized Species* 2021; pp 118–127. https://doi.org/10.4324/9781003044802-10/.

Vázquez Yanes, C.; Batis Muñoz, A. I.; Alcocer Silva, M. I.; Gual Díaz, M.; Sánchez Dirzo, C. *Árboles y Arbustos Nativos Potencialmente Valiosos Para La Restauración Ecológica y La Reforestación*; México, D.F, 1999. CONABIO (accessed May 06, 2023).

Vidal-Lezama, E.; Villegas-Monter, Á.; Vaquera-Huerta, H.; Robledo-Paz, A.; Martínez-Palacios, A. Annona Purpurea Moc. & Sessé Ex Dunal Especie Nativa de México, Subutilizada. *Agro Product.* **2019,** *12* (3), 9–15. https://doi.org/10.32854/agrop.v0i0.1326.

Villar-Herrera, S. Evaluación Nutrimental Del Fruto *Parmentiera edulis* Para Su Posible Utilización En La Industria Alimentaria., Universidad Autónoma Agraria Antonio Narro, 2011.

Wang, S.; Zhu, F. Tamarillo (*Solanum betaceum*): Chemical Composition, Biological Properties, and Product Innovation. *Trends Food Sci. Technol.* **2020,** *95*, 45–58. https://doi.org/10.1016/j.tifs.2019.11.004.

Wen, W.; Lin, Y.; Ti, Z. Antidiabetic, Antihyperlipidemic, Antioxidant, Anti-Inflammatory Activities of Ethanolic Seed Extract of *Annona reticulata* L. in Streptozotocin Induced Diabetic Rats. *Front. Endocrinol. (Lausanne).* **2019,** *10* (October), 1–15. https://doi.org/10.3389/fendo.2019.00716.

Zeng, S.; Cao, J.; Wei, C.; Chen, Y.; Liu, Q.; Li, C.; Zhang, Y.; Zhu, K.; Wu, G.; Tan, L. Polysaccharides from *Artocarpus heterophyllus* Lam. (Jackfruit) Pulp Alleviate Obesity by Modulating Gut Microbiota in High Fat Diet-Induced Rats. *Food Hydrocoll.* **2023,** *139*, 108521. https://doi.org/10.1016/j.foodhyd.2023.108521.

CHAPTER 7

Trichoderma Roles in Sustainable Agriculture

ERIKA SIRENI RODRÍGUEZ-MARTÍNEZ[1], EDEL TORRES-TORRES[2], CÉSAR GUIGÓN-LÓPEZ[2], and MÓNICA ALVARADO-GONZALEZ[1*]

[1]*Centro de Investigación en Alimentación y Desarrollo, A.C. Coordinación Delicias, Fracc, Vencedores del desierto. Delicias, Chihuahua, México*

[2]*Universidad Autónoma de Chihuahua, Facultad de Ciencias Agrícolas y Forestales, Delicias, Chihuahua, México*

ABSTRACT

A food system that conveys food security and nutrition for people implies guaranteeing sustainability throughout the agri-food chain. It starts with healthy soil, the development of a disease-free crop, and good post-harvest control that extends the shelf life of the food produced. However, currently achieving good development in food cultivation is a challenge. The soils are increasingly degraded and contaminated. As a result, many applications of chemical compounds are required to keep crops free of pests, affecting the quality and shelf life of the final product, the food. Moreover, the immoderate application of chemical fertilizers and pesticides causes a negative impact on the environment and human health. Undoubtedly, sustainable agriculture is an alternative that proposes to restore the holistic balance of soil composition through the biological mechanisms of native microorganisms. Microorganisms are valuable natural resources for sustainable agriculture. A lot of bacterial and fungal strains have been applied as plant growth promoter microorganisms. *Trichoderma* is a proposed biocontrol agent used in crops showing effectiveness in plant diseases. *Trichoderma* presents additional biocontrol properties as a biofertilizer and biostimulant at the rhizosphere

Sustainable Engineering and Agro-Food Processing: Renewable and Nonrenewable Resources.
José Juan Buenrostro-Figueroa, Julio César Tafolla-Arellano, Mónica Chávez-González,
Pedro Aguilar-Zárate, A. K. Haghi (Eds.)

and foliar level, enhancing microbial activities, bioremediation, and biosynthesis of nanoparticles. In addition, *Trichoderma* is a cosmopolitan fungus, it tolerates adverse conditions with a great ability to adapt, characteristics that give it potential applicability in sustainable agriculture.

7.1 INTRODUCTION

A food system that delivers food security and nutrition for all, implies guaranteeing sustainability throughout the agri-food chain; however, currently achieving a good development of the crops and quality foods is a challenge. Growing high-quality field crops, fruit, and vegetables which contribute to our global diet, is vital for current agriculture in order to provide the needs and demands of the population. In addition, the sustainability of food supply requires the development of new agricultural practices that focus on minimizing soil degradation, environmental contamination, and the adverse effects of climate change.

Modern agriculture urgently demands alternatives to synthetic chemical fertilizers and pesticides to respond to changing international regulations, as well as consumer demands for food without potentially toxic residues. Currently, agriculture is highly dependent on the use of chemical-based fertilizers and pesticides that affect the nutritional quality, health status, and productivity of crops. The continuous release of these chemical inputs causes various toxic compounds to accumulate in the soil and move to plants with prolonged exposure, ultimately affecting human health. Therefore, it is necessary to generate alternatives to chemical pesticides/fertilizers to improve agricultural production.

Other problems facing agriculture are various abiotic stresses, drought, salinity, extreme temperatures, which reduce their productivity and are most of the times are related to global warming, and which affect the ability of agricultural regions to sustain adequate food production. For this reason, it is necessary to have a comprehensive view of the true impact of climate change on our global food supply, which requires understanding the effects of climate change on the nutritional value of crops. Abiotic stresses have an unusual effect on plant growth, minimize the yield of all the crops, and cause considerable losses (Bulgari et al., 2019).

Research indicates that soil microbiome can be of great help because it has great potential as a driver of soil functions and increases the productivity of agricultural crops. Microorganisms perform basic functions such as the supply of nitrogen, phosphorus, potassium, sulfur, and micronutrients, and also act as a functional extension in plant defense against biotic and abiotic stress. Thus, plant-microbiome interactions are crucial for plant health, as microorganisms can affect plant growth, development, and productivity. Therefore, sustainable agriculture proposes to restore the holistic balance of the composition of agricultural soils through biological mechanisms of native microorganisms, which represent valuable tools for this purpose.

The success of different microorganisms as biological control agents, biofertilizers, and biostimulants is known, with positive effects on plant development modulating different physiological processes (Lombardi et al., 2020; Rouphael and Colla, 2020; Sánchez-Montesinos et al., 2020).

Climate change is considered to be very dangerous for food production and crops have to adapt to the changing conditions of the environment in which they grow, therefore improving the stress resistance of plants is essential for agricultural productivity and also for environmental sustainability. Scientific knowledge supports that *Trichoderma* is a fungus with a great capacity for adaptation and has the potential to be incorporated into the development of sustainable agriculture. Knowing the mechanisms involved in the multiple beneficial capacities of *Trichoderma* species will allow a better use of these beneficial microorganisms to produce healthy food and nutritional quality, and to feed the populations of the different regions of the world in these difficult times of great challenges.

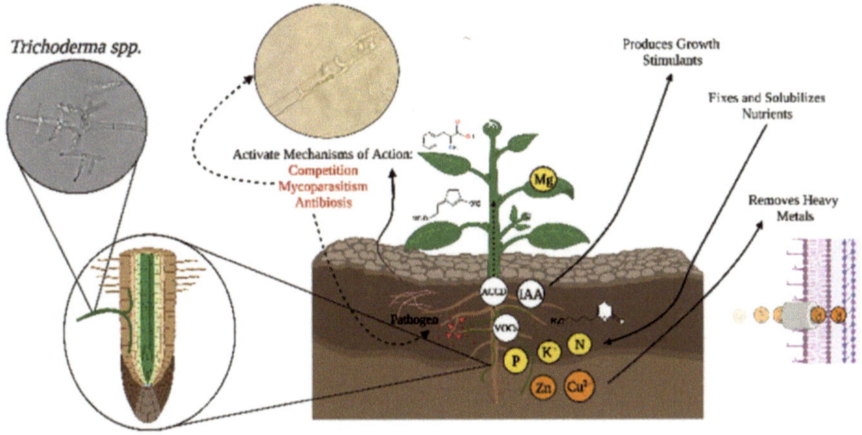

FIGURE 7.1 Beneficial relationships between *Trichoderma* species and agricultural crops.

7.2 TRICHODERMA SOIL INHABITANT AND ROOT COLONIZER

Trichoderma is commonly found in the soil and rhizosphere, to such an extent that its presence has been proposed as a marker of healthy soils (Zachow et al., 2009; Woo et al., 2022). It has a large number of ecological niches, from very alkaline to very acidic soils and in a wide variety of climates, even surviving in conditions that are considered unfavorable due to its excellent adaptability and the speed with which it manages to take advantage of available nutrients (Macías-Rodríguez et al., 2020). One of their lifestyles is saprophytic, in which they take advantage of their enzymatic mechanism to degrade highly complex substrates that may be composed mainly of cellulose, chitin, and lignin (Harman and Uphoff, 2019).

The applications of *Trichoderma* have been shown to help recover or maintain the diversity of beneficial plant microorganisms in soils that had previously been affected by high doses of fertilizer (Woo et al., 2022; Illescas et al., 2020). In addition, it has been able to increase and stabilize the presence of beneficial soil microorganisms after agrochemical spraying (Wu et al., 2022).

Trichoderma also seeks to colonize the roots of plants as it is attracted by the carbohydrates they secrete, which when detected by the fungus act as a signaling molecule and cause *Trichoderma* to respond by using its mycelium to move towards the root. These carbohydrates can include arabinose, xylose, monoinositol, fructose, and glucose, this last being the one that Trichoderma can absorb more easily (Macías-Rodríguez et al., 2018). Once *Trichoderma* reaches the roots, it uses small hydrophobic proteins to adhere to the root surface (Hermosa et al., 2012). Some species can live as endophytes because they have the ability to penetrate the cell wall of the plant and resist the toxic metabolites it produces to defend itself (Harman and Uphoff, 2019).

7.3 TRICHODERMA AS BIOCONTROL AGENT

Currently, plant disease control is mainly carried out using chemical fungicides; however, this has several disadvantages since they have a negative impact on the environment and, due to their frequent use, pathogenic microorganisms can develop resistance. On the other hand, the use of biological control microorganisms is an ideal strategy to ensure a more sustainable and environmentally friendly agricultural production (Dutta et al., 2022).

Biological control is defined as the use of organisms or their products that are natural enemies, competitors, or antagonists to disease-causing organisms, that also support plant development, and that have no or minimal environmental impact compared to chemical control methods (Köhl et al., 2019; Sood et al., 2020). It must directly or indirectly target the pathogenic organism and it is important to determine the mechanisms involved in its control over the pathogen (Stenberg et al., 2021).

Most *Trichoderma* species have proven biological control capabilities against plant pathogenic microorganisms, including fungi, oomycetes, bacteria, and even nematodes (Harman, G. E.; Uphoff, 2019; Singh et al., 2018).

One of the important aspects to consider when it is used as biological control is the compatibility that *Trichoderma* has with the medium to which it is going to be applied, that is why preference should be given to the use of native strains that present a greater adaptability to the medium, the temperature, the humidity, and the availability of nutrient (Ferreira and Musumeci, 2021).

The strains that are usually useful for agriculture are mainly distributed among the following species: *T. afroharzianum*, *T. guizhouenense, T. virens, T. atroviride, T. gamsii, T. viride, T. asperellum*, *T. asperelloides*, among others (Woo et al., 2022).

Among the microorganisms that *Trichoderma* has been shown to control are *Botrytis cinerea, Alternaria alternata, Sclerotinia sclerotium, Rhizoctonia solani, Fusarium* spp, *Pythium* spp, and *Phytophthora* capsici (Macías-Rodríguez et al., 2020).

7.3.1 MECHANISM OF ACTION

Trichoderma has several mechanisms of action for the direct control of phytopathogenic microorganisms such as competition for space, mycoparasitism, and antibiosis. The capacity to carry out these mechanisms depends on the species and not all of them present the same mechanisms or are equally efficient. Nor do they respond in the same way in all crops or geographical locations (Woo et al., 2022).

7.3.2 COMPETITION

Soil is a complex ecosystem with limited space and nutrients where a large number of microorganisms live. The rhizosphere tends to be an environment richer in nutrients from plant roots, which is why there is greater competition to colonize this space, favoring microorganisms with better capabilities to metabolize the available nutrients and take advantage of the space (Guzmán-Guzmán et al., 2023). *Trichoderma* is considered a fungus of aggressive growth as it is excellent in terms of mobility and nutrient uptake compared to other soil microorganisms, which is why competition for space and nutrients is considered one of the most important methods of biological control of this fungus (Sood et al., 2020). Another important feature that makes *Trichoderma* excellent in the competition for space is that it can resist toxins produced by other microorganisms in the rhizosphere (Dutta et al., 2022).

A 2018 study tested the ability of *T. atroviride* to compete for space and nutrients in the rhizosphere of tomato plants compared to *Phytophthora cinnamomi*. Inoculation of *T. atroviride* inhibited *P. cinnamomi* mycelial growth and zoospore production. On the other hand, *T. atroviride* increased its growth rate and spore production (Macías-Rodríguez et al., 2018).

7.3.3 MYCOPARASITISM

Mycoparasitism is one of the main mechanisms of control of pathogens, and consist that during the interaction Trichoderma-pathogen, the first presents a parasitic action on the second; previously, it was considered only as a necrotophic parasitism in which *Trichoderma* completely destroys the host to feed on it. However, it has been discovered that it penetrates through holes in the cell wall and coexists with the pathogen for a significant time, so it was proposed to consider it as a hemibiotrophic parasite (Woo et al., 2022; Lace et al., 2015; Mukherjee et al., 2022).

This interaction between fungi consists of a sequence of events that starts with the recognition of the phytopathogenic microorganism due to diffusible signals such as chitin secretion or the detection of ligands on the surface of the host, which causes the activation of genes to increase the production of chitinases and other cell wall degrading enzymes by *Trichoderma* while directing its hyphae toward the phytopathogen (Guzmán-Guzmán et al., 2019; Alfiky and Weisskopf, 2021). Then, the formation of papillae or appressoria-like structures begins just before it starts to coil over the hyphae of the phytopathogenic microorganism and starts to degrade the fungal cell wall consisting mainly of chitin and glucans or cellulose in the case of oomycetes. This is why cell wall degrading enzymes such as cellulases, hemicellulases, chitinases, protease, and β-1,3-glucanase are generated. During this interaction, the host also secretes metabolites as a defense mechanism, which causes *Trichoderma* to activate stress resistance and detoxification genes (Guzmán-Guzmán et al., 2023; Fang et al., 2021; Viterbo and Horwitz, 2010). During this process, *Trichoderma* not only degrades the cell wall of the pathogenic fungus, but also inactivates its enzyme production, which limits its ability to colonize and penetrate plant tissues (Sood et al., 2020). Signaling pathways that trigger *Trichoderma* mycoparasitism include the G protein and MAP kinase pathway (Sood et al., 2020; Karlsson et al., 2017). The secretion of lytic enzymes causes a series of physiological changes in the host microorganism which facilitates the flow of nutrients to *Trichoderma*, while continuing to destroy the target microorganism (Dutta et al., 2022).

7.3.4 ANTIBIOSIS

The different species of *Trichoderma* are able to produce a wide variety of secondary metabolites, some of them with antibiotic capacity that inhibit or limit the growth of pathogens. The activation of genes encoding metabolites with antibiotic capacity increases when *Trichoderma* is in the presence of pathogens and plant-secreted products (Guzmán-Guzmán et al., 2023).

More than 180 secondary metabolites have been described, which can be of the volatile or non-volatile type. Depending on the biosynthetic origin they are also classified as peptaibol, polyketide, and terpene. Several species of *Trichoderma* produce peptaibols, non-proteinogenic amino acid compounds with a molecular weight between 500 and 2200 Da; these are inserted in the membrane to form ionic channels depending on the voltage, an important characteristic for their capacity of antibiosis. They are synthesized by non-ribosomal peptide synthetases (Sood et al., 2020; Khan et al., 2020). Peiptaboles act during the mycoparasitism process by combining with cell wall lytic enzymes (Alfiky and Weisskopf, 2021). The microorganism that acts as host of *Trichoderma* can impact the metabolism of the peptaibols; in studies carried out in vitro, it was verified that the mixtures of peptaibols produced by *T. asperellum* and *T. longibrachiatum* presented variations depending on the pathogen with which it is confronted (Rahimi Tamandegani et al., 2020).

Among the wide variety of antibiotics that can be produced by *Trichoderma* spp. are: trichotoxins A and *B. trichodeceins*, trichorovins, viridins and gliovirins, produced by *T. virens*. Harzianic acid, trichorzines and trichorzianins were isolated from *T. harzianum*, among many others (Sood et al., 2020; Reino et al., 2008).

Viridin and gliovirin, two metabolites produced by *T. virens*, reduced germination and inhibited the growth of several oomycetes of the genus

Phytophthora (Pakora et al., 2018). Harzianic acid produced by *T. harzianum* showed antibiotic activity against *Rhizoctonia solani* in a study conducted on infected tomato plants and also induced in the plant the overexpression of genes involved in the defense response related to cuticle biosynthesis and the production of toxic metabolites (Manganiello et al., 2018).

The secondary metabolite 6-pentyl-α-pyrone (6-PP) is the most studied and has been shown to inhibit several phytopathogenic fungi, but it also has the ability to promote plant growth and modify the root inducing development of lateral roots (Alfiky and Weisskopf, 2021).

7.3.5 CONTROLLING PLANT DISEASES PROSPECT

Trichoderma has been widely used in agriculture to control numerous plant pathogenic organisms and the diseases they cause, focusing generally but not exclusively on diseases related to soil microorganisms (Dutta et al., 2022). Some examples of these diseases are root, stem, crown rots, and wilt, which can be caused by *Phytophthora spp, Pythium spp, Rhizoctonia solani, Fusarium spp, Sclerotium rolfsii, Macrophomina phaseolina*, among many others (Ketta and Hewedy, 2021; Sánchez-Montesinos et al., 2019; Sridharan et al., 2020).

7.3.6 REDUCTION OF DISEASES THROW FOLIAR APPLICATIONS

The method of application depends on the mode of action of the pathogen, *Trichoderma* is usually highly effective against pathogens that attack areas of the plant in contact with the soil; however, the edaphic application is not effective for pathogens that present foliar phases for what has been studied the possibility of realizing foliar applications taking advantage of the capacity of *Trichoderma* to colonize and to survive in the surface of the leaves (Ferreira and Musumeci, 2021).

The combination of soil applications followed by monthly foliar applications of *T. harzianum* proved efficient in controlling the different stages of various diseases that attack bentgrass (Lo et al., 1997). Other foliar application protocols have also been studied in combination with other biological control organisms or fungicides, and it was found that the foliar application of *T. asperellum* spores alone or in combination with *Bacillus amyloliquefaciens* significantly reduced the final severity of bacterial spot disease in tomato caused by *Xanthomonas perforans*, although when combined with applications to the substrate increased its potential to control the disease (Chien and Huang, 2020). Another study conducted on rice plants showed that the combination of native Trichoderma species mixed with the fungicide Trivia 73 WP reduced brown leaf spot disease caused by *Drechslera oryzae* by 27–36% and increased grain yield per plant (Mau et al., 2022). Foliar spraying of *T. asperellum* spores also reduced the severity of rice sheath blight damage caused by *Rhizoctonia solani*, matching the effectiveness of the typically used fungicide (de França et al., 2015). Most of these studies suggest evaluating the best protocol for the frequency of application to increase the efficiency of biological control of *Trichoderma* spp. by foliar application, since this method of application usually reduces the damage caused by pathogens, but does not eradicate or control them completely, which is why foliar applications of *Trichoderma* are recommended as part of an integrated disease management program complemented with other disease control tools.

7.4 TRICHODERMA AS BIOSTIMULANT

Plant biostimulants are defined as products obtained from different organic or inorganic substances and/or microorganisms, that are able to improve plant growth, productivity, and alleviate the negative effects of abiotic stresses (Bulgari et al., 2019; Rouphael and Colla, 2020). They can act directly on plant physiology and metabolism by improving soil conditions and are able to modify some molecular processes that allow to improve water and nutrient use efficiency of crops, stimulate plant development, and counteract abiotic stresses by enhancing primary and secondary metabolism. Plant biostimulants are often included in agricultural management practices to improve crop production and nutritional quality of agrifood products, reducing chemical inputs, increasing productivity, and recovering the natural equilibrium in agroecosystems (Woo and Pepe, 2018).

Microorganisms biostimulants includes bacteria, yeast, filamentous fungi, and micro algae that are isolated from soil, plants, water, and composted manures or other organic materials and are applied to soil to increase crop productivity through metabolic activities. A good example might be, for example, taking endophytes from desert plants, which grow in low-nutrient, high-stress environments, and transferring them to crop plants. Since desert endophytes have the ability to increase nutrient uptake and increase plant resistance to drought, heat stress, salt stress, and pathogen attack; they may help alleviate some of the challenges facing the agricultural industry, such as increased frequency of droughts and growing opposition to the use of chemicals in agriculture (Zhang and White, 2021).

Among the most studied endophytic fungi is *Trichoderma* spp. which exerts a wide range of effects on root activity and specific metabolic pathways of plants (Visconti et al., 2020). The contributions of *Trichoderma* species with the purpose of limiting the use of the external chemicals in agroecosystems and avoiding environmental and health risks have been reported in the scientific literature.

7.4.1 EFFECT ON PLANT GROWTH

7.4.1.1 EFFECTS OF TRICHODERMA ACTING ALONE

Some strains of species of *Trichoderma* trigger mechanisms entailing multifaceted cross-talk communication between root and shoot systems. In strains of *T. harzianum, T. virens,* and *T. atroviride*, the process includes the transfer of auxins (i.e., indole-3-acetaldehyde, indole-3-carboxaldehyde, and indole-3-ethanol), oligopeptides, volatile, and soluble secondary metabolites to the rhizosphere, which augment root branching and nutrient uptake efficiency that consequently improve crop performance (Visconti et al., 2020). The direct and indirect benefits to the plants depend upon this *Trichoderma*-plant molecular crosstalk and the reorganization of gene expression patterns in shoots and roots with significant changes in the metabolism of the plant and consequently improvement in growth and yield (Lombardi et al., 2020; Rouphael et al., 2020).

7.4.2 ENHANCING ABIOTIC STRESS TOLERANCE OF CROPS

Abiotic stresses strongly affect the growth, development, and quality of plant production, among which drought stress is perhaps the most significant abiotic stress limiting plant growth and development (Bulgari et al., 2019). However, drought and heat waves often occur together. Photosynthesis is limited by water and heat stress, so droughts and heat waves reduce crop yields (Zhang and White, 2021). Water stress results in a variety of biochemical and physiological disorders, including stomatal closure, impaired enzyme activity, and membrane damage. Heat stress causes similar alterations in CO_2 assimilation and photosynthesis, but also increases numbers of reactive oxygen species and alters chloroplast metabolism. Soil salinity is a global problem in crop productivity due to land clearing and unsustainable irrigation practices. Moreover, a constant increase in soil salinity may mean that cropping becomes impossible (Munns and Gilliham, 2015). To avoid the osmotic damage caused by salt in their roots, and to be able to absorb the water present in the soil, plants must pay a high energy cost.

There are alternatives to increase the tolerance of crops to stress, such as through genetic improvements, which require a lot of time and different cultivation environments. For this reason, biostimulants have been proposed since they have a beneficial effect on plants and improve their ability to cope with adverse environmental conditions, acting on primary or secondary metabolism.

Fungi can enhance the adaptability and tolerance of their host plants to drought stress after infecting plant roots and establishing a symbiotic association with their host plant. Among these fungi, the species of the genus *Trichoderma* stimulates plant growth and resistance to a wide range of adverse environmental conditions, through mechanisms that involve the antioxidant defense system and expression and increase of endogenous hormones in stressed plants (Boamah et al., 2021). This gives *Trichoderma* the potential to be used to improve the growth of cultivated plants under abiotic stress and can be considered a beneficial and sustainable way to improve crop survival (Yusnawan et al., 2021). However, for the biostimulant capacity of these fungi to take place, several factors, such as the availability of nutrients in the soil and environmental conditions, greatly influence these responses. These influences need to be studied, analyzed and considered when using these biostimulants, in order to adapt and modify strategies to generate sustainable systems that can reduce inputs while maintaining yield and increasing stress tolerance (Silletti et al., 2021).

7.4.2.1 CHANGES IN HORMONE REGULATION

Abscisic acid (ABA) and ethylen (ET) participates in an important way in tolerance to abiotic stress (Poveda, 2020). ABA possibly due to physiological mechanisms of the plant in response to it, such as the regulation of stomatal opening, a key mechanism to limit the loss of water from the leaves in drought conditions, and to induce the synthesis of osmoprotective compounds. ET, due to an increase in the expression of genes that alter Na+/

K+ ratio and selectively activate salt-tolerance genes in plants. In addition, ascorbate participates, which is an important antioxidant that is involved in the ascorbate-glutathione cycle (Brotman et al., 2013).

Plants treated with *Trichoderma* in salt stress imposition affect the expression of several genes related to osmo-protection and general oxidative stress in the roots of plants. For example, *T. parareesei* increases rapeseed productivity under drought or salinity conditions by a reduction in oxidative stress and the regulation of the expression of hormone genes related to abiotic stresses in the plant (Poveda, 2020). Besides this, the activation of other genes that have a protective role against salinity and osmotic stress has also been observed, such as the upregulation of two members of the aquaporin gene family AT2G34390 and AT2G29870 and high expression of phenylalanine ammonia lyase (PAL) (Boamah et al., 2021).

The enzyme chorismate mutase plays an important role in the ability of *T. parareesei* to promote tolerance to salinity and drought in plants by increasing the expression of genes related to the hormonal pathways of ABA under drought stress, and ethylene ET under salt stress (Poveda, 2020). The activity of the enzyme 1-aminocyclopropane-1-carboxylate deaminase (ACCD) reduces the availability of ACC necessary for ethylene biosynthesis. *Trichoderma* produces ACCD that enhances plant growth under abiotic stress conditions, reducing increases in ethylene levels and promoting high antioxidant capacity. In turn, ethylene reductions promote plant growth by increasing the involvement of gibberellins through the degradation of DELLA proteins that are repressors of gibberellin signaling. Furthermore, gibberellins can control the initiation of plant JA- and SA-dependent defense responses (Hermosa et al., 2012).

In the same way, the expression of genes known to be involved in the response of wheat to abiotic stress LEA2 and ABA8 increased during drought stress. The LEA2 gene is a key regulator of drought stress responses and was highly expressed in *T. harzianum* inoculated plants under severe water stress (Silletti et al., 2021). Besides that, it seems that the activation of the antioxidant machinery to recycle oxidized ascorbate is a mechanism activated by different *Trichoderma* strains in different plant species to improve tolerance to a range of abiotic stresses (Brotman et al., 2013). High levels of ascorbate in its reduced form can be found in plants treated with *Trichoderma*. In the regeneration of reduced ascorbate, the enzyme involved is monodehydroascorbate reductase (MDAR), whose gene is significantly upregulated and was found to increase the reduced ascorbic acid pool.

7.4.2.2 PHYSIOLOGICAL ADAPTATION

The *Trichoderma* treated plants accumulate secondary metabolites in order to protect them from salinity and drought damage, increased tolerance to oxidative stress and membrane protection during stress by increasing the amount of carotenoids, enhanced photosynthesis, and favored plant development by increasing indole acetic acid synthase, since this plant hormone is responsible for the regulation of a series of cellular mechanisms related to cell elongation, organ development, and plant growth. Proline levels are higher under severe drought and low N_2. With *Trichoderma*, levels of proline have been associated to stress adaptation (Silletti et al., 2021). In plants growing under heat stress, proline accumulation has been shown to upregulate the antioxidant defense system, elevated levels of enzymatic and non-enzymatic antioxidants leading to less oxidative damage, less membrane damage, and increased chlorophyll and water content and photosynthesis (Kaushal et al., 2011; Sehar et al., 2022). A significant improvement in the activities of carbon metabolism enzymes was also noted. *Trichoderma pseudokoningii* SMF2 produces trichokonin antimicrobial peptaibols that induce systemic resistance against gray mold caused by *Botrytis cinerea* on the orchid *Phalaenopsis*, but also promotes growth, since after treatment with trichokonin, both aboveground and belowground biomass increased, particularly root length, surface area, volume and branching (Zhao et al., 2018). In heat-stressed Lanzhou lily plants, trichokonins application conferred improved thermotolerance, where the activity of antioxidant enzyme systems such as SOD, CAT, and POD, the level of hormones, such as ABA, acids salicylic and jasmonic, relative water content, chlorophyll content, and net photosynthetic rate were promoted by trichokonins treatment.

Some conditions that are favored in plants treated with *T. harzianum* are stomatal conductance and net photosynthetic rate (Silletti et al., 2021). The stomatal conductance was higher at full irrigation regime under suboptimal fertilization an under optimal fertilization. The NH4+ and NO3- contents in roots, were higher under optimal nitrogen and water regimes and were always higher under optimal and suboptimal nitrogen. Highest NO3-contents were observed under optimal fertilization and 50% water. The application of *T. longibrachiatum* TG1 in plants improved seedling emergence and vigor and reduced the deleterious effects of NaCl stress and the pathogens on wheat seedling growth, increasing shoot and root length and enhancing root hair formation, which is beneficial for stressed plants (Boamah et al., 2021). The groundnut seeds that were treated with *T. virens* and *T. asperellum* increase shoots and biomass weight. These *Trichoderma* species increased the phenolics and flavonoids content, the levels of chlorophyll a, chlorophyll b, carotenoids, and the accumulation of IAA synthase in plants. The effects of increase varied depending on the stage of development of the plants and among leaves and roots (Yusnawan et al., 2021). Wheat seedlings with *T. longibrachiatum* treatments under different NaCl stress levels decreased diseases, increased the superoxide dismutase (SOD), peroxidase (POD), and catalase (CAT) activity and decreased malondialdehyde (MDA) and hydrogen peroxide (H2O2) content (Boamah et al., 2021).

Also, under salinity conditions *Trichoderma* spp. displayed changes in the compositions and relative amounts of volatile organic compounds that were produced (Yusnawan et al., 2021). *T. virens* produced mainly the VOCs isoprophyl-1-methyl, ledene oxide, and acetone, while *T. asperellum* produced mainly gurjunene, butanol, and himachalene. Apparently, these VOCs possess some properties among which antioxidant properties, with roles in cellular repair. Example of this is himachalene that acts as an antioxidant that is capable of reducing singlet oxygen activity in plants that are exposed to abiotic stresses and is involved in repairing cell damage (Rinkel and Dickschat, 2019).

7.5 TRICHODERMA AS BIOFERTILIZER: A NUTRITIONAL ENRICHER

Trichoderma is an organism with great capacities, known by different investigations over time that have determined its various functions (Woo et al., 2022). This fungus, as mentioned earlier in this chapter, acts in the protection of plants directly or indirectly against other pathogenic microorganisms, participates in the stimulation of vegetative growth, and is also attributed the ability to provide nutrients. Taking this last point into account, we will talk about the use of *Trichoderma* as a biofertilizer.

Biofertilizer is defined as a product based on microorganisms that have the ability to benefit plants, mainly due to the availability of various nutrients, so that it also contributes to improving some aspects of agricultural soil and increasing microbial activity (Bargaz et al., 2018; Daniel et al., 2022). There is a relationship that is positively established in the presence of *Trichoderma spp.* with agricultural land and plants, for example, we can talk about what was commented on by Zhang et al. (2019) in their research, where the filamentous fungus increased and developed alterations in the chemical compounds 1-hexacosene and 1-heptadecanol from the rhizosphere, as well as established a higher concentration of phosphorus (P) in the form of absorbable molecules for the plants. Derived from these situations, alfalfa cultivation improved, which is attributed to chemical compounds that increasing the activity of other microorganisms, point that will be discussed in a later subtopic. It is understandable that the availability of an element such as P has had an influence on alfalfa, being one of the essentials for the plant's activity. Another case is that of Bedine et al. (2022) who speak in their study about the ability of *T. gamsii* It-62 to transform P into a soluble molecule, demonstrating it first in in vitro tests and later in bean plants with P-deficient soil. Among the results appeared the use of this element for the characteristics favored in the development of the plants. The aforementioned activity has the element for the various species of *Trichoderma* based on the production of organic acids, production of enzymes such as phytase, or a relationship of both, as mentioned by Aishwarya et al. (2020). Under favorable conditions, various species can solubilize P, but in cases of abiotic stress, there must be an adequate selection of the species to guarantee its permanence under the conditions presented and have an effective activity on the element, as mentioned by Tandon et al. (2020). Within his research, *T. koningiopsis* NBRI-PR5 was tested under unfavorable conditions where an effect on nutrient availability is shown, the production of organic acids in an alkaline state provided HPO_4^{2-} in an absorbable form by plant roots. They also mention that it is a good species for use in degraded soils, looking for the opportunity to recover them. The study by Bononi et al. (2020) corroborates that soil conditions are a determining point for the P solubilization level, in addition to the form of application of the microorganism and its concentration. The same study mentions the high production of the enzyme such as phosphatase or phytase by *Trichoderma*, in addition to organic acids that are produced at the time of nutrient solubilization, such as lactic acid, fumaric acid, ascorbic acid, gluconic acid, d-malic acid, d-isocitric acid, citric acid, and phytic acid. For the effect of transforming P into an absorbable nutrient, the conditions that the soil presents are important, speaking about physical and chemical properties, hence the advances that today have been developed with biofertilizers based on *Trichoderma spp.*

Fu et al. (2021) make a proposal based on the results found in their study of the cultivation of maize, they observed the capacities by applications of *T. asperellum* spores, the pH values and especially the salinity in the rhizosphere are significantly reduced, which can be a factor determinant negative in the development of the crop. Emphasis is placed on an increase in permeability and the balance of water-salt in the saline-alkaline soil, being a high-activity effect of the fungus enzymes so that the plants tolerate the stress caused by the sodic soil conditions. In the same way, it is highlighted that not only P is modified to a soluble form; by following the same investigation it is mentioned that in the soil analyzed after the applications of the fungus, a higher concentration of Ca^{2+}, Mg^{2+}, K^{+}, and N was found. Continuing with the availability of other elements, talk about the role of *T. virens* for the efficiency in the disposition of nitrogen in absorbable form for the roots of lettuce and arugula in natural conditions and by fertilization effect. Another case is due to the participation of *T. asperellum* that improved the absorption of N (NO_3^- / NH_4^+) by increasing and improving the root system of the plant, thus being one of the factors that define the best form of use of the nutrient (Singh et al., 2018).

Potassium (K) is another important macronutrient for plants. The possible solubilization of this nutrient by various species such as *T. asperelloides*, *T. harzianum*, *T. afroharzianum*, *T. viride,* and *T. atrobrunneum* has been reported in vitro in the Aleksandrov Agar medium. The characteristic that was presented was the change of the medium due to the production of acids by *Trichoderma*, the solubilization of Zn is reported in the same study in the same way by the previous species. It is mentioned that these elements can be present in an available and absorbable form in the rhizosphere when any of the already identified *Trichoderma* species are present (Hewedy et al., 2022). Similarly, organic acids participate in the solubilization of the Zn, but the extrusion of protons and the production of chelating agents have also been reported to make the nutrient available to the root system of the plant (Anuradha and Syed, 2016). Likewise, Trichoderma has activity on iron (Fe) in low concentrations in the soil, providing it in an assimilable form through the production of siderophores, in addition to contributing by maintaining an adequate balance of properties in the rhizosphere (Ansari et al., 2017). *T. asperellum* has been reported to be active in media with iron deficiency, through the production of siderophores it allowed a better development of the plants. Species such as *T. viride*, *T. harzianum,* and *T. longibrachiatum* produce two types of siderophores known as hydroxymate and carboxylate (Zhao et al., 2020; Ghosh et al., 2017). In the corn crop, from the seed germination stage to the continuous growth, significant results were manifested by the application of *Trichoderma*, before in vitro evaluations were carried out to determine various activities of which the production of siderophores was important. Among the isolated strains, *T. harzianum*, *T. koningiopsis*, and *T. asperellum* stood out for their activity (Kumar et al., 2017).

The availability of some nutrients that Trichoderma induces has been observed from in vitro tests to field experiments, where the composition of the soil shows very low or absent amounts of elements without the action of the fungus. In this regard, once the various minerals are assimilable, it is easier for the roots of the plant to take them, so that when studying the composition of flowers, roots, shoots, seeds, etc., we can find the elements in them. Awad-Allah et al. (2011), during their study, found in tomato leaf the content of N, P, K, Mg, Ca, Fe, Zn, and Cu, mentioning that the characteristics of the plants inside the greenhouse were benefited by *T. viride* and *T. harzianum* in individual preparations or consortium. In the zucchini crop, a higher concentration of P and Mg has also been reported by Formisano et al. (2021), which were significantly different from the control, but among them were Ca, S, and K without any significant difference with another treatment and therefore for the researchers it was not very relevant. In the case of the roots, they presented better characteristics when being in endophytic interaction with *Trichoderma*, the authors mention that the availability of P had a role in the morphological changes. The pistachio fertilized with *T. harzianum*, *T. viride* alone, or in consortium had interesting effects on the leaves, branches, and in general leaf area, where K, Fe, P, Ca, Mg, and Zn were present. Within the comments in the research, it is expected that both species will be produced with the purpose of biofertilizer for agronomic activity within the crop and surely on another (Hosseinzeynali et al., 2020).

The diversity of *Trichoderma* generates a viable opportunity for its use for crop improvement purposes, by having macronutrients and micronutrients available in rhizospheric soils in conditions that can be assimilated mainly by plant roots, creating a very interesting fungus-plant-soil interaction. In addition to providing an improvement to soil conditions to maintain a state of greater nutrient productivity and optimal properties such as pH and salinity. However, it must be taken into account that it is a priority to ensure the development of their activities for each species since each one adapts differently to the conditions in which they are evaluated. Among other aspects to consider are that each species will transport and have different minerals that can be assimilated by plants, in addition to other mechanisms that, when they interact with other organisms, it will be difficult to control the specific activity and each species will determine if it will use one or more functions to provide the best conditions for the crop. The form, solid or liquid, of the inoculation and the times of application will also determine its effectiveness (Abdul-Halim et al., 2023).

7.5.1 INCREASING THE EFFICIENCY OF CHEMICAL FERTILIZERS

The participation of chemical fertilizers has been of great help in the benefits granted to crops, but mainly the misuse of them leads to environmental problems that take their toll today. However, even in the face of the problems generated by their applications, it is currently difficult to put them aside and look for other alternatives, although their benefits in short times in agricultural systems make them highly desirable (Abdul-Halim et al., 2023; Sharma and Singhvi, 2017; Craswell, 2021). Given the situation presented of the problems and the search for a sustainable agriculture, it has been found that *Trichoderma* can work in collaboration with established fertilizers, in order to further reduce the concentrations of chemical products and increase biological solutions. Said activity will surely seek to maintain the optimal results of the crops and mainly reduce the contamination of the environment, water, soil, and the preservation of beneficial microorganisms. Studies speak of the possible relationship between fertilization doses with bioproducts based on *Trichoderma*. Khan et al. (2017) during his research, obtained important results from a *T. harzianum*/composting applied individually on tomato plants, but also in concentrations of 75% with 25% N, 50% with 50% N, 25% with 75% N, recommended dose of NPK (100%) and the control (without any application). The results indicate that BioF/compost (100%) increased the minerals K, Cu, Fe, and Zn in the roots of the plant, but it worked the same in relation to the different doses of N. In the case of K and Zn, the best significant result appeared in the treatment of 50% BioF/Compost and 50% N. Similar results were presented in the shoots and fruit, as a final conclusion the authors mention that the 100% bioorganic fertilizer generates efficient results that facilitates the substitution of 100% inorganic fertilizers, but although sometimes 50% both also improve the availability of assimilable nutrients. Khan (2018) reports in his research on fertilization applications in rice at 100% N produces excellent results, but in doses of 75% N, deficiencies in the crop begin to be noticed. However, when applying 75% N + *Trichoderma* and 75% N + *Azospirillium* + *Trichoderma*, similar results are maintained at a dose of 100% of the chemical product. As a highlight of the fungus, it is the possibility that it is producing other components to maintain a balanced level with the fertilizer at 100%. The physiological improvements of the plant will be derived from the availability of nutrients from the rhizosphere, the best combinations of chemical products with biofertilizers will increase crop yields. Based on this, the perfect interaction between the two must be presented, biochar + *Trichoderma* + 50% dose of

NPK is a perfect equation for the absorption of N, P, K, Fe, and Zn by plants and these move to different sites inside of her. The aforementioned formula outperformed fertilization-based treatments (standard NPK dose, 50% NPK dose [control]), proposing a viable alternative to reduce the use of chemical products (Sani et al., 2020). The purpose of the combinations of biofertilizers with inorganic fertilizers is to reduce the excessive use of the latter and increase the activity and presence of microorganisms, from there are derived other no less important ones such as lowering the input costs, reducing contamination, and increasing the soil fertility. The plant, even reducing the concentration of the inorganic product a little or to a greater extent, but with a good proportion of the bioorganic alternative, will maintain or increase the results in final yield, mainly due to the availability of nutrients that are generated in the consortium between both, one of them provides in the rhizosphere and another transforms them into assimilable molecules for root systems (Barua et al., 2018). It is an advance to improve the quality of life of organisms and the environment. This is also pointed out by Pang et al. (2017) in their research where emphasis is placed on the following points: synthetic fertilization undoubtedly helps to promote benefits to the crop under proper use, the combination of chemical/organic fertilizer is also an acceptable option in terms of results but it is even better indicated the preparation of chemical fertilization + organic fertilization enriched with some species of *Trichoderma*. The latter, apart from providing nutrients, is a composition of attractants for other microorganisms that contribute to activities to promote vegetative growth and antagonism, demonstrated in the experiment they carried out. Within the investigations, the effectiveness of the interaction of *Trichoderma* with fertilizers of chemical or organic origin or both is demonstrated, allowing more adequate alternatives for the needs that today we are asked to solve to maintain a sustainable and safe environment for future generations.

7.5.2 ENHANCE MICROBIAL ACTIVITIES

Trichoderma inhabits soils and roots where it seeks its survival and in them, it performs beneficial functions that improve its physical-chemical conditions. During its establishment when applied as a biofertilizer, it attracts other microorganisms which interact with each other, promoting an increase in both fungi and bacteria. Mainly these microscopic organisms can participate as growth promoters or with antagonistic activity and in lower concentrations those called undesirables (pathogens) appear. It is difficult to determine which characteristic of *Trichoderma* spp. it is what attracts other organisms, but the combination of factors is what can promote this microbial contribution. For example, the availability of nutrients in the rhizosphere will attract new organisms by finding a better for their metabolic activities, less damaged soils would be another factor, and the exudates of the roots that had better growth and possibly a better availability of food if the biofertilizer was contributed with organic matter (Zhang et al., 2018). After *Trichoderma* inoculations, various microorganisms have been reported coexisting without presenting risks, including fungi of the phyla Glomeromycota, Chytridiomycota, and Zygomycota, as well as acidobacteria, bacteroidetes, firmicutes, and proteobacteria (Pang et al., 2017). Both beneficial and pathogenic bacteria and fungi of the above phyla may be present. Applications of *T. harzianum* reduced the presence of *Gibberella, Paraphoma,* and *Fusarium*, most of which are known to cause plant diseases. *Chaetomium, Tetracladium, Alternaria,* and *Mortierella* increased with the application of the bioproduct. Among the bacteria, mainly those that are beneficial in agriculture (*Rhizobium, Streptomyces, Steroidobacter,* and *Gaiella*) increased. Research by Liu et al. (2022) is a contribution to understand another function of *Trichoderma*, its attributions as an excellent functional microbiota that is surely reflected in the growth of crops. Fungi such as *Penicillium*, *Chaetomium*, *Aspergillus,* and *Acremonium* can be affected by *T. asperellum* but bacteria such as *Sphingomonas*, *Pseudomonas*, *Actinomadura,* and *Rhodanobacter* are the opposite case, they increase their participation. Apparently, *Trichoderma* tends to co-inhabit more easily with bacteria than with some fungi, it could be considered that there is more competition between the two for space and nutrients, causing less presence of the fungi kingdo (Zhang et al., 2020). However, Hang et al. (2022) observed a contrasting situation. After three years of *Trichoderma* applications, they attribute the most important role in promoting plant growth to fungal communities, and not to bacterial ones. The fungal community of the soils specifically showed increases in resident *Aspergillus* species. Interestingly, *Aspergillus tamarii* did not promote plant growth by itself but did so strongly after co-inoculation with *Trichoderma*.

When presenting a diversity of bacteria and fungi, they will be able to work in consortium and improve their activities, for example, *T. viride* and *P. fluorescens* when developing a biofertilizer based on both, the effects were positive on the availability of nutrients and consequently the physiological state of the plants. The same fungal species and *Azotobacter chroococcum* attributed benefits such as the production of organic acids which we know participate in the solubilization of minerals, exudates that increase the diversity of microorganisms, enzymes, organic matter, and physiological improvements of the plant (Velmourougane et al., 2017; Vij et al., 2022). *Serendipita indica* is a fungus that in the same way can be attracted by the conditions that are established by the induction of the biofertilizer based on the *Trichoderma* organism. In the reports, *T. simmonsii* is found to have good activity in said relationship, attributing important characteristics such as the control of *Phytopthtora capsici* and promoting nutrients between both for the cultivation of chili (Rokni et al., 2021). The constitution of a biofertilizer with any of the various known species of *Trichoderma* will attract other microorganisms that together will improve their beneficial capacities and provide an attribute for plants and soils alike, which will be a large solid consortium, rich in beneficial microbiota.

7.6 TRICHODERMA AS SOIL BIOREMEDIATOR

7.6.1 SYNTHETIC AGROCHEMICALS DEGRADER

Environment represents an equilibrium between free and bonded chemical elements. However, this equilibrium is often disrupted by population growth and demand for more yield and quality agricultural products. Often, the continuous applications of chemical compounds required to keep crops free of pests, leads to excessive, and uncontrolled use of chemical fertilizers and pesticides. These actions cause accumulation of pesticides and heavy metals in the soil negatively impacting on the environment and resulting in detrimental effects on the shelf life of agricultural products and human health quality.

An alternative to this global problem is bioremediation, a biological strategy that has been used in recent years. Endophytic fungi have attracted the attention of researchers for being a microorganism that adapts to adverse habitat conditions, capable of coexist in plant tissue without causing apparent damage (Khan and Mohiddin, 2018; Pietro-Souza et al., 2020). The *Trichoderma* genus is known as a potential bioremediator. Defining bioremediation as the degradation of pollutant agents present in the soil by using biological agents capable of exerting a crucial role in the restoration of soil health (Lorito et al., 2010). *Trichoderma* is a ubiquitous fungus, hence a wide variety of species of the genus is distributed worldwide. *Trichoderma* is a fungal endophyte that colonizes the first outer layer of the root (Poveda, 2020), activates defense systems through signaling to initiate the production of degrading enzymes of the cell wall (typical of mycoparasitism), and also promotes plant growth (Sood et al., 2020; Guzmán-Guzmán et al., 2019). *Trichoderma* genus have been applied as biocontrol agents with great benefits in sustainable agriculture. The antagonistic capacity of *Trichoderma* against pathogens has been known since the beginning of the 19th century (Harman, 2006). This genus is distinguished by rapid growth and colonization of the rhizosphere of the plant. Also is a potential antagonistic even in polluted environments. Representing all this, the main characteristics of a remedial microorganism (Hu et al., 2020). *Trichoderma* is a fungus with biotechnological potential for sustainable agriculture. It has a multifaceted role as a biofertilizer, biocontrol, and bioremediation. Therefore, bioremediation process with *Trichoderma* is economical (Argumedo-Delira et al., 2012). Also, *Trichoderma* fungus is adaptable to various substrates where it can develop and grow even in adverse conditions (Tripathi et al., 2013). There is a great diversity of *Trichoderma* strains with different characteristics, therefore, the ability to degrade pesticides or contaminants is specie specific but can be combined with other strains to offer a broader spectrum of decontamination (Bhandari et al., 2021).

7.6.1.1 DEGRADATION MECHANISMS

Trichoderma has demonstrated excellent qualities as a biological control microorganism even when constantly exposed to chemical pesticides, demonstrating effectiveness in degrading pesticides and tolerance (Tripathi et al., 2013). *Trichoderma* metabolizes polycyclic aromatic hydrocarbon compounds in soil, minimizing their molecular weight, and decreasing toxicity (Yao et al., 2015). Other mechanism used by *Trichoderma* for soil bioremediation is the production of enzymes that break down harmful chemicals resulting in a non-toxic compounds, restoring soil health by means of hydrolytic enzymes as amylases, cellulases, 1,3 β-glucanases, and chitinases (Kumar et al., 2012). Moreover, organic acids produced by *Trichoderma*, such as citric acid, fumaric acid and gluconic acid cause a decrease in soil pH, contributing to dissolution of manganese, iron, and magnesium that are metabolized by the plant. Mechanisms used by *Trichoderma* to degrade pesticides in the soil are mainly enzymatic (Li et al., 2016) coordinated metabolism and biosorption (Lu et al., 2016). Dichlorvos (2,2-dichlorovinyldimethyl phosphate) is a synthetic organophosphate pesticide commonly used as a broad-spectrum pesticide in agricultural industries (Asemoloye et al., 2019). Its residues in crop soils contaminate food chains impacting negatively on human health by inhibiting acetylcholinesterase (Fu et al., 2022). Previous studies have evidence of the ability of *Trichoderma atroviridae* T23 to degrade dichlorvos related to cytochrome P450, paraoxonase 1 (PON1), and ABC transporter system (Sun et al., 2022). Recent research finds that growth, sporulation, and pigmentation of T23 are inhibited in the presence of dichlorvos as an initial induction effect. Afterward, *Trichoderma* T23 synthesizes intracellular enzymes to degrade dichlorvos, possibly related energy, and antioxidant capacity in dichlorvos stress, concluding that some T23 dichlorvos degradation-inducing enzymes are also involved in dichlorvos degradation (Solanki et al., 2019). Additionally, it has been found that *Trichoderma* have the ability to degrade pesticides such as carbendazim and mancozeb, insecticides like chlorpyrifos and chlordane, and herbicides such as sulfonylurea (Vázquez et al., 2015). Given that, *Trichoderma* has great remedial potential and may be able to substitute synthetic chemical pesticides in a few years.

7.6.2 HEAVY METALS REMOVAL

Population growth and human activity have led to heavy metal accumulation. Combustion, emission of gases, and industries of waste, tanning, and dying are mainly the origin of heavy metal pollution. The accumulation of heavy metals in the soil also causes contamination in bodies of water, thus impacting the environment. Furthermore, it affects the metabolism of plants and animal health altering the food chain (Saba et al., 2012; Cai et al., 2019; Adimalla et al., 2020). Heavy metal is a metallic element whose density exceeds 5 g/cm^3, so it easily penetrates the soil and plants. The homeostasis system deteriorates and affects the transport of minerals from the soil to the roots. Soil structure, pH, and components are modified, affecting soil productivity and fertility (Wagh et al., 2018). A high concentration of these metals in soil triggers accumulation in all plant sections, causing phytotoxicity. And the biggest drawback of heavy metals is that they are not biodegradable (Olawoyin et al., 2018). The traditional methods to treat metal ionic contaminated soils are ion exchange, electrochemical destruction, biological stabilization, chemical precipitation, ozonation, chemical reduction or oxidation reverse osmosis, membrane processes, coagulation, and electro remediation, however, they are expensive and not efficient (Sharma et al., 2018; Piccolo et al., 2021). In addition, it presents high energy demand, sludge generation and requires periodic sorbent replacement (Kiwaan et al., 2021; AlHazmi et al., 2022).

7.6.2.1 BIOSORPTION

In recent years, sorption is a method of heavy metal decontamination, is efficient, easy to access, low cost, and low sludge production (Yaseen, 2021). Development of methods using indigenous microorganisms is promising, especially biological sorption is an eco-friendly way that allows the removal of heavy metals with high-efficiency (El-Bindary et al., 2022; Thakare et al., 2021). Biosorption is the ability of biomass to bind metal ions from soil, employing indirect metabolic activities or physicochemical adsorption methods without producing toxic substances. The biomass may contain living or dead microbial from microalgae, bacteria, yeasts, or fungi (Thakare et al., 2021; Şenol et al., 2021). Initially, functional groups from the cell wall surface interact with metal ions and are absorbed through chelation, precipitation, ion exchange, and physical absorption mechanisms (passive transport), metals remaining on the external sorbent surface. Another mechanism is that metal ions go through the cell wall (active transport), such as binding of metals with organic acids, peptides, polyphosphates, and sulfurs pass into the cell by intracellular uptake (Khan et al., 2021; Emara et al., 2020).

Trichoderma is beneficial fungi with great adaptability to grow in almost all soils, widely used as a biocontrol agent (Oliveira et al., 2021). *Trichoderma* has been used as a metal biosorbent for its fast growth even in stressful biotic and abiotic conditions (Ferreira and Musumeci, 2021; Dusengemungu et al., 2020). *Trichoderma* cell wall composition has suitable metal-binding properties in living and dead mycelium. However, living mycelium has an intracellular metal uptake that includes dealkylation, oxidation-reduction, methylation, and precipitation mechanisms, storing ions in vesicles and vacuoles (Wong et al., 2018; Alothman et al., 2020). The advantages of using *Trichoderma* as a biosorbent of heavy metal in soil are its easy availability, low cost, and chemical stability. The intracellular mechanisms of *Trichoderma* neutralize the hurt effects of ionic metals in the cytosol to protect their cells and those of the host plant (Bellion et al., 2006). Recent studies based *Trichoderma* genus have shown the remarkable capacity of the fungi to remove heavy metals from soil. Nazir et al. (2022) reported strains of *T. harzianum* and *T. pseudokoningii* autochthonous collected from tannery solid waste of Kasur with heavy metal resistance potential. While damage to potato crops *Trichoderma asperellum* Cu tolerant in combination with a lower dose of fungicide (copper) showed effectivity potato late blight disease caused by *Phytophthora infestans* (Ladi et al., 2020). Some strains of species of *Trichoderma* can adsorb metal ions like *Trichoderma* sp. and *T. atrobrunneum* that were isolated from contaminated soil of a tannery agency. Both strains could adsorb more metal ions when grown in high metal concentrations has been effective multi-metal biosorbent that can be used for bio-treatment of heavy metal polluted (Mushtaq et al., 2023). However, parameters like pH, temperature, and time of contact of fungi and mechanism of the interactions between autochthonous *Trichoderma* species and heavy metals need to be investigated.

7.6.3 NANOPARTICLES BIOSYNTHESIZER

Nowadays, a great global food demand corresponding to growth in population, makes necessary the increase the production of agricultural products. But it is also required to maintain healthy soils with the necessary nutrients, and for this a sustainable strategy must be employed to protect crops from possible damage caused by the pathogens. The excessive use of herbicides, insecticides, and chemical pesticides has caused crops' bioaccumulation of toxic chemical residues. Lately, nanostructured agrochemicals are an alternative to sustainable agriculture (Singh et al., 2021). Nanoparticles (NPs) reduce nutrient exhaustion, are economical, improve crop yields, and are eco-friendly (Salem and Fouda, 2021; Husen and Iqbal, 2019). It also reduces the use of conventional agrochemicals and fertilizer. Among the most used nanoparticles in agriculture are silver (AgNPs), oxide zinc (ZnONPs), and copper oxide (CuONPs) highly toxic elements at high concentrations and cumulative in crops and soil.

NPs can be obtained in two ways, physical-chemical or biosynthesis. The first option uses methods such as laser ablation, ultraviolet radiation, and ultrasonic fields that are very expensive and generate toxic waste that impacts the environment (Fraceto et al., 2018; Guilger-Casagrande and Lima, 2019), while biosynthesis is a straightforward, eco-friendly, and cost-effective method (Guilger-Casagrande and Lima, 2019; Mishra et al., 2014). Moreover, these novel eco-friendly synthesis methods are controllable and biocompatible for use in agriculture (Kareem et al., 2020). The biosynthesis of NPs can be achieved by using bacteria, algae, yeasts, fungi, or their bioproducts (Elegbede et al., 2020). This biological system is often known as nanofactories or nanoparticle producers, from which their cellular machinery, may be controlled to start the synthesis of NPs (Mahawar and Prasanna, 2018). Currently, NPs biosynthesis in agriculture is in the initial period of research where the first findings of microorganism-NPs-plant interaction have been at the laboratory level and in greenhouses (Alghuthaymi et al., 2021).

The biosynthesis of NPs based on filamentous fungi is potentially effective due to the generation of diverse metabolites, easy large-scale cultivation, biomass generator, tolerates adverse conditions such as high metals concentration, high pressure, and agitation (Guilger-Casagrande and Lima, 2019; Narayanan and Sakthivel, 2010; Salvadori et al., 2014). Furthermore, they have a high production of antimicrobial compounds and extracellular proteins involved in the capping mechanism (Guilger-Casagrande and Lima, 2019; Chauhan et al., 2022; Grasso et al., 2021). Among the fungi of interest for the biosynthesis of NPs are *Trichoderma, Fusarium, Penicillium,* and *Aspergillus* (Rai et al., 2021).

7.6.3.1 BIOSYNTHESIS OF NANOPARTICLES BY TRICHODERMA GENUS

Trichoderma genus has been extensively studied for its effectiveness as biocontrol agent. Its species are widely distributed in soil. Their potential activity against pathogenic microorganisms, plant growth promoter, increased yield output, fungicide, and biofertilizer makes them good candidates (Hermosa et al., 2012; Zin and Badaluddin, 2020). Furthermore, recent research has demonstrated that *Trichoderma* species tolerate environments with high concentrations of metals and can accumulate metals in their tissues. Mainly, these fungi produce a large variety of intracellular and extracellular bioactive metabolites such as proteins and enzymes that function as reducing agents, making fungal emerge as more suited microorganism for the biosynthesis of nanoparticles (Zhao et al., 2018; Ayad et al., 2018; Misra et al., 2022).

Trichoderma has a high tolerance to metals, a characteristic that gives it potential bioremediation agent for contaminated environments (Tripathi et al., 2013). Recently, their resistance to nano compounds has been reported; however, there is scarce information about the synthesis mechanism (Ramírez-Valdespino and Orrantia-Borunda, 2021).

Nanoparticle synthesis by *Trichoderma* can be performed by proteins, enzymes, and secondary metabolites (Elegbede et al., 2020), which can provide stability to the nanostructures formed (Mishra et al., 2014). Enzymes such as reductases take a reducing effect in the biosynthesis of NPs, which can provide additional biological activity, and is considered a key mechanism in capping (Mishra et al., 2014; Elegbede et al., 2020).

Enzymes and other secondary metabolites released by *Trichoderma* act as a protective mechanism in NPs biosynthesis by converting ions into elemental metal (Fraceto et al., 2018). Thus, extracellular metabolites can reduce and serve as stabilize agents favoring the biosynthesis of metallic NPs (Guilger-Casagrande and Lima, 2019). These stabilizing or protective effects, known as capping, mediate the physicochemical interaction of NPs with their surrounding environment (Grasso et al., 2021). In addition, NPs synthesized by *Trichoderma* species provide antimicrobial activity against different phytopathogenic microorganisms such as *Pseudomonas*, *Fusarium*, *Xanthomonas,* and *Aspergillus* (Shobha et al., 2020; Bilesky-Jose et al., 2021; Boruah and Dutta, 2021).

Among the main NPs biosynthesized by *Trichoderma* are silver (AgNPs), gold (AuNPs), zinc ZnONPs, copper (CuNPs), CuONPs, and selenium (SeNPs). The first finding of biosynthesis of silver nanoparticles was from *Trichoderma asperellum.* Biosynthesis was performed under regulated culture conditions supplemented with $AgNO_3$ 1 mM at 25 °C for five days (Mukherjee et al., 2008). Silver NPs have been highly studied due to their high bioactivity and antimicrobial power (Tomah et al., 2020). Furthermore, secondary metabolites secreted by *Trichoderma* reduce $AgNO_3$ salt, generating AgNPs (Guilger-Casagrande and Lima, 2019). The ZnONPs have been widely applied in agriculture for their usefulness in plants and soil (Shobha et al., 2020). It also presents antifungal activities against soil pathogens such as *Fusarium* spp., *Rhizhopus stolonifera*, *A. niger*, *Penicillium expansum*, and *Botrytis cinérea* and antibacterial activities of broad spectrum (Consolo et al., 2020). SeNPs synthesized by *Trichoderma* can be an option to use as Se salts. Previous studies indicate that SeNPs generate amides, which function as stabilizers in the capping process. However, regarding SeNPs, is neccesary more information, and more research. Gold NPs can be utilized to catalyze reactions, but also as antimicrobial agents, nanodiagnosis, drug nanocarrier, and nanodetection (do Nascimento et al., 2021). Research reports indicate that AuNPs possess antimicrobial activity against *Aspergillus* species, *E. coli, S, aureus*, *P. aeruginosa*, *K. granulomatis,* and *P. aeruginosa* (Mishra et al., 2014; Elegbede et al., 2020). Besides, AuNPs biosynthesis is economical and gold can be recovered. Research suggests that under metallic stress conditions, *T. harzianum* release cysteine encapsulates and stabilizes gold ions reducing gold ion toxicity (Tripathi et al., 2014). Especially, CuNPs have been widely used in the agricultural industry due to their antimicrobial selectivity at low concentrations presenting low toxicity. On the other hand, the biosynthesis of CuNPs is cheaper than AuNPs and AgNPs (Maqsood et al., 2020). Its antifungal efficiency is dependent on the concentration and size of the nanoparticle, being effective against *Fusarium* and *Aspergillus*. CuNPs modify the morphology and structural surface of the mycelium, therefore, the germination of conidia becomes weak and is inhibited. Subsequently, the cell wall is damaged and the intracellular content is released, causing cell death. The synthesis of structural components of the cell wall is inhibited, favoring the entry of CuNPs (Pariona et al., 2019).

7.6.3.2 EFFECTS OF NPS ON TRICHODERMA

Trichoderma and plants communicate bidirectionally, establishing the symbiotic effect between them. This relationship is generated when *Trichoderma* secondary metabolites are released into the rhizosphere, creating a favorable relationship (Madbouly, 2021). Then *Trichoderma* colonizes the roots of the plants, by biostimulating their response, growth promotion, enhanced yield and ability to alleviate adverse abiotic effects (Gupta and Bar, 2020). *Trichoderma* activates its response mechanism when the extracellular presence of ROS produced by metals is detected. However, research that describes the effect of *Trichoderma* is relatively recent and there is little information about it. Some defense mechanisms that *Trichoderma* activates in the presence of NPs are described below. The presence of AgNPs and CuNPs raises the enzymatic activity of peroxidase and catalase to neutralize the ROS species, which can be a mechanism of tolerance (Mahdizadeh et al., 2015). Faced with various types of stress, T*richoderma* activates proton transport systems to regulate pH and normalize growth (Ali and Hashem, 2007). This tolerance mechanism can occur in the presence of AuNPs where *Trichoderma* activates the synthesis of proton transporters to neutralize the effect, thus generating resistance to the nanocomposite. In the case of high concentrations of lead nanoparticles (up to 4000 mg/L), *Trichoderma* produces melanin in the chlamydospores activating the synthesis of enzymes to fix and convert lead into a less harmful molecule (Povedano-Priego et al., 2017). This detoxification mechanism of lead-contaminated soils tends to be more effective at pH with an acid tendency. Possibly this effect can be due to the balance of charges in the *Trichoderma* cell wall facilitating biosorption of metal ions (Errasquın and Vazquez, 2003). In the presence of Zinc, *Trichoderma* activates the detoxification mechanism when carbon sources are available, otherwise, if carbon sources are deficient, autolysis occurs. The copper absorption can be due to the negatively charged molecules on the cell surface, which have an affinity for metal ions.

Trichoderma is a key factor in safe biocontrol strategies. This fungus is closely in contact with plants, soil, and heavy metals that may be present in the ecosystem, where *Trichoderma* can activate its tolerance mechanism and form nanoparticles. However, the tolerance of *Trichoderma* to NPs needs to be better understood. It is well known that *Trichoderma* is a controllable biosynthesis source of NPs (Kareem et al., 2020). Recent investigations describe that biosynthesis depends on heavy metal concentration, species of *Trichoderm*a, fungal growth conditions, reagent concentration, reaction time, initial pH, and temperature (Ramírez-Valdespino and Orrantia-Borunda, 2021; Gemishev et al., 2019). Nevertheless, more research is needed to define optimal conditions to attain NPs with suitable characteristics.

7.7 FUTURE TRENDS/PERPECTIVES

Sustainable agriculture proposes restoring soil health through native microorganisms. Beneficial microbes include *Trichoderma,* a fungus characterized by its strong competitive and reproductive potential, as it adapts to unfavorable abiotic stress conditions, such as salinity, water stress, or the presence of various toxic chemicals. The species of the genus *Trichoderma* have multiple beneficial capacities for plants that positively influence the productivity and quality of crops and/or confer tolerance to biotic or abiotic stress. Currently, agriculture face various abiotic stresses derived from climate change, which reduce their productivity and the adequate food production. These conditions make agriculture highly dependent on the use of chemical fertilizers and pesticides that affect the environment and the nutritional quality, health status, and productivity of crops. Therefore, modern agriculture urgently requires alternatives to synthetic agrochemicals. The alternative of a *Trichoderma*-based fertilizer is present today in various places with different conditions, and even if one represents a good benefit in one place, it does not mean that it will be the same in another place, let alone if it presents different conditions. For this reason, before the applications, studies are carried out that guarantee their activity under the conditions presented by the site. Therefore, many species are already known today, but others that possibly have the same potential still need to be investigated. Although there are numerous studies of *Trichoderma* biological control, these are generally carried out in laboratory conditions where the substrate, temperature, pH, and humidity are usually very controlled. That is why it is necessary to increase the number of studies at the field level, where all the conditions can vary and therefore affect the behavior of *Trichoderma*. The innovation in biodegradation, biosorption, and nanosynthesis of particle processes generated by *Trichoderma* fungus and the evaluation of results are still developing. Therefore, researchers must overcome some inconveniences to scale agriculture toward green agrochemicals nano-structured with potential bioremediation that fulfill the demands of the actual global, preventing the development of fungal pathogens.

7.8 CONCLUDING REMARKS

The use of *Trichoderma* as a biological control has proven to be effective against a wide variety of phytopathogenic agents, however, methods must be sought to increase its efficiency, reliability, and persistence. Either using its wide variety of metabolites or developing new strategies of integrated management of diseases that involve microbial consortia, a combination of application methods, or even the combination with chemical fungicides seeking to reduce the current doses. Furthermore, the relationship of the three components planta-*Trichoderma*-soil, has characterized an agricultural benefit that comes in potential growth; The various species of *Trichoderma* reported in use as a biofertilizer, which have generated a promotion of nutrients in the rhizosphere, in such a way that they would be considered nutrient facilitators for the roots of the plants and with this they have been improving their characteristics during the physiological process. As a biostimulant, *Trichoderma* can act directly on plant physiology and metabolism, either by improving soil conditions, modifying molecular processes that improve the efficiency of crop water and nutrient use, stimulating plant development and counteracting abiotic stress by improving primary and secondary metabolism. However, since the positive effects depend on different factors, that interact with each other, it is necessary to continue evaluating in each agricultural region the behavior of the native strains of *Trichoderma* in the different cultivation systems, which include factors such as edaphoclimatic conditions, plant species, fertilizer management, and other biostimulants. It should be taken into account that the greatest benefits of these fungal species can only be obtained through an adequate combination of strategies defined by studies that consider the factors mentioned throughout this chapter.

KEYWORDS

- **biostimulant**
- **healthy soil**
- **bioremediation**
- **sustainability**
- **biocontrol**
- **biosynthesis**

REFERENCES

Abdul-Halim, A. M. A.-A.; Shivanand, P.; Krishnamoorthy, S.; Taha, H. A Review on the Biological Properties of *Trichoderma* spp. as a Prospective Biocontrol Agent and Biofertilizer. *J. Appl. Biol. Biotechnol.* **2023**.

Adimalla, N.; Chen, J.; Qian, H. Spatial Characteristics of Heavy Metal Contamination and Potential Human Health Risk Assessment of Urban Soils: A Case Study from an Urban Region of South India. *Ecotoxicol. Environ. Safety* **2020,** *194*, 110406.

Aishwarya, S.; Viswanath, H.; Singh, A.; Singh, R. Biosolubilization of Different Nutrients by *Trichoderma* spp. and their Mechanisms Involved: A Review. *Int. J. Adv. Agric. Sci. Technol.* **2020,** *7* (6), 34–39.

Alfiky, A.; Weisskopf, L. Deciphering *Trichoderma*–Plant–Pathogen Interactions for Better Development of Biocontrol Applications. *J. Fungi* **2021,** *7* (1), 61.

Alghuthaymi, M. A.; Kalia, A.; Bhardwaj, K.; Bhardwaj, P.; Abd-Elsalam, K. A.; Valis, M.; Kuca, K. Nanohybrid Antifungals for Control of Plant Diseases: Current Status and Future Perspectives. *J. Fungi* **2021,** *7* (1), 48.

AlHazmi, G. A.; AbouMelha, K. S.; El-Desouky, M. G.; El-Bindary, A. A. Effective Adsorption of Doxorubicin Hydrochloride on Zirconium Metal-Organic Framework: Equilibrium, Kinetic and Thermodynamic Studies. *J. Mol. Struct.* **2022,** *1258*, 132679.

Ali, E. H.; Hashem, M. Removal Efficiency of the Heavy Metals Zn (II), Pb (II) and Cd (II) by Saprolegnia delica and *Trichoderma viride* at Different pH Values and Temperature Degrees. *Mycobiology* **2007,** *35* (3), 135–144.

Alothman, Z. A.; Bahkali, A. H.; Khiyami, M. A.; Alfadul, S. M.; Wabaidur, S. M.; Alam, M.; Alfarhan, B. Z. Low Cost Biosorbents from Fungi for Heavy Metals Removal from Wastewater. *Sep. Sci. Technol.* **2020,** *55* (10), 1766–1775.

Ansari, R. A.; Mahmood, I.; Rizvi, R.; Sumbul, A.; Ansari, S. Siderophores: Augmentation of Soil Health and Crop Productivity. In *Probiotics in agroecosystem*, 2017; pp 291–312.

Anuradha, P.; Syed, I. Evaluation of Zinc Solubilization Capacity of Different Microbial Strains and their Effect on Bt Cotton. *Int. J. Agric. Sci.* **2016,** 0975–3710.

Argumedo-Delira, R.; Alarcón, A.; Ferrera-Cerrato, R.; Almaraz, J. J.; Peña-Cabriales, J. J. Tolerance and Growth of 11 Trichoderma Strains to Crude Oil, Naphthalene, Phenanthrene and Benzo [a] Pyrene. *J. Environ. Manag.* **2012,** *95*, S291–S299.

Asemoloye, M. D.; Jonathan, S. G.; Ahmad, R. Degradation of 2, 2-Dichlorovinyl Dimethyl Phosphate (dichlorvos) through the Rhizosphere Interaction between *Panicum Maximum* Jacq and Some Selected Fungi. *Chemosphere* **2019,** *221*, 403–411.

Awad-Allah, E. F.; Shams, A. H.; Helaly, A. A.; Ragheb, E. I. Effective Applications of *Trichoderma* spp. as Biofertilizers and Biocontrol Agents Mitigate Tomato Fusarium Wilt Disease. *Agriculture* **2022,** *12* (11), 1950.

Ayad, F.; Matallah-Boutiba, A.; Rouane-Hacene, O.; Bouderbala, M.; Boutiba, Z. Tolerance of *Trichoderma* sp. to Heavy Metals and its Antifungal Activity in Algerian Marine Environment. *J. Mar. Biol. Oceanogr.* **2018,** *7* (3), 2.

Bargaz, A.; Lyamlouli, K.; Chtouki, M.; Zeroual, Y.; Dhiba, D. Soil Microbial Resources for Improving Fertilizers Efficiency in an Integrated Plant Nutrient Management System. *Front. Microbiol.* **2018,** *9*, 1606.

Barua, S.; Molla, A.; Haque, M. M.; Alam, M. Performance of *Trichoderma*-Enriched Bio-organic Fertilizer in N Supplementation and Bottle Gourd Production in Field Condition. *Hortic. Int. J.* **2018,** *2* (3), 106–114.

Bedine, M. A. B.; Iacomi, B.; Tchameni, S. N.; Sameza, M. L.; Fekam, F. B. Harnessing the Phosphate-Solubilizing Ability of *Trichoderma* Strains to Improve Plant Growth, Phosphorus Uptake and Photosynthetic Pigment Contents in Common Bean (*Phaseolus vulgaris*). *Biocatal. Agric. Biotechnol.* **2022,** *45*, 102510.

Bellion, M.; Courbot, M.; Jacob, C.; Blaudez, D.; Chalot, M. Extracellular and Cellular Mechanisms Sustaining Metal Tolerance in Ectomycorrhizal Fungi. *FEMS Microbiol. Lett.* **2006,** *254* (2), 173–181.

Bhandari, S.; Pandey, K. R.; Joshi, Y. R.; Lamichhane, S. K. An Overview of Multifaceted Role of *Trichoderma* spp. for Sustainable Agriculture. *Arch. Agric. Environ. Sci.* **2021,** *6* (1), 72–79.

Bilesky-Jose, N.; Maruyama, C.; Germano-Costa, T.; Campos, E.; Carvalho, L.; Grillo, R.; Fraceto, L. F.; De Lima, R. Biogenic α-Fe_2O_3 Nanoparticles Enhance the Biological Activity of *Trichoderma* against the Plant Pathogen Sclerotinia sclerotiorum. *ACS Sustain. Chem. Eng.* **2021,** *9* (4), 1669–1683.

Boamah, S.; Zhang, S.; Xu, B.; Li, T.; Calderón-Urrea, A. *Trichoderma longibrachiatum* (TG1) Enhances Wheat Seedlings Tolerance to Salt Stress and Resistance to *Fusarium pseudograminearum*. *Front. Plant Sci.* **2021,** *12*, 741231.

Bononi, L.; Chiaramonte, J. B.; Pansa, C. C.; Moitinho, M. A.; Melo, I. S. Phosphorus-Solubilizing *Trichoderma* spp. from Amazon Soils Improve Soybean Plant Growth. *Sci. Rep.* **2020,** *10* (1), 2858.

Boruah, S.; Dutta, P. Fungus Mediated Biogenic Synthesis and Characterization of Chitosan Nanoparticles and its Combine Effect with *Trichoderma asperellum* against *Fusarium oxysporum, Sclerotium rolfsii* and *Rhizoctonia solani*. *Indian Phytopathol.* **2021,** *74* (1), 81–93.

Brotman, Y.; Landau, U.; Cuadros-Inostroza, A.; Takayuki, T.; Fernie, A. R.; Chet, I.; Viterbo, A.; Willmitzer, L. *Trichoderma*-Plant Root Colonization: Escaping Early Plant Defense Responses and Activation of the Antioxidant Machinery for Saline Stress Tolerance. *PLoS Pathog.* **2013,** *9* (3), e1003221.

Bulgari, R.; Franzoni, G.; Ferrante, A. Biostimulants Application in Horticultural Crops under Abiotic Stress Conditions. *Agronomy* **2019,** *9* (6), 306.

Cai, L.-M.; Wang, Q.-S.; Luo, J.; Chen, L.-G.; Zhu, R.-L.; Wang, S.; Tang, C.-H. Heavy Metal Contamination and Health Risk Assessment for Children Near a Large Cu-smelter in Central China. *Sci. Total Environ.* **2019,** *650*, 725–733.

Caruso, G.; El-Nakhel, C.; Rouphael, Y.; Comite, E.; Lombardi, N.; Cuciniello, A.; Woo, S. L. *Diplotaxis tenuifolia* (L.) DC. Yield and Quality as Influenced by Cropping Season, Protein Hydrolysates, and *Trichoderma* Applications. *Plants* **2020,** *9* (6), 697.

Chauhan, A.; Anand, J.; Parkash, V.; Rai, N. Biogenic Synthesis: A Sustainable Approach for Nanoparticles Synthesis Mediated by Fungi. In *Inorganic and Nano-Metal Chemistry,* 2022; pp 1–14.

Chien, Y.-C.; Huang, C.-H. Biocontrol of Bacterial Spot on Tomato by Foliar Spray and Growth Medium Application of *Bacillus amyloliquefaciens* and *Trichoderma asperellum. Eur. J. Plant Pathol.* **2020,** *156* (4), 995–1003.

Consolo, V. F.; Torres-Nicolini, A.; Alvarez, V. A. Mycosinthetized Ag, CuO and ZnO Nanoparticles from a Promising *Trichoderma harzianum* Strain and their Antifungal Potential against Important Phytopathogens. *Sci. Rep.* **2020,** *10* (1), 1–9.

Craswell, E. Fertilizers and Nitrate Pollution of Surface and Ground Water: An Increasingly Pervasive Global Problem. *SN Appl. Sci.* **2021,** *3* (4), 518.

Daniel, A. I.; Fadaka, A. O.; Gokul, A.; Bakare, O. O.; Aina, O.; Fisher, S.; Burt, A. F.; Mavumengwana, V.; Keyster, M.; Klein, A. Biofertilizer: The Future of Food Security and Food Safety. *Microorganisms* **2022,** *10* (6), 1220.

de França, S. K. S.; Cardoso, A. F.; Lustosa, D. C.; Ramos, E. M. L. S.; de Filippi, M. C. C.; da Silva, G. B. Biocontrol of Sheath Blight by *Trichoderma asperellum* in Tropical Lowland Rice. *Agron. Sustain. Dev.* **2015,** *35* (1), 317–324.

do Nascimento, J. M.; Cruz, N. D.; de Oliveira, G. R.; Sa, W. S.; de Oliveira, J. D.; Ribeiro, P. R. S.; Leite, S. G. Evaluation of the Kinetics of Gold Biosorption Processes and Consequent Biogenic Synthesis of AuNPs Mediated by the Fungus *Trichoderma harzianum. Environ. Technol. Innov.* **2021,** *21*, 101238.

Dusengemungu, L.; Kasali, G.; Gwanama, C.; Ouma, K. O. Recent Advances in Biosorption of Copper and Cobalt by Filamentous Fungi. *Front. Microbiol.* **2020,** *11*, 582016.

Dutta, P.; Deb, L.; Pandey, A. K. *Trichoderma*-from Lab Bench to Field Application: Looking Back over 50 Years. *Front. Agron.* **2022,** *4*, 932839.

El-Bindary, M.; El-Desouky, M.; El-Bindary, A. Adsorption of Industrial Dye from Aqueous Solutions onto Thermally Treated Green Adsorbent: A Complete Batch System Evaluation. *J. Mol. Liq.* **2022,** *346*, 117082.

Elegbede, J.; Lateef, A.; Azeez, M.; Asafa, T.; Yekeen, T.; Oladipo, I.; Aina, D.; Beukes, L.; Gueguim-Kana, E. Biofabrication of Gold Nanoparticles using Xylanases through Valorization of Corncob by *Aspergillus niger* and *Trichoderma longibrachiatum*: Antimicrobial, Antioxidant, Anticoagulant and Thrombolytic Activities. *Waste Biomass Valori.* **2020,** *11*, 781–791.

Emara, N.; Amin, R. M.; Youssef, A. F.; Elfeky, S. A. Recycling of Steel Industry Waste Acid in the Preparation of Fe_3O_4 Nanocomposite for Heavy Metals Remediation from Wastewater. *Rev. Chim.* **2020,** *71*, 34–46.

Errasquın, E. L.; Vazquez, C. Tolerance and Uptake of Heavy Metals by *Trichoderma atroviride* Isolated from Sludge. *Chemosphere* **2003,** *50* (1), 137–143.

Fang, H.; Li, C.; Zhao, J.; Zhao, C. Biotechnological Advances and Trends in Engineering *Trichoderma reesei* Towards Cellulase Hyperproducer. *Biotechnol. Bioproc. Eng.* **2021,** *26*, 517–528.

Ferreira, F. V.; Musumeci, M. A. *Trichoderma* as Biological Control Agent: Scope and Prospects to Improve Efficacy. *World J. Microbiol. Biotechnol.* **2021,** *37* (5), 90.

Formisano, L.; Miras-Moreno, B.; Ciriello, M.; El-Nakhel, C.; Corrado, G.; Lucini, L.; Colla, G.; Rouphael, Y. *Trichoderma* and Phosphite Elicited Distinctive Secondary Metabolite Signatures in Zucchini Squash Plants. *Agronomy* **2021,** *11* (6), 1205.

Fraceto, L. F.; Maruyama, C. R.; Guilger, M.; Mishra, S.; Keswani, C.; Singh, H. B.; de Lima, R. *Trichoderma harzianum*-Based Novel Formulations: Potential Applications for Management of Next-Gen Agricultural Challenges. *J. Chem. Technol. Biotechnol.* **2018,** *93* (8), 2056–2063.

Fu, H.; Tan, P.; Wang, R.; Li, S.; Liu, H.; Yang, Y.; Wu, Z. Advances in Organophosphorus Pesticides Pollution: Current Status and Challenges in Ecotoxicological, Sustainable Agriculture, and Degradation Strategies. *J. Hazard. Mater.* **2022,** *424*, 127494.

Fu, J.; Xiao, Y.; Wang, Y.-F.; Liu, Z.-H.; Yang, K. Saline–Alkaline Stress in Growing Maize Seedlings is Alleviated by *Trichoderma asperellum* through Regulation of the Soil Environment. *Sci. Rep.* **2021,** *11* (1), 11152.

Gemishev, O.; Panayotova, M.; Mintcheva, N.; Djerahov, L.; Tyuliev, G.; Gicheva, G. A Green Approach for Silver Nanoparticles Preparation by Cell-Free Extract from *Trichoderma reesei* Fungi and their Characterization. *Mater. Res. Express* **2019,** *6* (9), 095040.

Ghosh, S. K.; Banerjee, S.; Sengupta, C. Bioassay, Characterization and Estimation of Siderophores from Some Important Antagonistic Fungi. *J. Biopestic.* **2017,** *10* (2), 105–112.

Grasso, G.; Zane, D.; Dragone, R. Precision Microbial Nanobiosynthesis: Knowledge, Issues, and Potentiality for the In Vivo Tuning of Microbial Nanomaterials. In *Microbial Nanobiotechnology: Principles and Applications,* 2021; pp 75–112.

Guilger-Casagrande, M.; Lima, R. d. Synthesis of Silver Nanoparticles Mediated by Fungi: A Review. *Front. Bioeng. Biotechnol.* **2019,** *7*, 287.

Guzmán-Guzmán, P.; Kumar, A.; de los Santos-Villalobos, S.; Parra-Cota, F. I.; Orozco-Mosqueda, M. d. C.; Fadiji, A. E.; Hyder, S.; Babalola, O. O.; Santoyo, G. *Trichoderma* Species: Our Best Fungal Allies in the Biocontrol of Plant Diseases—A Review. *Plants* **2023,** *12* (3), 432.

Guzmán-Guzmán, P.; Porras-Troncoso, M. D.; Olmedo-Monfil, V.; Herrera-Estrella, A. *Trichoderma* Species: Versatile Plant Symbionts. *Phytopathology* **2019,** *109* (1), 6–16.

Hang, X.; Meng, L.; Ou, Y.; Shao, C.; Xiong, W.; Zhang, N.; Liu, H.; Li, R.; Shen, Q.; Kowalchuk, G. A. *Trichoderma*-Amended Biofertilizer Stimulates Soil Resident *Aspergillus* Population for Joint Plant Growth Promotion. *NPJ Biofilms Microbiom.* **2022,** *8* (1), 57.

Harman, G. E. Overview of Mechanisms and Uses of *Trichoderma* spp. *Phytopathology* **2006,** *96* (2), 190–194.

Harman, G. E.; Uphoff, N. Symbiotic Root-Endophytic Soil Microbes Improve Crop Productivity and Provide Environmental Benefits. *Scientifica* **2019,** *2019.*

Hermosa, R.; Viterbo, A.; Chet, I.; Monte, E. Plant-Beneficial Effects of *Trichoderma* and of its Genes. *Microbiology* **2012,** *158* (1), 17–25.

Hewedy, O. A.; Mansour, A.; Ali, M. G.; Lateif, K. S. A.; Ismaiel, M. H.; El-Meihy, R. M. Comprehensive Characterization and Screening of Different *Trichoderma* Isolates as Plant Growth Promoters: Insight to Metal Solubilization, Enzymatic Activity, and Antagonistic Effect, 2022.

Hosseinzeynali, A.; Abbaszadeh Dahaji, P.; Alaei, H.; Hosseinifard, J.; Akhgar, A. Effect of *Trichoderma* on Growth and Nutrition of Pistachio Trees under Common Garden Condition. *J. Sol Biol.* **2020,** *8* (2), 115–128.

Hu, J.; Chen, K.; Li, J.; Wei, Y.; Wang, Y.; Wu, Y.; Yang, H.; Zhou, Y.; Ryder, M.; Denton, M. D. Large-Scale *Trichoderma* Diversity was Associated with Ecosystem, Climate and Geographic Location. *Environ. Microbiol.* **2020,** *22* (3), 1011–1024.

Husen, A.; Iqbal, M. In *Nanomaterials and Plant Potential: An Overview*; Springer, 2019.

Illescas, M.; Rubio, M. B.; Hernández-Ruiz, V.; Morán-Diez, M. E.; Martínez de Alba, A. E.; Nicolás, C.; Monte, E.; Hermosa, R. Effect of Inorganic N Top Dressing and *Trichoderma harzianum* Seed-Inoculation on Crop Yield and the Shaping of Root Microbial Communities of Wheat Plants Cultivated under High Basal N Fertilization. *Front. Plant Sci.* **2020,** *11*, 575861.

Kareem, S.; Adeleye, T.; Ojo, R. In *Effects of pH, Temperature and Agitation on the Biosynthesis of Iron Nanoparticles Produced by Trichoderma Species*, IOP Conference Series: Materials Science and Engineering, IOP Publishing, 2020; p 012036.

Karlsson, M.; Atanasova, L.; Jensen, D. F.; Zeilinger, S. Necrotrophic Mycoparasites and their Genomes. *Microbiol. Spectr.* **2017,** *5* (2).

Kaushal, N.; Gupta, K.; Bhandhari, K.; Kumar, S.; Thakur, P.; Nayyar, H. Proline Induces Heat Tolerance in Chickpea (*Cicer arietinum* L.) Plants by Protecting Vital Enzymes of Carbon and Antioxidative Metabolism. *Physiol. Mol. Biol. Plants* **2011,** *17*, 203–213.

Ketta, H. A.; Hewedy, O. A. E.-R. Biological Control of *Phaseolus vulgaris* and *Pisum sativum* Root Rot Disease using *Trichoderma* Species. *Egypt. J. Biol. Pest Control* **2021,** *31*, 1–9.

Khan, H. I. Appraisal of Biofertilizers in Rice: To Supplement Inorganic Chemical Fertilizer. *Rice Sci.* **2018,** *25* (6), 357–362.

Khan, M. Y.; Haque, M. M.; Molla, A. H.; Rahman, M. M.; Alam, M. Z. Antioxidant Compounds and Minerals in Tomatoes by *Trichoderma*-Enriched Biofertilizer and their Relationship with the Soil Environments. *J. Integr. Agric.* **2017,** *16* (3), 691–703.

Khan, A. H. A.; Kiyani, A.; Mirza, C. R.; Butt, T. A.; Barros, R.; Ali, B.; Iqbal, M.; Yousaf, S. Ornamental Plants for the Phytoremediation of Heavy Metals: Present Knowledge and Future Perspectives. *Environ. Res.* **2021,** *195*, 110780.

Khan, M. R.; Mohiddin, F. A. *Trichoderma*: Its Multifarious Utility in Crop Improvement. In *Crop Improvement Through Microbial Biotechnology*; Elsevier, 2018; pp 263–291.

Khan, R. A. A.; Najeeb, S.; Hussain, S.; Xie, B.; Li, Y. Bioactive Secondary Metabolites from *Trichoderma* spp. against Phytopathogenic Fungi. *Microorganisms* **2020,** *8* (6), 817.

Kiwaan, H. A.; Mohamed, F. S.; El-Bindary, A. A.; El-Ghamaz, N. A.; Abo-Yassin, H. R.; El-Bindary, M. A. Synthesis, Identification and Application of Metal Organic Framework for Removal of Industrial Cationic Dyes. *J. Mol. Liq.* **2021,** *342*, 117435.

Köhl, J.; Kolnaar, R.; Ravensberg, W. J. Mode of Action of Microbial Biological Control Agents Against Plant Diseases: Relevance Beyond Efficacy. *Front. Plant Sci.* **2019,** *845*.

Kumar, K.; Manigundan, K.; Amaresan, N. Influence of Salt Tolerant *Trichoderma* spp. on Growth of Maize (*Zea mays*) Under Different Salinity Conditions. *J. Basic Microbiol.* **2017,** *57* (2), 141–150.

Kumar, D. P.; Singh, R. K.; Anupama, P.; Solanki, M. K.; Kumar, S.; Srivastava, A. K.; Singhal, P. K.; Arora, D. K. Studies on Exo-chitinase Production from *Trichoderma asperellum* UTP-16 and its Characterization. *Indian J. Microbiol.* **2012,** *52*, 388–395.

Lace, B.; Genre, A.; Woo, S.; Faccio, A.; Lorito, M.; Bonfante, P. Gate Crashing Arbuscular Mycorrhizas: *In Vivo* Imaging Shows the Extensive Colonization of Both Symbionts by *Trichoderma atroviride*. *Environ. Microbiol. Rep.* **2015,** *7* (1), 64–77.

Ladi, E.; Shukla, N.; Bohra, Y.; Tiwari, A. K.; Kumar, J. Copper *Trichoderma asperellum* Increases Bio-Efficacy of Copper Against *Phytophthora infestans* in Dual Combination. *Phytoparasitica* **2020,** *48*, 357–370.

Li, Y.; Sun, R.; Yu, J.; Saravanakumar, K.; Chen, J. Antagonistic and Biocontrol Potential of *Trichoderma asperellum* ZJSX5003 against the Maize Stalk Rot Pathogen *Fusarium graminearum*. *Indian J. Microbiol.* **2016,** *56*, 318–327.

Liu, L.; Xu, Y.; Cao, H.; Fan, Y.; Du, K.; Bu, X.; Gao, D. Effects of *Trichoderma harzianum* Biofertilizer on Growth, Yield, and Quality of Bupleurum Chinense. *Plant Direct* **2022,** *6* (11), e461.

Lo, C.-T.; Nelson, E.; Harman, G. Improved Biocontrol Efficacy of *Trichoderma harzianum* 1295-22 for Foliar Phases of Turf Diseases by use of Spray Applications. *Plant Dis.* **1997,** *81* (10), 1132–1138.

Lombardi, N.; Caira, S.; Troise, A. D.; Scaloni, A.; Vitaglione, P.; Vinale, F.; Marra, R.; Salzano, A. M.; Lorito, M.; Woo, S. L. *Trichoderma* Applications on Strawberry Plants Modulate the Physiological Processes Positively Affecting Fruit Production and Quality. *Front. Microbiol.* **2020,** *11*, 1364.

Lorito, M.; Woo, S. L.; Harman, G. E.; Monte, E. Translational Research on *Trichoderma*: From'Omics to the Field. *Ann. Rev. Phytopathol.* **2010,** *48*, 395–417.

Lu, T.; Zhang, Q.-L.; Yao, S.-J. Application of Biosorption and Biodegradation Functions of Fungi in Wastewater and Sludge Treatment. In *Fungal Applications in Sustainable Environmental Biotechnology,* 2016; pp 65–90.

Macías-Rodríguez, L.; Contreras-Cornejo, H. A.; Adame-Garnica, S. G.; Del-Val, E.; Larsen, J. The Interactions of *Trichoderma* at Multiple Trophic Levels: Inter-Kingdom Communication. *Microbiol. Res.* **2020,** *240*, 126552.

Macías-Rodríguez, L.; Guzmán-Gómez, A.; García-Juárez, P.; Contreras-Cornejo, H. A. *Trichoderma atroviride* Promotes Tomato Development and Alters the Root Exudation of Carbohydrates, which Stimulates Fungal Growth and the Biocontrol of the Phytopathogen Phytophthora Cinnamomi in a Tripartite Interaction System. *FEMS Microbiol. Ecol.* **2018,** *94* (9), fiy137.

Madbouly, A. K. Biodiversity of Genus *Trichoderma* and Their Potential Applications. In *Industrially Important Fungi for Sustainable Development: Volume 1: Biodiversity and Ecological Perspectives*; Springer, 2021; pp 429–460.

Mahawar, H.; Prasanna, R. Prospecting the Interactions of Nanoparticles with Beneficial Microorganisms for Developing Green Technologies for Agriculture. *Environ. Nanotechnol. Monit. Manag.* **2018,** *10*, 477–485.

Mahdizadeh, V.; Safaie, N.; Khelghatibana, F. Evaluation of Antifungal Activity of Silver Nanoparticles against Some Phytopathogenic Fungi and *Trichoderma harzianum. J. Crop Prot.* **2015,** *4* (3), 291–300.

Manganiello, G.; Sacco, A.; Ercolano, M. R.; Vinale, F.; Lanzuise, S.; Pascale, A.; Napolitano, M.; Lombardi, N.; Lorito, M.; Woo, S. L. Modulation of Tomato Response to *Rhizoctonia solani* by *Trichoderma harzianum* and its Secondary Metabolite Harzianic Acid. *Front. Microbiol.* **2018,** *9*, 1966.

Maqsood, S.; Qadir, S.; Hussain, A.; Asghar, A.; Saleem, R.; Zaheer, S.; Nayyar, N. Antifungal Properties of Copper Nanoparticles Against Aspergillus Niger. *Sch. Int. J. Biochem.* **2020,** *3*, 87–91.

Mau, Y. S.; Prayetno, R. S.; Kaka, H.; Naat, K. D.; Henuk, J. B. D.; Hahuly, M. V.; Widinugraheni, S.; Gandut, Y. R. Y. Efficacy of Indigenous *Trichoderma* Isolates of West

Timor, Indonesia, as Biocontrol Agents of Brown Spot (*Drechslera oryzae*) on Two Upland Rice Varieties. *Egypt. J. Biol. Pest Control* **2022,** *32* (1), 62.

Mishra, A.; Kumari, M.; Pandey, S.; Chaudhry, V.; Gupta, K.; Nautiyal, C. Biocatalytic and Antimicrobial Activities of Gold Nanoparticles Synthesized by *Trichoderma* sp. *Bioresour. Technol.* **2014,** *166*, 235–242.

Misra, M.; Chattopadhyay, S.; Sachan, A.; Sachan, S. G. Microbially Synthesized Nanoparticles and their Applications in Environmental Clean-up. *Environ. Technol. Rev.* **2022,** *11* (1), 18–32.

Mukherjee, P. K.; Mendoza-Mendoza, A.; Zeilinger, S.; Horwitz, B. A. Mycoparasitism as a Mechanism of *Trichoderma*-Mediated Suppression of Plant Diseases. *Fungal Biol. Rev.* **2022,** *39*, 15–33.

Mukherjee, P.; Roy, M.; Mandal, B.; Dey, G.; Mukherjee, P.; Ghatak, J.; Tyagi, A.; Kale, S. Green Synthesis of Highly Stabilized Nanocrystalline Silver Particles by a Non-pathogenic and Agriculturally Important Fungus *T. asperellum*. *Nanotechnology* **2008,** *19* (7), 075103.

Munns, R.; Gilliham, M. Salinity Tolerance of Crops–What is the Cost? *New Phytol.* **2015,** *208* (3), 668–673.

Mushtaq, S.; Bareen, F. e.; Tayyeb, A. Equilibrium Kinetics and Thermodynamic Studies on Biosorption of Heavy Metals by Metal-Resistant Strains of *Trichoderma* Isolated from Tannery Solid Waste. *Environ. Sci. Pollut. Res.* **2023,** *30* (4), 10925–10954.

Narayanan, K. B.; Sakthivel, N. Biological Synthesis of Metal Nanoparticles by Microbes. *Adv. Colloid Interface Sci.* **2010,** *156* (1-2), 1–13.

Nazir, A.; Shafiq, M.; Bareen, F.-e. Fungal Biostimulant-Driven Phytoextraction of Heavy Metals from Tannery Solid Waste Contaminated Soils. *Int. J. Phytoremediat.* **2022,** *24* (1), 47–58.

Olawoyin, R.; Schweitzer, L.; Zhang, K.; Okareh, O.; Slates, K. Index Analysis and Human Health Risk Model Application for Evaluating Ambient Air-Heavy Metal Contamination in Chemical Valley Sarnia. *Ecotoxicol. Environ. Safety* **2018,** *148*, 72–81.

Oliveira, A. F.; Florentino, A. C.; Ferreira, A. M.; Ferreira, I. M.; Costa, I. F.; Carvalho, J. T. Potential Absorption of Mercury-Contaminated Substrate by *Trichoderma* sp Isolated from Brazil Nuts and Amazon Soil. *Ciência e Natura* **2021,** *43*, e29–e29.

Pakora, G.-A.; Mpika, J.; Kone, D.; Ducamp, M.; Kebe, I.; Nay, B.; Buisson, D. Inhibition of Phytophthora Species, Agents of Cocoa Black Pod Disease, by Secondary Metabolites of *Trichoderma* Species. *Environ. Sci. Pollut. Res.* **2018,** *25*, 29901–29909.

Pang, G.; Cai, F.; Li, R.; Zhao, Z.; Li, R.; Gu, X.; Shen, Q.; Chen, W. Trichoderma-Enriched Organic Fertilizer can Mitigate Microbiome Degeneration of Monocropped Soil to Maintain Better Plant Growth. *Plant Soil* **2017,** *416*, 181–192.

Pariona, N.; Mtz-Enriquez, A. I.; Sánchez-Rangel, D.; Carrión, G.; Paraguay-Delgado, F.; Rosas-Saito, G. Green-Synthesized Copper Nanoparticles as a Potential Antifungal Against Plant Pathogens. *RSC Adv.* **2019,** *9* (33), 18835–18843.

Piccolo, A.; De Martino, A.; Scognamiglio, F.; Ricci, R.; Spaccini, R. Efficient Simultaneous Removal of Heavy Metals and Polychlorobiphenyls from a Polluted Industrial Site by Washing the Soil with Natural Humic Surfactants. *Environ. Sci. Pollut. Res.* **2021,** *28*, 25748–25757.

Pietro-Souza, W.; de Campos Pereira, F.; Mello, I. S.; Stachack, F. F. F.; Terezo, A. J.; da Cunha, C. N.; White, J. F.; Li, H.; Soares, M. A. Mercury Resistance and Bioremediation Mediated by Endophytic Fungi. *Chemosphere* **2020,** *240*, 124874.

Poveda, J. *Trichoderma parareesei* Favors the Tolerance of Rapeseed (*Brassica napus* L.) to Salinity and Drought Due to a Chorismate Mutase. *Agronomy* **2020,** *10* (1), 118.
-Priego, C.; Martín-Sánchez, I.; Jroundi, F.; Sánchez-Castro, I.; Merroun, M. L. Fungal Biomineralization of Lead Phosphates on the Surface of Lead Metal. *Miner. Eng.* **2017,** *106*, 46–54.
Rahimi Tamandegani, P.; Marik, T.; Zafari, D.; Balázs, D.; Vágvölgyi, C.; Szekeres, A.; Kredics, L. Changes in Peptaibol Production of Trichoderma Species during in Vitro Antagonistic Interactions with Fungal Plant Pathogens. *Biomolecules* **2020,** *10* (5), 730.
Rai, M.; Bonde, S.; Golinska, P.; Trzcińska-Wencel, J.; Gade, A.; Abd-Elsalam, K. A.; Shende, S.; Gaikwad, S.; Ingle, A. P. *Fusarium* as a Novel Fungus for the Synthesis of Nanoparticles: Mechanism and Applications. *J. Fungi* **2021,** *7* (2), 139.
Ramírez-Valdespino, C. A.; Orrantia-Borunda, E. *Trichoderma* and Nanotechnology in Sustainable Agriculture: A Review. *Front. Fungal Biol.* **2021,** 61.
Reino, J. L.; Guerrero, R. F.; Hernandez-Galan, R.; Collado, I. G. Secondary Metabolites from Species of the Biocontrol Agent *Trichoderma. Phytochem. Rev.* **2008,** *7*, 89–123.
Rinkel, J.; Dickschat, J. S. Stereochemical Investigations on the Biosynthesis of Achiral (Z)-γ-bisabolene in Cryptosporangium Arvum. *Beilstein J. Organic Chem.* **2019,** *15* (1), 789–794.
Rokni, N.; Alizadeh, H. S.; Bazgir, E.; Darvishnia, M.; Mirzaei-Najafgholi, H. The Tripartite Consortium of *Serendipita indica, Trichoderma simmonsii*, and Bell Pepper (*Capsicum annum*). *Biol. Control* **2021,** *158*, 104608.
Rouphael, Y.; Carillo, P.; Colla, G.; Fiorentino, N.; Sabatino, L.; El-Nakhel, C.; Giordano, M.; Pannico, A.; Cirillo, V.; Shabani, E. Appraisal of Combined Applications of *Trichoderma virens* and a Biopolymer-Based Biostimulant on Lettuce Agronomical, Physiological, and Qualitative Properties under Variable N Regimes. *Agronomy* **2020,** *10* (2), 196.
Rouphael, Y.; Colla, G. Toward a Sustainable Agriculture through Plant Biostimulants: From Experimental Data to Practical Applications. *MDPI* **2020,** *10*, 1461.
Rouphael, Y.; Lucini, L.; Miras-Moreno, B.; Colla, G.; Bonini, P.; Cardarelli, M. Metabolomic Responses of Maize Shoots and Roots Elicited by Combinatorial Seed Treatments with Microbial and Non-microbial Biostimulants. *Front. Microbiol.* **2020,** *11*, 664.
Saba, H.; Vibhash, D.; Manisha, M.; Prashant, K.; Farhan, H.; Tauseef, A. *Trichoderma*–a Promising Plant Growth Stimulator and Biocontrol Agent. *Mycosphere* **2012,** *3* (4), 524–531.
Salem, S. S.; Fouda, A. Green Synthesis of Metallic Nanoparticles and their Prospective Biotechnological Applications: An Overview. *Biol. Trace Element Res.* **2021,** *199*, 344–370.
Salvadori, M. R.; Ando, R. A.; Oller Do Nascimento, C. A.; Corrêa, B. Bioremediation from Wastewater and Extracellular Synthesis of Copper Nanoparticles by the Fungus *Trichoderma koningiopsis. J. Environ. Sci. Health A* **2014,** *49* (11), 1286–1295.
Sánchez-Montesinos, B.; Diánez, F.; Moreno-Gavira, A.; Gea, F. J.; Santos, M. Plant Growth Promotion and Biocontrol of Pythium Ultimum by Saline Tolerant Trichoderma Isolates under Salinity Stress. *Int. J. Environ. Res. Public Health* **2019,** *16* (11), 2053.
Sánchez-Montesinos, B.; Diánez, F.; Moreno-Gavíra, A.; Gea, F. J.; Santos, M. Role of *Trichoderma aggressivum* f. europaeum as Plant-Growth Promoter in Horticulture. *Agronomy* **2020,** *10* (7), 1004.
Sani, M. N. H.; Hasan, M.; Uddain, J.; Subramaniam, S. Impact of Application of *Trichoderma* and Biochar on Growth, Productivity and Nutritional Quality of Tomato under Reduced NPK Fertilization. *Ann. Agric. Sci.* **2020,** *65* (1), 107–115.

Sehar, Z.; Gautam, H.; Masood, A.; Khan, N. A. Ethylene-and Proline-Dependent Regulation of Antioxidant Enzymes to Mitigate Heat Stress and Boost Photosynthetic Efficacy in Wheat Plants. *J. Plant Growth Regul.* **2022,** 1–15.

Şenol, Z. M.; Gül, Ü. D.; Gurbanov, R.; Şimşek, S. Optimization the Removal of Lead Ions by Fungi: Explanation of the Mycosorption Mechanism. *J. Environ. Chem. Eng.* **2021,** *9* (2), 104760.

Şesan, T. E.; Oancea, A. O.; Ştefan, L. M.; Mănoiu, V. S.; Ghiurea, M.; Răut, I.; Constantinescu-Aruxandei, D.; Toma, A.; Savin, S.; Bira, A. F. Effects of Foliar Treatment with a *Trichoderma* Plant Biostimulant Consortium on *Passiflora caerulea* L. Yield and Quality. *Microorganisms* **2020,** *8* (1), 123.

Sharma, N.; Singhvi, R. Effects of Chemical Fertilizers and Pesticides on Human Health and Environment: A Review. *Int. J. Agric. Environ. Biotechnol.* **2017,** *10* (6), 675–680.

Sharma, S.; Tiwari, S.; Hasan, A.; Saxena, V.; Pandey, L. M. Recent Advances in Conventional and Contemporary Methods for Remediation of Heavy Metal-Contaminated Soils. *3 Biotech* **2018,** *8*, 1–18.

Shobha, B.; Lakshmeesha, T. R.; Ansari, M. A.; Almatroudi, A.; Alzohairy, M. A.; Basavaraju, S.; Alurappa, R.; Niranjana, S. R.; Chowdappa, S. Mycosynthesis of ZnO Nanoparticles using *Trichoderma* spp. Isolated from Rhizosphere Soils and its Synergistic Antibacterial Effect against *Xanthomonas oryzae* pv. oryzae. *J. Fungi* **2020,** *6* (3), 181.

Silletti, S.; Di Stasio, E.; Van Oosten, M. J.; Ventorino, V.; Pepe, O.; Napolitano, M.; Marra, R.; Woo, S. L.; Cirillo, V.; Maggio, A. Biostimulant Activity of *Azotobacter chroococcum* and *Trichoderma harzianum* in Durum Wheat under Water and Nitrogen Deficiency. *Agronomy* **2021,** *11* (2), 380.

Singh, R. P.; Handa, R.; Manchanda, G. Nanoparticles in Sustainable Agriculture: An Emerging Opportunity. *J. Controll. Release* **2021,** *329*, 1234–1248.

Singh, A.; Shukla, N.; Kabadwal, B.; Tewari, A.; Kumar, J. Review on Plant-*Trichoderma*-Pathogen Interaction. *Int. J. Curr. Microbiol. Appl. Sci.* **2018,** *7* (2), 2382–2397.

Solanki, M. K.; Kashyap, B. K.; Solanki, A. C.; Malviya, M. K.; Surapathrudu, K. Helpful Linkages of *Trichoderma* s in the Process of Mycoremediation and Mycorestoration. In *Plant Health Under Biotic Stress: Volume 2: Microbial Interactions,* 2019; pp 51–64.

Sood, M.; Kapoor, D.; Kumar, V.; Sheteiwy, M. S.; Ramakrishnan, M.; Landi, M.; Araniti, F.; Sharma, A. *Trichoderma*: The "Secrets" of a Multitalented Biocontrol Agent. *Plants* **2020,** *9* (6), 762.

Sridharan, A.; Thankappan, S.; Karthikeyan, G.; Uthandi, S. Comprehensive Profiling of the VOCs of *Trichoderma longibrachiatum* EF5 while Interacting with *Sclerotium rolfsii* and *Macrophomina phaseolina*. *Microbiol. Res.* **2020,** *236*, 126436.

Stenberg, J. A.; Sundh, I.; Becher, P. G.; Björkman, C.; Dubey, M.; Egan, P. A.; Friberg, H.; Gil, J. F.; Jensen, D. F.; Jonsson, M. When is it Biological Control? A Framework of Definitions, Mechanisms, and Classifications. *J. Pest Sci.* **2021,** *94* (3), 665–676.

Sun, J.; Karuppiah, V.; Li, Y.; Pandian, S.; Kumaran, S.; Chen, J. Role of Cytochrome P450 Genes of *Trichoderma atroviride* T23 on the Resistance and Degradation of Dichlorvos. *Chemosphere* **2022,** *290*, 133173.

Tandon, A.; Fatima, T.; Shukla, D.; Tripathi, P.; Srivastava, S.; Singh, P. C. Phosphate Solubilization by *Trichoderma koningiopsis* (NBRI-PR5) under Abiotic Stress Conditions. *J. King Saud Univ. Sci.* **2020,** *32* (1), 791–798.

Thakare, M.; Sarma, H.; Datar, S.; Roy, A.; Pawar, P.; Gupta, K.; Pandit, S.; Prasad, R. Understanding the Holistic Approach to Plant-Microbe Remediation Technologies for

Removing Heavy Metals and Radionuclides from Soil. *Curr. Res. Biotechnol.* **2021,** *3*, 84–98.

Tomah, A. A.; Alamer, I. S. A.; Li, B.; Zhang, J.-Z. Mycosynthesis of Silver Nanoparticles using Screened Trichoderma Isolates and their Antifungal Activity against *Sclerotinia sclerotiorum*. *Nanomaterials* **2020,** *10* (10), 1955.

Tripathi, R.; Gupta, R. K.; Singh, P.; Bhadwal, A. S.; Shrivastav, A.; Kumar, N.; Shrivastav, B. Ultra-sensitive Detection of Mercury (II) Ions in Water Sample using Gold Nanoparticles Synthesized by *Trichoderma harzianum* and their Mechanistic Approach. *Sens. Actuators B Chem.* **2014,** *204*, 637–646.

Tripathi, P.; Singh, P. C.; Mishra, A.; Chauhan, P. S.; Dwivedi, S.; Bais, R. T.; Tripathi, R. D. *Trichoderma*: A Potential Bioremediator for Environmental Clean Up. *Clean Technol. Environ. Policy* **2013,** *15*, 541–550.

Wong, C.; Tan, L.; Mujahid, A.; Lihan, S.; Wee, J.; Ting, L.; Müller, M. Biosorption of Copper by Endophytic Fungi Isolated from *Nepenthes ampullaria*. *Lett. Appl. Microbiol.* **2018,** *67* (4), 384–391.

Woo, S. L.; Hermosa, R.; Lorito, M.; Monte, E. *Trichoderma*: A Multipurpose, Plant-Beneficial Microorganism for Eco-sustainable Agriculture. *Nat. Rev. Microbiol.* **2022,** 1–15.

Woo, S. L.; Pepe, O. Microbial Consortia: Promising Probiotics as Plant Biostimulants for Sustainable Agriculture. *Front. Plant Sci.* **2018,** *9*, 1801.

Wu, J.; Zhu, J.; Zhang, D.; Cheng, H.; Hao, B.; Cao, A.; Yan, D.; Wang, Q.; Li, Y. Beneficial Effect on the Soil Microenvironment of Trichoderma Applied after Fumigation for Cucumber Production. *PLos One* **2022,** *17* (8), e0266347.

Vázquez, M. B.; Barrera, V.; Bianchinotti, V. Molecular Identification of Three Isolates of *Trichoderma harzianum* Isolated from Agricultural Soils in Argentina, and their Abilities to Detoxify In Vitro Metsulfuron Methyl. *Botany* **2015,** *93* (11), 793–800.

Velmourougane, K.; Prasanna, R.; Singh, S.; Chawla, G.; Kumar, A.; Saxena, A. K. Modulating Rhizosphere Colonisation, Plant Growth, Soil Nutrient Availability and Plant Defense Enzyme Activity through *Trichoderma viride-Azotobacter chroococcum* Biofilm Inoculation in Chickpea. *Plant Soil* **2017,** *421*, 157–174.

Vij, S.; Sharma, N.; Sharma, M.; Mohanta, T. K.; Kaushik, P. Application of Trichoderma viride and Pseudomonas Fluorescens to Cabbage (*Brassica oleracea* L.) Improves both its Seedling Quality and Field Performance. *Sustainability* **2022,** *14* (13), 7583.

Visconti, D.; Fiorentino, N.; Cozzolino, E.; Woo, S. L.; Fagnano, M.; Rouphael, Y. Can Trichoderma-Based Biostimulants Optimize N Use Efficiency and Stimulate Growth of Leafy Vegetables in Greenhouse Intensive Cropping Systems? *Agronomy* **2020,** *10* (1), 121.

Viterbo, A.; Horwitz, B. A. Mycoparasitism. In *Cellular and Molecular Biology of Filamentous Fungi,* 2010; pp 676–693.

Wagh, V. M.; Panaskar, D. B.; Mukate, S. V.; Gaikwad, S. K.; Muley, A. A.; Varade, A. M. Health Risk Assessment of Heavy Metal Contamination in Groundwater of Kadava River Basin, Nashik, India. *Model. Earth Syst. Environ.* **2018,** *4*, 969–980.

Yao, L.; Teng, Y.; Luo, Y.; Christie, P.; Ma, W.; Liu, F.; Wu, Y.; Luo, Y.; Li, Z. Biodegradation of Polycyclic Aromatic Hydrocarbons (PAHs) by *Trichoderma reesei* FS10-C and Effect of Bioaugmentation on an Aged PAH-Contaminated Soil. *Bioremediat. J.* **2015,** *19* (1), 9–17.

Yaseen, Z. M. An Insight into Machine Learning Models Era in Simulating Soil, Water Bodies and Adsorption Heavy Metals: Review, Challenges and Solutions. *Chemosphere* **2021,** *277*, 130126.

Yusnawan, E.; Taufiq, A.; Wijanarko, A.; Susilowati, D. N.; Praptana, R. H.; Chandra-Hioe, M. V.; Supriyo, A.; Inayati, A. Changes in Volatile Organic Compounds from Salt-Tolerant *Trichoderma* and the Biochemical Response and Growth Performance in Saline-Stressed Groundnut. *Sustainability* **2021,** *13* (23), 13226.

Zachow, C.; Berg, C.; Müller, H.; Meincke, R.; Komon-Zelazowska, M.; Druzhinina, I. S.; Kubicek, C. P.; Berg, G. Fungal Diversity in the Rhizosphere of Endemic Plant Species of Tenerife (Canary Islands): Relationship to Vegetation Zones and Environmental Factors. *ISME J.* **2009,** *3* (1), 79–92.

Zin, N. A.; Badaluddin, N. A. Biological Functions of *Trichoderma* spp. for Agriculture Applications. *Ann. Agric. Sci.* **2020,** *65* (2), 168–178.

Zhang, F.; Huo, Y.; Cobb, A. B.; Luo, G.; Zhou, J.; Yang, G.; Wilson, G. W.; Zhang, Y. Trichoderma Biofertilizer Links to Altered Soil Chemistry, Altered Microbial Communities, and Improved Grassland Biomass. *Front. Microbiol.* **2018,** *9*, 848.

Zhang, Y.; Tian, C.; Xiao, J.; Wei, L.; Tian, Y.; Liang, Z. Soil Inoculation of *Trichoderma asperellum* M45a Regulates Rhizosphere Microbes and Triggers Watermelon Resistance to Fusarium wilt. *AMB Express* **2020,** *10*, 1–13.

Zhang, Q.; White, J. F. Bioprospecting Desert Plants for Endophytic and Biostimulant Microbes: A Strategy for Enhancing Agricultural Production in a Hotter, Drier Future. *Biology* **2021,** *10* (10), 961.

Zhang, F.; Xu, X.; Huo, Y.; Xiao, Y. *Trichoderma*-Inoculation and Mowing Synergistically Altered Soil Available Nutrients, Rhizosphere Chemical Compounds and Soil Microbial Community, Potentially Driving Alfalfa Growth. *Front. Microbiol.* **2019,** *9*, 3241.

Zhao, P.; Ren, A.; Dong, P.; Sheng, Y.; Li, D. Antimicrobial Peptaibols, Trichokonins, Inhibit Mycelial Growth and Sporulation and Induce Cell Apoptosis in the Pathogenic Fungus *Botrytis cinerea*. *Appl. Biochem. Microbiol.* **2018,** *54* (4), 396–403.

Zhao, L.; Wang, Y.; Kong, S. Effects of *Trichoderma asperellum* and Its Siderophores on Endogenous Auxin in *Arabidopsis thaliana* under Iron-Deficiency Stress. *Int. Microbiol.* **2020,** *23* (4), 501–509.

Zhao, X.; Zhou, L.; Riaz Rajoka, M. S.; Yan, L.; Jiang, C.; Shao, D.; Zhu, J.; Shi, J.; Huang, Q.; Yang, H. Fungal Silver Nanoparticles: Synthesis, Application and Challenges. *Crit. Rev. Biotechnol.* **2018,** *38* (6), 817–835.

CHAPTER 8

Use of Oregano as a Sustainable Perspective

YAHAIRA J. NAVARRO-MELGAREJO[1], MÓNICA L. CHÁVEZ-GONZALEZ[1*], CRISTÓBAL N. AGUILAR[1], CRISTIAN TORRES-LÉON[2], LEONARDO SEPULVEDA-TORRE[2], and ANNA ILINÁ[2]

[1]*Food Research Department, School of Chemistry, Universidad Autónoma de Coahuila, Saltillo, Coahuila, México*

[2]*Bioprocesses and Bioproducts Department, School of Chemistry, Universidad Autónoma de Coahuila, Saltillo, Coahuila, México*

[3]*Ethnobiological Garden and Research Center-UAdeC (CIJE), Universidad Autónoma de Coahuila, Saltillo Mexico*

ABSTRACT

The popularity of oregano in recent years has increased owing to the nutraceutical properties that have been found, including antioxidant, antiseptic antiviral, anti-inflammatory, and, very importantly, anti-chronic degenerative and neurodegenerative diseases. This potential is attributed to certain compounds known as phytochemicals. This plant includes different species cultivated worldwide, but mainly in Europe and America in areas generally with arid or semi-arid climates. However, the great commercial demand gives rise to irrational management of oregano, as most of the oregano that exists is wild species and there needs to be more organization for the management and agronomic improvement of this plant. Apart from the collection of oregano for use as a condiment, essential oils are extracted, which have industrial importance, such as pharmaceutical, cosmetic, and food. Although essential oil is well known for its bioactive compounds, there is very little information

Sustainable Engineering and Agro-Food Processing: Renewable and Nonrenewable Resources.
José Juan Buenrostro-Figueroa, Julio César Tafolla-Arellano, Mónica Chávez-González, Pedro Aguilar-Zárate, A. K. Haghi (Eds.)

regarding the residues that it generates; although there are different methods to extract these compounds, solid-state fermentation (SSF) is a methodology that has been explored for a few years, but which has shown promising results and the most important advantage of this is how environmentally friendly it is.

8.1 INTRODUCTION

In recent years, interest in aromatic plants has increased—oregano is one of them—as some research has shown its nutraceutical properties, among which we can also find antioxidant, antiseptic antiviral, anti-inflammatory properties, and very importantly, it combats chronic degenerative and neuro-degenerative diseases, this potential is attributed to certain compounds known as phytochemicals. Therefore, its value in the market is very important, in 2012, it was estimated world production of 15,000 tons being Türkiye the first producer, followed by Mexico, where the species *Lippia* is the most significant commercial production thanks to the quality of its essential oil contained in its leaves (García-Pérez et al., 2012), but current data report Mexico as the first producer contributing about 40% of world production leaving Türkiye with 30% and in third place Greece contributing 22.5% (Orona-Castillo et al., 2017).

This increase in production is important, mainly because the harvesting of this plant is carried out by low-income farmers, representing the only source of income and, as has been reported, commercialization is done mainly of the leaf, which is used as a condiment for a variety of dishes and also, considering its importance as a raw material in food and pharmaceutical industries both in Mexico and other countries (INIFAP, 2012). Foreign companies generally use oregano to extract essential oil from it, thus benefiting economically, leaving a very small economic margin for the farmers.

However, a large amount of waste is generated from oil extraction without being previously used. As Fríaz-Zepeda (2021) indicates, it is an important source of phenolic compounds with a high antioxidant potential, which is reported using a method of extraction by mixing ethanol with water.

Likewise, the interest in finding new extraction techniques for this type of residue is growing. Some examples are the conventional extraction methodologies, such as maceration, reflux heating, etc. An example of the use of these is the following research in which polyphenolic compounds were extracted from coffee residues by ultrasound-assisted extraction, obtaining a maximum of polyphenolic compounds of 1429.09 mg AG/g SCG (Marcelo-Díaz, 2017). Another research used common mango (*Mangifera indica* L.) residues to extract phenolic compounds using a direct technique with ultrasound, determining that this technique is an efficient process for obtaining these compounds (Estrada-Jaramillo et al., 2012). Thus, the use of conventional and emerging technologies has proven to be able to obtain good yields of polyphenolic compounds, however, these can be harmful to the environment and even expensive; however, in recent years, a technology has been studied which presents a wide variety of uses giving added value to waste (Martínez-Vera and Mendoza Camejo, 2018) and we refer to solid-state fermentation (SSF), with advantages such as simple techniques, higher yields, more economical, environmentally friendly and requires less energy (Soccol et al., 2017). Likewise, Bautista-Hernández et al. (2022), reported an increase in the content of polyphenolic compounds in fermented extracts using Mexican oregano residues as substrate, however the ideal parameters for the production of secondary metabolites from oregano essential oil extraction residues have not been explored, thus suggesting the importance of the Foundation for Ecological Security (FES) as a tool to increase the bioactive potential of the extracts.

In this context, the main objective of this chapter is to address the use of oregano residues generated in the industry, being SSF an alternative for their treatment, as well as contributing as an economic source for farmers.

8.2 OREGANO

8.2.1 BOTANICAL CHARACTERISTICS

Although there are different types of oregano in the world, the best-known genera are *Origanum* and *Lippia*. The *Origanum* genus is more common in Mediterranean countries, being native to the islands of Crete, southern Greece, while the *Lippia* genus is usually found in Mexico and other Latin American countries (Marieschi et al., 2009), this herbaceous perennial plant has a high concentration of essential oil between 1 and 8% m/v depending on the habitat (Karamanos and Sotiropoulou, 2013); for example, with *Lippia* genus possess stronger flavor and aromas compared to *Origanum* genus and this is due to the percentage of essential oil. In the case of oregano leaves for *Lippia* genus presents 3–4% of oil, while the percentage inside the Mediterranean oregano leaves is around 2–2.5% (Connor Black et al., 2016).

In general, this plant usually reaches a maximum height of up to 2 m and has woody stems that are highly branched from the base, its leaves are finely crenulate and tomentose, oblong or elliptic in shape, 1.5-cm wide and 1–1.5 cm long. The flowers are arranged in subglobose spikes, with white zygomorphic corollas. The fruits are dehiscent and dry (Mata-González and Meléndez-González, 2005). However, these specifications may vary depending on the location and type of species.

8.2.2 CHEMICAL COMPOSITION

Among the main components in oregano leaves, carvacrol, and thymol have been reported, the latter in higher percentages in the essential oil of oregano. However, in various investigations, other components have been identified as polyphenolic compounds, as shown in Table 8.1, which presents the different compounds that can be found depending on the species of oregano.

TABLE 8.1 Chemical Composition of the Different Oregano Species.

Oregano species	Chemical composition	References
Origanum vulgare L.	α-Pinene; *cis*-β-Terpineol; 1,6-Ocatien-3-ol,3,7,7-dimethyl; (+)-4-Carene; Benzene,1-methyl-4-(1-methyl; β-Myrcene	(Sánchez Tito and Layme Huanca, 2022)
Origanum spp.	α-Terpinene; Limonene; Linalol; Thymol; Carvacrol	(Flores et al., 2019)
O. vulgare subsp. *Hirtum* (Greek oregano)	Carvacrol; Thymol; p-Cymeno; γ-Terpinene linalol; para-Cymen; α-Terpinene	(Solís, 2011; Rodriguez et al., 2021)

TABLE 8.1 *(Continued)*

Oregano species	Chemical composition	References
Origanum majorana (Sicilian oregano)	Menthol; L-Mentone; Pulegona; Linalol; DL-limonene; 1,8-Cineole	(Barón-Malaver and García Rincón, 2019)
Origanum onites	Ferulic acid; Caffeic acid p-Hydroxybenzoic acid; Vanillic acid	(Gerothanassis et al., 1998)
Lippia graveolens (Mexican oregano)	Terpinene; Limonene; Linalol; Thymol; Carvacrol; Loganine; 8-*epi*-Loganic acid	(Gallegos-Flores et al., 2019; Martinez, 2003)
Lippia origanoides Kunth	p-Cymene; 1,8-Cineole; Thymol; and p-Cymene; Thymyl acetate; *trans*-β-Caryophyllene; α-Humulene	(Vazquez-Carreño, 2012)
Lippia alba	Borneol; Camphor; 1,8-Cineole; p-Cymene; Limonene, Linaool, Piperitone, Carvone, Thymol; α-Muurolene; *trans*-β-Caryophyllene; β-Cubebene	(Stashenko et al., 2003)
Lippia palmeri W.	α-Pinene; p-Cymene; Limonene; Carvacrol; Thymol; Eugenol; Borneol; α-Felnadrene; α-Terpinene; Linalol; Caryophyllene; α-Myrcene	(Corella-Bernal and Ortega-Nieblas, 2013)

In this way, it is observed that among the main compounds that can be found in the different species of oregano are carvacrol, thymol, cymonene, and linalool. These components provide antibacterial, antioxidant, antiviral, antifungal, and insecticidal activity (Aguirre, 2000). However, some studies have also reported components such as caffeic acid and catechin, lignan floretin (Gutierrez-Crijalva, 2017; Irakli, 2021; Cortés-Chital et al., 2021), and it has also been found to possess hydroxybenzoic acids and some conjugated amino acids (Bautista-Hernández et al., 2022).

All these compounds have been classified in several researches as volatile compounds and phenolic compounds, where the volatile compounds can be defined as those responsible for the sensory characteristics that, depending on their concentration, modify the aroma and flavor of oregano, while the phenolic compounds represent the bioactive compounds that the plant presents, giving it all the bioactivities mentioned above (Arias et al., 2020; Machado et al., 2010).

8.2.3 LOCATION IN MEXICO

Mexico is a country valued as a center of origin and domestication of various aromatic plants (Calvo-Irabien, 2018). Therefore, oregano is considered a nontimber forest resource in Mexico, occupying second place as a world producer of oregano with species of the *Lippia* family (Mastelić et al., 2008), which is regulated by the General Law of Sustainable Forest Development (SEMARNAT, 2021) and the NOM-007-RECNAT-1997 (SEMARNAT, 1997). It is a species distributed in states of the Mexican Republic with arid and semi-arid zones (Aguirre, 2000) such as Chihuahua, Tamaulipas, Hidalgo, Zacatecas, Jalisco, Durango, Coahuila, Oaxaca, and Baja California Sur (Villavicencio-Gutierrez and Martinez-Burciaga, 2007; SIAP, 2010). However, the main producers at the national level are Durango, Chihuahua, Baja California, Baja California Sur, State of Mexico, and Oaxaca, with a cultivated area of 57.16 ha and a production of 117.37 t (SIAP, 2018).

8.2.4 OREGANO SPECIES

Globally, there are four families of plants that are known under the name of "oregano," which are Lamiaceae, Verbenaceae, Asteraceae, and Fabaceae; however, Lamiaceae and Verbenaceae are two families that are the most recognized as there is more information, but there are still species that is studied its composition and certain biological properties (Brown et al., 2006).

Although oregano is well known around the world, there are four species of great economic importance: Turkish oregano (*O. onites* L.), Mexican oregano (*L. graveolens* Kunth), European oregano (*O. vulgare*), and Spanish oregano (*Coridohymus capitatu* (L) Hoffmanns and Links). For example, in the case of the genus Origanum, 38 species with 6 subspecies and 17 hybrids have been reported (Arcila-Lozano et al., 2004). In Mexico, the genus *Lippia* has about 40 known species of oregano, some endemic, distributed in several states of the Mexican Republic (see Fig. 8.1).

FIGURE 8.1 Mexican oregano (*L. graveolens*).

8.2.5 GLOBAL IMPORTANCE OF OREGANO

Interest in aromatic plants has been expanding for some years; in the case of oregano, its value in the market is based on its aroma, so it is widely used as a spice and condiment (Ait-Ouazzou et al., 2011). World production is estimated at around 15,000 tons, Türkiye being the leading producer, followed by Mexico, where the genus *Lippia* has the highest commercial production because of the quality of the essential oil contained in its leaves (García-Pérez et al., 2012). However, current data report Mexico as the leading producer, contributing around 40% of world production, leaving Türkiye with 30% and Greece in third place with 22.5% (Orona-Castillo et al., 2017). So, the oregano produced in Mexico is mainly destined for the United States, to which around 85% of national production is exported; 10% goes to the domestic market and 5% to European and Asian countries (CONABIO, 2005). On the contrary, in world production, it generates a value of approximately 22.5 million dollars, although according to data from the National Forestry Commission (CONAFOR), in 2005 oregano sales added up to more than 75 billion euros (CONAFOR, 2009; Koksal et al., 2010).

8.2.6 MAJOR OREGANO PRODUCING COUNTRIES

Among the largest producers of oregano worldwide are the countries of Peru, Türkiye, Greece, Argentina, Mexico, Spain, Syria, Morocco, Egypt, and Israel; however, the origin of this plant is in the Mediterranean area (UNC, 2013). Thereby, the most common species are *O. vulgare*, being native to Europe, and *L. graveolens*, of Mexican origin (Arcila et al., 2004). In the case of *Origanum vulgare*, it is usually very abundant in northern Spain, growing on the shores of some streams, sunny slopes, and rocky areas, on the contrary, some subspecies extend from countries in Europe and Asia to areas of northeastern Africa (Di Fabio et al., 2005). And in the case of the species *L. graveolens*, it extends over most of the Mexican Republic, but mainly in arid and semi-arid areas of the country. However, even though the main producers of oregano are countries in Europe and Mexico, there are now countries in South America such as Peru, Argentina, and Colombia, among others, which have located areas of oregano production and are considered to be one of the most profitable cultivations (Condori-Tintaya and Fernández Cutire, 2022).

8.3 APPLICATIONS

8.3.1 CONSUMPTION IN FRESH

Oregano leaves are generally part of the plant that is most used; for example, the consumption of this culinary herb is distinguished in different dishes such as pizza, for sauces, or some Mediterranean dishes, also oregano infusions have been used in traditional so for the treatment of digestive diseases and not only in that but also, to make cosmetics, drugs, liquors, and the essence obtained by distillation is commonly used for perfumery and some soaps (Muños, 1978); the reason why it makes it a successful export product (Ait-Ouazzou et al., 2011). Some authors suggest that using these plants began with preserving food or reducing unpleasant odors (Simpson and Gorzaly, 2001). Figure 8.2 shows different uses that we can commonly find for oregano.

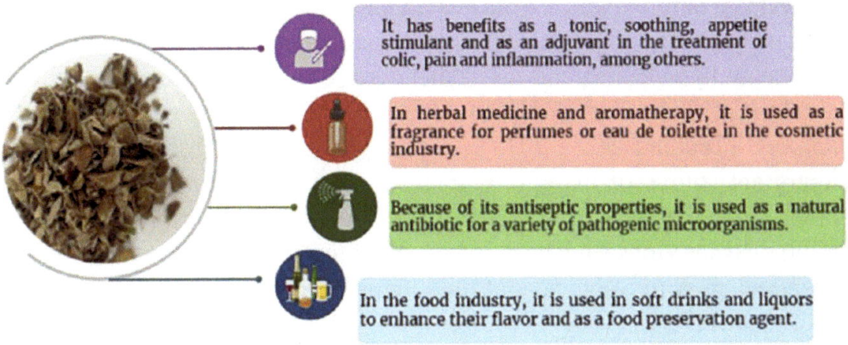

FIGURE 8.2 Some common uses of oregano (De la Torre et al., 2008; Sharamon and Bagiński, 1998; Punina-Caiza et al., 2022).

8.3.2 OREGANO ESSENTIAL OIL (OEO)

It is common for aromatic plants to be characterized by their special aroma and flavor; this is due to the essential oils they possess and something important to note is that they are not only produced in the leaves of the plants but also found in parts such as stems, roots, bark, flowers, fruits, and seeds. These essential oils are complex mixtures of chemical compounds defined as volatile. Among them mainly terpenoids, carotenoids, terpenes, antioxidants, and some phenolic compounds, which provide different biological activities important for pharmaceutical, food, or cosmetics industries (Baser and Buchbauer, 2009; Dhifi et al., 2016; Stanner et al., 2004). In general, the essential oils of plants help to protect them from some microorganisms or certain pests, apart from the fact that the smell they give off helps to achieve pollination. However, research has found beneficial health effects (Leyva-López et al., 2017). Although a wide variety of essential oils are commercialized in the market, one of the most important is oregano essential oil (OEO). Its main metabolites are carvacrol and thymol (Calvo-Irabien et al., 2014). However, it is important to mention that more components can be found in different proportions, due to factors such as the species of oregano, growing conditions, harvesting season, and geographical location of the plant (Leyva-López et al., 2017).) For example, for *O. majorana* it was found that its most predominant component was terpinen-4-ol and also that it has moderate activity for bacteria such as *Staphylococcus aureus*, *Bacillus cereus*, and *Staphylococcus epidermidis*. However, the most prominent antimicrobial activity was with the fungal strains; *Aspergillus niger*, *Candida albicans*, *Trichophyton mentagrophytes*, *Aspergillus fumigatus*, and its antioxidant activity was proven (Paudel et al., 2022). In another study, using oregano essential oil from *O. vulgare*, apart from carvacrol and thymol, β-phenylethyl alcohol and γ-terpinene were also determined and showed good results in inhibiting the growth of *Listeria monocytogenes* and *Listeria innocua* (Teixeira et al., 2013). In this so, we can realize that OEO is an important source of natural compounds such as terpenes, flavonoids, and other phytochemical compounds; therefore, thanks to the studies different biological activities have been proved and thus, they can be used in medicine, cosmetic, pharmacological and food industries (Moghrovyan et al., 2019).

8.4 SUSTAINABLE USE OF OREGANO

8.4.1 OREGANO RESIDUES

Oregano is a nontimber forest resource considered of great economic importance. Therefore, in the industry, the marketing product is the essential oil, due to its valuable content of thymol and carvacrol (Rolando, 2007). The extraction of the OEO is done by foreign companies that sell it to different parts of the world (CONAFOR, 2005) which is intended for industries such as the food industry, soft drink, and liquor industry, in the manufacture of oil for aeronautics, cleaning of automotive parts and even aromatherapy products (SEMARNAT, 2015). However, the extraction of oregano essential oil generates a significant amount of by-product considered as "bagasse," in addition to the fact that the sale of oregano leaves as a condiment generates the stems as waste. Also, as reported in SEMARNAT (2001), it constitutes almost 95% dry weight of its initial weight. However, various authors have shown that residual oregano material is rich in bioactive compounds (Cuevas-Gonzalez, 2022). As reported by Frías-Zepeda and Rosales-Castro (2021), using the residues of Mexican oregano (*L. graveolens* Kunth) evaluated its nutraceutical properties since, in different investigations its antimutagenic and anticarcinogenic effects have been reported, which represents the potential for the cancer treatment (Arcila et al., 2004), on the contrary, it has been found to have antioxidant and antimicrobial capacity against pathogens such as *Salmonella, Escherichia coli, S. aureus, Staphylococcus epidermis*, which favors the food industry in regarding food safety and security (Asensio, 2013; Hernández, 2015).

8.4.2 OREGANO HARNESSING WORKS

Currently, there are very few works about the residual material that is obtained from the extraction of oregano essential oil, it is generally used as a matter for fermentation or composting processes and even the stems that are discarded in the collection of the leaves are it is used minimally as woody material for campfires (Cid et al., 2019). Even so, it has been possible to give it a revaluation, an example of this is the following investigation, where extracts of Mexican oregano (*L. graveolens*) were evaluated, being an important source of pinocembrin due to its higher content of this compound (0.766 mg/mL), also the fractionation and identification that was carried out allowed it to identify galangin and naringenin, compounds with the highest concentration in Mexican oregano (Cuevas-Gonzalez, 2022). In other research, oregano bagasse was found to be efficient as feed for fattening rabbits, which improved productive characteristics, carcass, and meat quality (Aquino-López et al., 2020). On the contrary, Falcón Peña and Velarde Franco (2022), determined the calorific value and level of emissions of briquettes using olive grove and oregano residues to obtain a renewable energy source; however, the briquettes obtained with oregano residues were more fragile with a maximum value of 31 ± 1.15 bar compared to olive grove residues, but in the sensory preference test, the oregano briquette was the one that obtained the best scores. Also, these residues have been evaluated, determining that they are viable for obtaining extracts so that they can be added to products with nutraceutical potential (Fríaz-Zepeda, 2021). Thus, the great potential of the generated oregano residue can be perceived, which can be put to various uses.

8.4.3 SOLID-STATE FERMENTATION AS AN ALTERNATIVE USE

In recent years, the interest in bioactive compounds has grown, generating an important economic source in the industries since many investigations have demonstrated their various biological activities (e.g., antioxidant activity, anti-inflammatory, antimicrobial); many of these compounds are secondary metabolites such as polyphenols, carotenoids, terpenes, vitamins, among others; which can be found from fruits, vegetables, cereals, herbs, tubers (Kamiloglu et al., 2021; Santos et al., 2019). However, the extraction of this type of compound is usually carried out with methodologies that require a lot of energy and solvents that could be more environmentally friendly; below is Table 8.2 with the different methodologies used and their respective advantages and disadvantages.

TABLE 8.2 Extraction Method of Bioactive Compounds.

Extraction methods	Principle	Advantages	Disadvantages	References
Hydrodistillation (HD)	In this process, a suspension of a mixture of water and plant material is boiled and the vapors are condensed and collected in a cold phase.	Low cost Easy to use	Long extraction times Degradation of volatile components High energy consumption	(Bampouli et al., 2014; Morsy, 2015; Stanojevic et al., 2015)
Steam distillation	Selective vaporization of the volatile component is carried out.	Older method It is simple Low Temperatures	Long extraction times	(Quintero- Chacón et al.,2004; Barba, 1993)
Organic solvents	Use of solvents such as petroleum ether, benzene, and alcohol and subsequent heating.	Simple	Causes damage to health	(Palma, 2003, González- Trujano et al., 2017).
Supercritical fluids	Chemicals are used under special conditions of temperature and pressure.	High performance Easy to remove solvent Environmentally friendly	Expensive	(Kerrola and Kallio, 1993)
Ultrasound extraction	The bubbles generated by the ultrasonic cavitation destroying the lipid cells of the cellular tissue of the plant	Ecological Low temperatures Short times	Expensive	(Barrazueta et al., 2015; Albu et al., 2004; Mason et al., 2011)
Microwave-assisted extraction	Electromagnetic waves are generated to initiate the extraction process and the cell wall of the plant interacts with the waves, thus resulting in a heating of the plant matrix, which swells due to the pressure exerted, and causes an increase in the mass transfer of the solute to the solvent	Reduces energy consumption Extraction takes less time Solvent expenditure is lower and generates better yields of the compounds of interest	Low selectivity Matrix-dependent and reduced solvent use Possible degradation of compounds	(Ameer et al., 2017; Sadeghi et al., , 2017)

Extraction methods for bioactive compounds using plant matrices have advanced in response to the challenges of sustainability, energy efficiency, resource optimization, and the principles of green chemistry, so the choice of extraction method is important if good yields and selectivity are sought or if biological activities are to be exploited (Calvo-Flores et al., 2018).

On the contrary, there are other alternatives for the extraction of these compounds, and they are fermentations, such as SSF; it is a process that has major advantages, such as higher productivity, lower investment cost, low energy consumption, simpler processes, less waste water and better efficiency in the recovery of the products. It is a potential technology for the production of food, fuels, industrial chemicals, and pharmaceuticals (Zhao, 2019; Chohan, 2020).

Its application offers advantages in bioprocesses such as bioleaching, biobeneficiation, and bioremediation (Romero and Vargas, 2017). It is also used for the recycling of agro-industrial waste and offers an alternative that allows, through simple technologies, to modify or alter fermentation processes with the addition of microbial preparations rich in lactic acid bacteria, thus improving the quality and safety of the process (Elías and Herrera, 2008; Díaz, 2014) For example, many research studies have been carried out using potato peel, pineapple peel, pomegranate residues, including aromatic plant residues such as bagasse generated after the extraction of essential oils, stems, flowers, and leaves; and these studies have shown that polyphenols are more readily available through FES, because a higher concentration of polyphenols is obtained in the extracts of fermented residues (Sadh et al., 2018; Bautista-Hernández et al., 2022).

This availability is due to the microorganism used in the SSF produces enzymes that hydrolyze the plant matrix, causing the polyphenols and other bioactive compounds that are bound to cellulose, hemicellulose, pectin, and structural proteins to be released and can be extracted by the solvent (Torres-León et al., 2019). On the contrary, this technique represents a good option, as compared to conventional methods and emerging technologies. It does not require solvents, is less energy-intensive, and uses minimal amounts of water, which makes it a good eco-friendly and economical choice.

8.5 CONCLUSIONS

It is important to mention that in the case of oregano, SSF represents an option concerning the recovery of secondary metabolites, because, the oregano plant is overexploited in various countries around the world, so there is no regulation of it, so, the marketing of oregano leaves comes from poor collectors where they are sold at a very low value per kilo. As mentioned earlier in the chapter, some studies demonstrate the bioactive potential of this bagasse, so this SSF is an option in terms of revaluing the waste and generating an extra source of income for these farmers.

KEYWORDS

- **oregano leaves**
- **phytochemicals**
- **essential oil**
- **antioxidants**
- **solid-state fermentation**

REFERENCES

Aguirre, M. El aprovechamiento forestal de productos forestales no maderables en la Región Lagunera de Durango. [Tesis de Licenciatura, Universidad Autónoma de Chapingo], Unidad Regional Universitaria de Zonas Áridas (URUZA), Bermejillo, Durango, México, 2000.

Ait-Ouazzou, A.; Cherrat, L.; Espina, L.; Lorán, S.; Rota, C.; Pagán, R. La actividad antimicrobiana de los constituyentes de los aceites esenciales hidrófobos actuando solos o en procesos combinados de conservación de alimentos. *Ciencia de los alimentos innovadores y tecnologías emergentes* **2011,** *12* (3), 320–329.

Albu, S.; Joyce, E.; Paniwnyk, L.; Lorimer, J. P.; Mason, T. J. Potential for the Use of Ultrasound in the Extraction of Antioxidants from *Rosmarinus officinal* is for the Food and Pharmaceutical Industry. *Ultrason. Sonochem.* **2004,** *11* (3–4), 261–265. https://doi. org/10.1016/j.ultsonch.2004.01.015

Ameer, K.; Shahbaz, H. M.; Kwon, J. Green Extraction Methods for Polyphenols from Plant Matrices and their by Products: A Review. *Compr. Rev. Food Sci. Food Saf.* **2017,** *16,* 295–315. https://doi. org/10.1111/1541-4337.12253

Aquino-López, J. L.; Chávez-Martínez, A.; García-Macías, J. A.; Méndez-Zamora, G.; Rentería-Monterrubio, A. L.; Dalle-Zotte, A.; García-Flores, L. R. El aceite esencial y bagazo de orégano (*Lippia berlandieri* Schauer) afectan el comportamiento productivo y la calidad de la carne de conejo. *Revista mexicana de ciencias pecuarias* **2020,** *11* (3), 701–717.

Arcila, C.; Loarca, G.; Lecona, S.; Gonzales, E. El orégano: Propiedades, composicion y actividad biologica de sus componentes. *Archivos Latinoamericanos de Nutricion* **2004,** *54* (1), 100–111.

Arcila-Lozano, C.; Loarca-Piña, G.; Lecona-Uribe, S.; Gonzáles de Mejía, E. El orégano: Propiedades, composición y actividad biológica de sus componentes. *Arch. Latinoam. Nutr.* **2004,** *54*, 100–111.

Arias, J.; Mejía, J.; Córdoba, Y.; Martínez, J. R.; Stashenko, E.; del Valle, J. M. Optimization of Flavonoids Extraction from *Lippia graveolens* and *Lippia origanoides* Chemotypes with Ethanol-Modified Supercritical CO_2 after Steam Distillation. *Ind. Crops Prod.* **2020,** *146*, 112170. doi:10.1016/j.indcrop.2020.112170

Bampouli, A.; Kyriakopoulou, K.; Papaefstathiou, G.; Louli, V.; Krokida, M.; Magoulas, K. Comparison of Different Extraction Methods of *Pistacia lentiscus* var. Chia Leaves: Yield, Antioxidant Activity and Essential Oil Chemical Composition. *J. Appl. Res. Med. Aromat. Plants* **2014,** *1* (3), 81–91. https://doi.org/10.1016/ j.jarmap.2014.07.001

Barba, J. M. In *Introducción al análisis de los productos naturales*; Universidad Autónoma Metropolitana, México, 1993.

Barba, F. J.; Alexandre Saraiva, J. M.; Cravotto, G.; Lorenzo, J. M. In *Innovative Thermal and Non-Thermal Processing, Bioaccessibility and Bioavailability of Nutrients and Bioactive Compounds*; Elsevier Inc: Amsterdam, The Netherlands; 2019, pp 23–54. ISBN 97810128141748

Barón-Malaver, L. J.; García-Rincón, A. N. Revisión documental de aceites esenciales de *Mentha piperita* y *Origanum majorana* como antimicrobianos, una solución alternativa al uso de agroquímicos; Universidad Colegio Mayor de Cundinamarca, 2019.

Barrazueta, S.; Mendoza, G.; Fornari, T. Uso de ultrasonidos en la extracción de plantas labiadas combinando etanol y CO2 supercrítico. *Alimentos, Ciencia e investigación* **2015,** *23* (1), 37–46.

Baser, K. H. C.; Buchbauer, G. In *Handbook of Essential Oils: Science, technology, and Applications*; CRC Press: Boca Ratón, FL, Estados Unidos, 2015; p 975. ISBN 978-1-4200-6315-8.

Bautista-Hernández, I.; Aguilar, C. N.; Martínez-Ávila, G. C.; Torres-León, C.; Ilina, A.; Flores-Gallegos, A. C.; Chávez-González, M. L. Mexican Oregano (*Lippia graveolens* Kunth) as Source of Bioactive Compounds: A Review. *Molecules* **2021,** *26* (17), 5156. https://doi.org/10.3390/molecules26175156

Bautista-Hernández, I.; Aguilar, C. N.; Martínez-Ávila, G. C.; Ilina, A.; Torres-León, C.; Verma, D. K.; Chávez-González, M. L. Compuestos fenólicos y actividad antioxidante de hojas residuales de *Lippia graveolens* Kunth fermentadas por dos cepas de hongos filamentosos en proceso de estado sólido. *Procesamiento de Alimentos y Bioproductos* **2022,** *136*, 24–35. https://doi.org/10.1016/j.fbp.2022.09.001

Brown, M.; Henderson, D.; Hunt, C. Comparison of Antioxidant Properties of Supercritical Fluid Extracts of Herbs and the Confirmation of Pinocembrin as a Principle Antioxidant in Mexican oregano (*Lippia graveolens*). *Electron. J. Environ. Agric. Food Chem.* **2006,** *5*, 1265–1277.

Buchbauer, G.; Baser, K. H. C. In *Handbook of Essential Oils: Science, Technology, and Applications*; CRC Press Taylor & Francis Group: Boca Raton, Florida, 2009. https://doi.org/10.1201/9781420063165

Calvo-Flores, F. G.; Monteagudo-Arrebola, J.; Dobado, J. A.; Isac-García, J. Green and Bio-Based Solvents. *Top. Curr. Chem.* **2018,** *376* (3), 18. 10.1007/s41061-018-0191-6

Calvo-Irabien, L. M. Native Mexican Aromatic Flora and Essential Oils: Current Research Status, Gaps in Knowledge and Agro-industrial Potential. *Ind. Crops Prod.* **2018,** *111*, 807–822. https://doi.org/10.1016/j.indcrop.2017.11.044

Calvo-Irabien, L. M.; Parra-Tabla, V.; Acosta-Arriola, V.; Escalante-Erosa, F.; Díaz-Vera, L.; Dzib, G. R.; Peña-Rodríguez, L. M. Phytochemical Diversity of Essential Oils of Mexican oregano (*Lippia graveolens* Kunth) Populations Along Edapho-Climatic Gradient. *Chem. Biodivers.* **2014,** *11*, 1010–1021. https://doi.org/10.1002/cbdv.201300389

Castillo-Herrera, G. A.; García-Fajardo, J. A.; Estarrón-Espinosa, M. Extraction Method that Enriches Phenolic Content in Oregano (*Lippia graveolens* H. B. K.) Essential Oil. *J. Food Proc. Eng.* **2007,** *30*, 661–669. https://doi.org/10.1111/j.1745-4530.2007.00134.x

Cid, T. S.; Ávila, R.; Ochoa, C. E.; Rivera, B. E.; Nevárez, G. V. Actividad Antioxidante y Antimicrobiana del Aceite Esencial, Hidrosol y Extractos de Residuos Sólidos de Orégano Mexicano (*Poliomintha longiflora*). *Plants* **2019,** *8* (22).

Comisión Nacional Forestal (CONAFOR). Fichas de información comercial de productos forestales. Coordinación General de Educación y Desarrollo Tecnológico, Gerencia de Desarrollo y Transferencia de Tecnología. Fondo Sectorial para la Investigación, el Desarrollo y la Innovación Tecnológica Forestal, SEMARNAT, México, D. F., 2009; pp 8–9.

CONABIO. Informe 1992-2004. Comisión Nacional para el Conocimiento y Uso de la Biodiversidad, México, 2005. http://www.conabio.gob.mx/institucion/conabio_espanol/doctos/informe_doce_anios.pdf.[Consultado el 23 de abril de 2023].

Condori Tintaya, F.; Fernández Cutire, O. Resultados económicos de producción de orégano (*Origanum vulgare* L.) en cuencas interandinas de la región Tacna, Perú. *Idesia (Arica)* **2022,** *40* (2), 17–26. https://dx.doi.org/10.4067/S0718-34292022000200017

Connor Black, S. A.; Haughey, O. P.; Chevallier, P. G. K.; Christopher, T. E. A Comprehensive Strategy to Detect the Fraudulent Adulteration of Herbs: The Oregano Approach. *Food Chem.* **2016,** *2* (10), 551–557. https://doi.org/10.1016/j.foodchem.2016.05.004.

Corella-Bernal, R. A.; Ortega-Nieblas, M. M. IMPORTANCIA DEL ACEITE ESENCIAL Y LA PRODUCCION DE ORÉGANO *Lippia palmeri* Watson EN EL ESTADO DE SONORA. *Biotecnia* **2013,** *15* (1), 57–64. https://doi.org/10.18633/bt.v15i1.137

Cortés-Chitala, M. D.; Flores-Martínez, H.; Orozco-Ávila, I.; León-Campos, C.; Suárez-Jacobo, Á.; Estarrón-Espinosa, M.; López-Muraira, I. G. Identification and Quantification of Phenolic Compounds from Mexican Oregano (*Lippia graveolens* HBK) Hydroethanolic Extracts and Evaluation of Its Antioxidant Capacity. *Molecules* **2021,** *26 (3), 702. https://doi.org/10.3390/molecules26030702*

Cuevas-Gonzalez, M. *Valorización de los residuos agroindustriales de orégano mexicano y limón mexicano mediante la generación de productos enriquecidos con flavonoides(pinocembrina y hesperidina) de potencial interés biológico.* Tesis de Maestría, Tecnológico Nacional de México, Repositorio Institucional del Tecnológico Nacional de México, 2022. https://rinacional.tecnm.mx/jspui/handle/TecNM/4643

De la Torre, L.; Navarrete, H.; Muriel, P.; Macía, M.; Baslev, H. Enciclopedia de las plantas útiles del Ecuador; Herbario QCA y Herbario AAU, 2008.

Dhifi, W.; Bellili, S.; Jazi, S.; Bahloul, N.; Mnif, W. Essential Oils' Chemical Characterization and Investigation of Some Biological Activities: A Critical Review. *Medicines* **2016,** *3*, 25. https://doi.org/10.3390/medicines3040025

Di Fabio, A. El cultivo y su efecto sobre la calidad del oregano, 2005.

Estrada-Jaramillo, R.; Ruiz Otalora, F.; Martínez Correa, H. A. Extracción de compuestos fenólicos totales de residuo de mango común (*Mangifera indica* L.) empleando ultrasonido. *Acta Agronómica* **2012,** *61* (5), 95–96. https://revistas.unal.edu.co/index.php/acta_agronomica/article/view/41477

Falcón Peña, R. S.; Velarde Franco, A. *Efectos del pretratamiento de biomasa sobre poder calorífico y nivel de emisiones de briquetas de residuos del olivar y orégano.* Tesis de ingeniería, Universidad Privada de Tacna, Repositorio Institucional de la Universidad Privada de Tacna, 2022. https://repositorio.upt.edu.pe/handle/20.500.12969/2635

Flores, P. G.; Ruiz, L. D.; Ibarra, E. E.; Valenzuela, R. B. Actividad antibacteriana de extractos de orégano (*Origanum* spp.) en bacterias GRAM negativas y positivas. *Ciencias Químicas* **2019,** *1* (1), 1–8.

Fríaz-Zepeda, M. E. Evaluación nutraceútica de los residuos de Orégano Mexicano y su aplicación en un modelo murino con Síndrome Metabólico. Tesis de Doctorado, Instituto Politécnico Nacional, Repositorio del Instituto Politécnico Nacional, 2021. http://repositoriodigital.ipn.mx/handle/123456789/26434

Frías-Zepeda, M. E.; Rosales-Castro, M. Efecto de las condiciones de extracción sobre la concentración de compuestos fenólicos en residuos de orégano mexicano (*Lippia graveolens* Kunth). *Revista Chapingo, Serie Ciencias Forestales del Ambiente* **2021,** *27* (3), 367–381. https://doi.org/10.5154/R.RCHSCFA.2020.10.066

Gallegos-Flores, P.; Bañuelos-Valenzuela, R.; Delgadillo-Ruiz, L.; Echavarría-Cháirez, F.; Meza-López, C.; Rodríguez-Tenorio, D. Differential Evaluation of Oregano Extracts in the Production of Volatile Fatty Acids and Methane During Ruminal Fermentation in Vitro. *Abanico veterinario* **2019,** *9* (1), 1–18.

García-Pérez, E.; Castro-Álvarez, F. F.; Gutiérrez-Uribe, J. A.; García-Lara, S. Revisión de la producción, composición fitoquímica y propiedades nutracéuticas del orégano mexicano. *Revista mexicana de ciencias agrícolas* **2012,** *3* (2), 339–353. http://www.scielo.org.mx/scielo.php?script=sci_arttext&pid=S2007-09342012000200010&lng=es&tlng=es.

Garcia-Perez, E.; Francisco, C. Á.; Gutiérrez-Uribe, J. A.; García-Lara, S.. Revisión de la producción, composición fitoquímica y propiedades nutracéuticas del orégano mexicano. *Revista mexicana de ciencias agrícolas* **2012,** *3*, 339–353. 10.29312/remexca.v3i2.1467.

Gerothanassis, I. P.; Exarchou, V.; Lagouri, V.; Troganis, A.; Tsimidou, M.; Boskou, D. Methodology for Identification of Phenolic Acids in Complex Phenolic Mixtures by High-Resolution Two-Dimensional Nuclear Magnetic Resonance. Application to Methanolic Extracts of Two Oregano Species. *J. Agric. Food Chem.* **1998,** *46*, 4185–4192.

González-Trujano, M. E.; Hernández-Sánchez, L. Y.; Ocotero, V. M.; Dorazco-González, A.; Fefer, P. G.; Aguirre-Hernández, E. Pharmacological Evaluation of the Anxiolytic-Like Effects of *Lippia graveolens* and Bioactive Compounds. *Pharm. Biol.* **2017,** *55*, 1569–1576.

Kamiloglu, S.; Tomas, M.; Ozdal, T.; Yolci-Omeroglu, P.; Capanoglu, E. Bioactive Component Analysis. In *Innovative Food Analysis*; En Galanakis, C. M., Elsevier Inc.: Amsterdam, The Netherlands, 2021; pp 41–65. ISBN 9780128194935.

Karamanos, A. J.; Sotiropoulou, D. E. K. Estudios de campo de la aplicación de nitrógeno en el oreganogriego (*Origanum vulgare* ssp. hirtum (Link)letswaart)aceote esencial urante dos temporadas de cultivo. *Cultivo Ind. Pinchar.* **2013,** *46* (0), 246–252.

Kerrola, K.; Kallio, H. Volatile Compounds and Odor Characteristics of Carbon Dioxide Extracts of Coriander (*Coriandrum sativum* L.) Fruits. *Agric. Food Chem.* **1993,** *41* (5), 785–790. https://doi.org/10. 1021/jf00029a021

Kintzios, S. E. Profile of the Multifaceted Prince of the Herbs. In *The Genera Origanum and Lippia*, 1st Edition; Kintzios, S. E., Taylor and Francis: New York, 2002; pp 3–8.

Koksal, O.; Gunes, E.; Orkan, O.; Ozden, M. Analysis of Effective Factor on Information Sources at Turkish Oregano Farms. *African J. Agric.* **2010,** *5*, 142–149.

Leyva-López, N.; Gutiérrez-Grijalva, E.; Vazquez-Olivo, G.; Heredia, J. Aceites esenciales de orégano: Actividad biológica más allá de sus propiedades antimicrobianas. *Moléculas* **2017,** *22* (6), 989. https://doi.org/10.3390/molecules22060989

Machado, M.; Dinis, A. M.; Salguerio, L.; Cavaleiro, C.; Custódio, J. B. A.; do Céu Sousa, M. Anti-Giardia Activity of Phenolic-Rich Essential Oils: Effects of Thymbra capitata, *Origanum virens*, *Thymus zygis* subsp. sylvestris, and *Lippia graveolens* on Trophozoites Growth, Viability, Adherence, and Ultrastructure. *Parasitol. Res.***2010,** *106*, 1205–1215

Marieschi, M.; Torelli, A.; Poli, F.; Sacchetti, G.; Bruni, R. Método basado en RAPD para el control de calidad del orégano mediterráneo y su contribución a las técnicas farmacognósticas. *J. Agric. Food Chem.* **2009,** *57*, 1835–1840. https://doi.org/10.1021/jf8032649

Martínez Vera, M. A.; Mendoza Camejo, E. A. Producción de alfa amilasa por fermentación en estado sólido de residuos agroindustriales (cáscaras de banano) utilizando *Bacillus subtilis*. Bachelor's Thesis, Universidad de Guayaquil, Facultad de Ingeniería Química, 2018.

Mason, T.; Chemat, F.; Vinatoru, M. The Extraction of Natural Products Using Ultrasound or Microwaves. *Curr. Org. Chem.* **2011,** *15*, 237–247. https://doi.org/10.2174/138527211793979871

Mastelić, J.; Jerković, I.; Blazević, I.; Poljak-Blazi, M.; Borović, S.; Ivancić-Baće, I.; Smrecki, V.; Zarković, N.; Brcić-Kostic, K.; Vikić-Topić, D.; Müller, N. Estudio comparativo sobre las actividades antioxidantes y biológicas de los derivados de carvacrol, timol y eugenol. *J. Agric. Food Chem.* **2008,** *56* (11), 3989–3996. DOI: 10.1021/JF073272V

Mata-González, R.; Meléndez-González, R. Growth Characteristics of Mexican Oregano (*Lippia berlandieri* schauer) Under Salt Stress. *Southwestern Nat.* **2005,** *50* (1), 1–6. https://doi.org/10.1894/0038- 4909(2005)0502.0.CO;2

Moghrovyan, A.; Sahakyan, N.; Babayan, A.; Chichoyan, N.; Petrosyan, M.; Trchounian, A. Aceite esencial y extracto de etanol de orégano (*Origanum vulgare* L.) de la flora armenia como fuente natural de terpenos, flavonoides y otros fitoquímicos con actividad antirradical, antioxidante, quelante de metales, tirosinasa inhibitoria y antibacteriana. *Diseño farmacéutico actual* **2019,** *25* (16), 1809–1816. https://doi.org/10.2174/13816128 25666190702095612

Morsy, N. F. S. A Short Extraction Time of High Quality Hydrodistilled Cardamom (Elettaria cardamomum L. Maton) Essential Oil Using Ultrasoundas a Pretreatment. *Ind. Crops Prod.* **2015,** *65* (3), 287–292 . https://doi.org/10.1016/j.indcrop.2014.12.012

Muños, F. Plantas Medicinales y Aromáticas, Estudio, cultivo y procesado. *Ediciones Mundi-Prensa*, Madrid, 1978.

Orona Castillo, I.; Salvador Almazán, A. J.; Espinoza Arellano, J. D.; Vázquez, C. Recoleccion y comercializacion del oregano (*Lippia* spp) en el semi-desierto mexicano, un caso de estudio: Reserva ecologica municipal sierra y cañon de jimulco, mexico. *Revista Mexicana de Agronegocios* **2017,** *41*, 684–695.

Paudel, P. N.; Satyal, P.; Satyal, R.; Setzer, W. N.; Gyawali, R. Composición química, distribución enantiomérica, actividades antimicrobianas y antioxidantes de *Origanum majorana* L. Aceite esencial de Nepal. *Moléculas* **2022,** *27* (18), 6136. MDPI AG. Extraído de http://dx.doi.org/10.3390/molecules27186136

Punina-Caiza, N. M.; Ulcuango-Tuquerres, R. B. Aplicación de una película comestible a base de mucílago de nopal y aceite esencial de orégano en la conservación poscosecha de la fresa (*Fragaria x ananassa*); UTC, Latacunga, 2022; p 130.

Quintero-Chacón, N.; González de Colmenares, A.; Staschenko, E. Aceite esencial de las hojas de Hyptis umbrosa salzm extraído por diferentes técnicas. *Acta cient. Venez.* **2004,** *55* (2), 181–187.

Rodríguez, E. A.; Rey, P. A. S.; Godoy, J. A. Efectividad de los aceites esenciales de Thymus vulgaris (Tomillo) y *Origanum vulgare* subsp. Hirtum (Orégano griego) probados contra Ralstonia solanacearum fitopatógeno del cultivo de *Solanum lycopersicum* (Tomate). *Revista Environ. Technol.* **2021,** *2* (1), 1–18. https://doi.org/10.56205/ret.2-1.1

Rosero-Delgado, E. A.; Dustet-Mendoza, J. C. Cinética de la fermentación en estado sólido de cascarilla de arroz y bagazo de caña con *Auricularia auricula*. ICIDCA. *Sobre los Derivados de la Caña de Azúcar* **2017,** *51* (1), 28–38.

Sadeghi, A.; Hakimzadeh, V.; Karimifar, B. Extracción asistida por microondas de compuestos bioactivos de los alimentos: Una revisión. *Revista internacional de ciencia de los alimentos e ingeniería nutricional* **2017,** *7* (1), 19–27. DOI: 10.5923/j.food.20170701.03

Sadh, P. K.; Duhan, S.; Duhan, J. S. Agro-Industrial Wastes and their Utilization using 46 Solid State Fermentation: A Review. *Bioresour. Bioprocess.* **2018,** *5* (1), 1–15. https://doi.org/10.1186/s40643-017-0187-z

Sánchez Tito, M.; Layme Huanca, M. Composición química y actividad antibacteriana del aceite esencial de *Origanum vulgare* L. frente a *Enterococcus faecalis*. *Rev. Cubana. Estomatol.* **2022,** *59* (1), e3457. https://revestomatologia.sld.cu/index.php/est/article/view/3457

Santos, D. I.; Saraiva, J. M. A.; Vicente, A. A.; Moldão-Martins, M. Métodos para determinar la biodisponibilidad y bioaccesibilidad de compuestos bioactivos y nutrientes. En Procesamiento térmico y no térmico innovador, bioaccesibilidad y biodisponibilidad de nutrientes y compuestos bioactivosEditorial Woodhead, 2019; pp 23–54.

SAS, I. I. SAS/STAT® 15.1.User´s Guide. Car, NC, 2018.

SEMARNAT. Norma Oficial Mexicana-OO5-SEMARNAT-1997 que establece los procedimientos, criterios y especificaciones para realizar el aprovechamiento, transporte y almacenamiento de corteza, tallos y plantas completas de vegetación forestal. *Diario Oficial de la Federación* 1997. (Consultado el 23 de abril de 2003).

SEMARNAT. Ley General de Desarrollo Forestal Sustentable. *Diario Oficial de la Federación*, **2021**. (Consultado el 26 de abril 2023)

Servicio de Información Agroalimentaria y Pesquera (SIAP). Cierre de la producción agricola por estado, cultivos perennes, 2010. http://www.siap.gob.mx/ [accesado 25 Abril 2023].

Sharamon, S.; Bagiński, B. J. El poder de la semilla de pomelo; MH Publishing House: Warszawa, Polonia, 1998. [Google Académico]

Simpson, B.; Gorzaly, M. In *Economic Botany*; Mc Graw Hill, 2001; p 544.

Soccol, C. R.; Scopel Ferreira da Costa, E.; Junior Letti, L. A.; Grace Karp, S.; Lorenci Woiciechowski, A.; Porto de Souza Vandenberghe, L. Recent Developments and Innovations in Solid State Fermentation. *Biotechnol. Res. Innov.* **2017,** *1* (1), 52–71. https://doi.org/10.1016/j.biori.2017.01.002

Stanner, S.; Hughes, J.; Buttriss J. A Review of the Epidemiological Evidence for the Antioxidant Hypothesis. *Public Health Nutr.* **2004,** *7* (3), 407–422.

Stanojevic, L.; Radulovic, N.; Djokic, T.; Stankovic, B.; Ilic, D.; Cakic, M.; Nikolic, V. The Yield, Composition and Hydrodistillation Kinetics of the Essentialoil of Dill Seeds (*Anethii fructus*) Obtained by Different Hydrodistillation Techniques. *Ind. Crops Prod.* **2015,** *65* (3), 429–436. https://doi.org/10. 1016/j.indcrop.2014.10.067

Stashenko, E. E.; Jaramillo, B. E.; Martinez, J. R. Comparación de la composición química y de la actividad antioxidante in Vitro de los metabolitos secundarios volátiles de plantas de la familia Verbenaceae. *Rev. Acad. Colomb. Cienc.* **2003,** *27* (105), 579–597.

Teixeira, B.; Marques, A.; Ramos, C.; Serrano, C.; Matos, O.; Neng, N. R.; Nogueira, J. M. F.; Saraiva, J. A.; Nunes, M. L. Chemical Composition and Bioactivity of Different Oregano (*Origanum vulgare*) Extracts and Essential Oil. *Food Agric.* **2013,** *93*, 2707–2714. https://doi.org/10.1002/jsfa.6089

Torres-León, C.; Ramírez-Guzmán, N.; Ascacio-Valdés, J.; Serna-Cock, L.; dos Santos Correia, M. T.; Contreras-Esquivel, J. C.; Aguilar, C. N. Fermentación en estado sólido con *Aspergillus niger* para mejorar el contenido fenólico y la actividad antioxidante de la semilla de mango mexicano: Una fuente prometedora de antioxidantes naturales. *LWE* **2019,** *112*, 108236. https://doi.org/10.1016/j.lwt.2019.06.003

Universidad Nacional de Córdoba (UNC). Apunte de Cátedra Sistemas de producción de cultivos intensivos. Ediciones Académicas UNC, Córdoba, Argentina, 2013.

Vásquez Carreño, D. El orégano de monte (*Lippia origanoides*) del Alto Patía: Efecto del método de obtención de sus extractos sobre la composición y la actividad antioxidante de los mismos. Tesis de posgrado, Universidad Nacional de Colombia, Repositorio institucional de la Universidad Nacional de Colombia, 2012. https://repositorio.unal.edu.co/bitstream/handle/unal/11580/197514.2012.pdf?sequence=1&isAllowed=y

Villavicencio-Gutierrez, E. E.; Martinez-Burciaga,O. U.; Cano-Pineda, A. Orégano recurso con alto potencial. *Revista Ciencia y Desarrollo* **2007,** *33* (211), 60–66. https://www.cyd.conacyt.gob.mx/archivo/211/Articulos/Oregano/Oregano00.htm [Consultado el 27 de Abril de 2023].

CHAPTER 9

Oleogels: Healthy and Sustainable Replacement for Saturated and Hydrogenated Fats

MAYRA AGUILAR-ZÁRATE[1*], LAURA ARACELI LÓPEZ-MARTÍNEZ[2], and JORGE FERNANDO TORO-VAZQUEZ[1]

[1]*Universidad Autónoma de San Luis Potosí, Facultad de Ciencias Químicas, San Luis Potosí, SLP, México*

[2]*Universidad Autónoma de San Luis Potosí, Coordinación Académica Región Altiplano Oeste, Salinas de Hidalgo, SLP, México*

ABSTRACT

Oleogels are materials developed as an alternative to substitute saturated and trans fats in food systems using low molecular weight gelators and vegetable oils, plant-based components, noninvasives, and nontoxics for the consumers and the environment. This chapter highlights a selection of basic considerations to engineer oleogels, using healthy and sustainable components, with the objective to conscientize the scientists and developers about the importance of achieving adequate molecular self-assembly that permits obtaining a material with adequate properties that can successfully be applied in a food system or even in some cosmetics or pharmaceuticals products. It includes a revision of some studies that have demonstrated the influence of the type of gelators and solvents in oleogelation process, establishing some considerations to select the adequate components and examples of how their properties can be modified to obtain the desired materials. Also, highlighted here are the recent challenges in oleogels' functionality when they are applied to food systems, the achievements obtained so far, and the

Sustainable Engineering and Agro-Food Processing: Renewable and Nonrenewable Resources.
José Juan Buenrostro-Figueroa, Julio César Tafolla-Arellano, Mónica Chávez-González, Pedro Aguilar-Zárate, A. K. Haghi (Eds.)

direction such materials are taking, whose main function is to replace saturated fats and trans fats in food products or to be used as plant-based fats for other potential applications.

9.1 INTRODUCTION

Oleogels are lipid-based solid-like materials composed of 95% or more liquid vegetable oil entrapped in a tridimensional network formed by food-grade gelator molecules. All these components are biomolecules, such as mono- and poly-unsaturated fatty acids contained in vegetable oils (Lesellier et al., 2021) and the gelators are low molecular weight molecules (<3000 Da), for example, monoglycerides (MG) (López-Martínez et al., 2014), phospholipids (Murashova et al., 2018; Martínez-Ávila et al., 2019) plant sterol derivatives (γ-oryzanol, β-sitosterol) (Bot et al., 2008; Sawalha et al., 2011), polysaccharides as ethylcellulose (Zetzl et al., 2014), and vegetable waxes (Blake et al., 2014; Wettlaufer et al., 2022). The oleogels structuration only requires applying physical factors to engineer these materials, mainly considering the solubilization temperature of the gelators in the solvent (i.e., vegetable oil), heating and cooling rates, shear rate, and the time for the application of each parameter. This permits that the chemical structure of the oleogels components does not suffer changes as in other processes for fat structuration (i.e., hydrogenation, chemical or enzymatic transesterification). Oleogels are very versatile materials, and their thermal and mechanical properties can be adapted to design temperature-resistant, viscoelastic, plastic, or thixotropic materials. This can be achieved by manipulating the molecular self-assembly of gelator–gelator (Avendaño-Vásquez et al., 2020; Sahu et al., 2021), gelator–solvent (Zhu and Dordick, 2006), modifying the solvent polarity (Gravelle et al., 2016; Gravelle et al., 2017), or adjusting some of the physical parameters during the preparation process (Alvarez-Mitre et al., 2012, 2013; Sawalha et al., 2011). All the aforementioned processes neither require any extra reagent nor produce waste or toxic residues. Recently, also the pharmaceutical and cosmetic industries are substituting the petroleum and animal-based fat materials from their products for new renewable vegetal-based materials as the oleogels. For these reasons, oleogels have been established as a sustainable and healthy alternative for saturated and hydrogenated fat substitution in food processed products (Dassanayake et al., 2011; Manzocco et al., 2012; Patel et al., 2014; Rogers et al., 2014). In this chapter, we are highlighting some recent advances developed in our lab group related to the oleogels engineering and about some proposes that other researchers have been done to include oleogels to formulate food products.

9.2 BASICS TO ENGINEER OLEOGELS

9.2.1 MOLECULAR SELF-ASSEMBLY

For oleogels structuration, the gelator molecules need to be dissolved in the vegetable oil by heating at its solubilization temperature and then a cooling process is applied to achieve the solidification of the material. This permits the spontaneous molecular assembly through short-range bonds (i.e., hydrogen bonds, van der Waals or London forces), favoring the development of a tridimensional network with microdomains (Fig. 9.1) in which the liquid vegetable oil is entrapped.

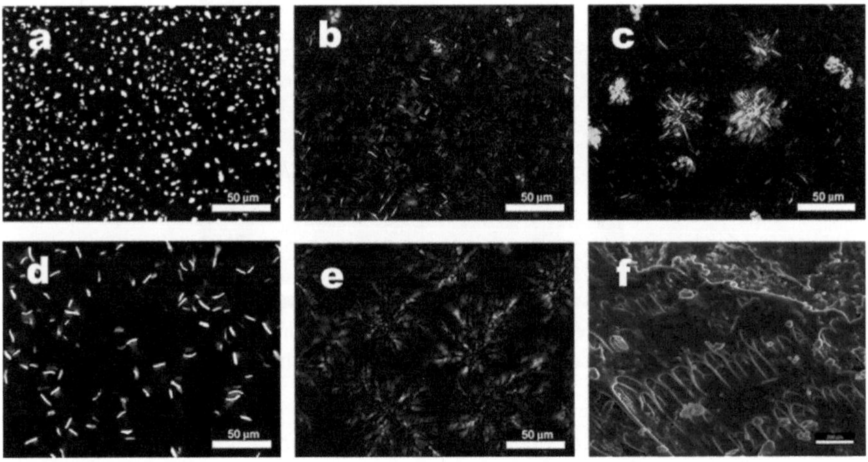

FIGURE 9.1 Comparison of the microstructure that constitutes the tridimensional network in different oleogels formulated in our group (some are unpublished data). (a) Lecithin in safflower oil (Springer ©) (Aguilar-Zárate et al., 2019), (b) candelilla wax in soybean oil oleogel, (c) carnauba wax in soybean oil oleogel, (d) MG in mineral oil Source: Reprinted with permission from Aguilar-Zárate et al., 2019. © Copyright. Springer, (e) 12-hydroxistearic acid in soybean oil, and (f) ethylcellulose in canola oil. All are polarized light microscopy (PLM) micrographs except aforementioned item (f), which is an SEM micrograph (bar: 200 µm)

These microdomains are formed for the molecular aggregates (i.e., solid phase) that have a characteristic microstructure depending on the type of oil, as can be distinguished in Figure 9.2, or gelator molecules as is shown in Figure 9.1. The result, at the macroscopic level (i.e., bulk fat), is a solid-like material with specific thermal and mechanical properties. Such physical properties provide the oleogels characteristics and functionality to be like a solid fat (Fig. 9.3).

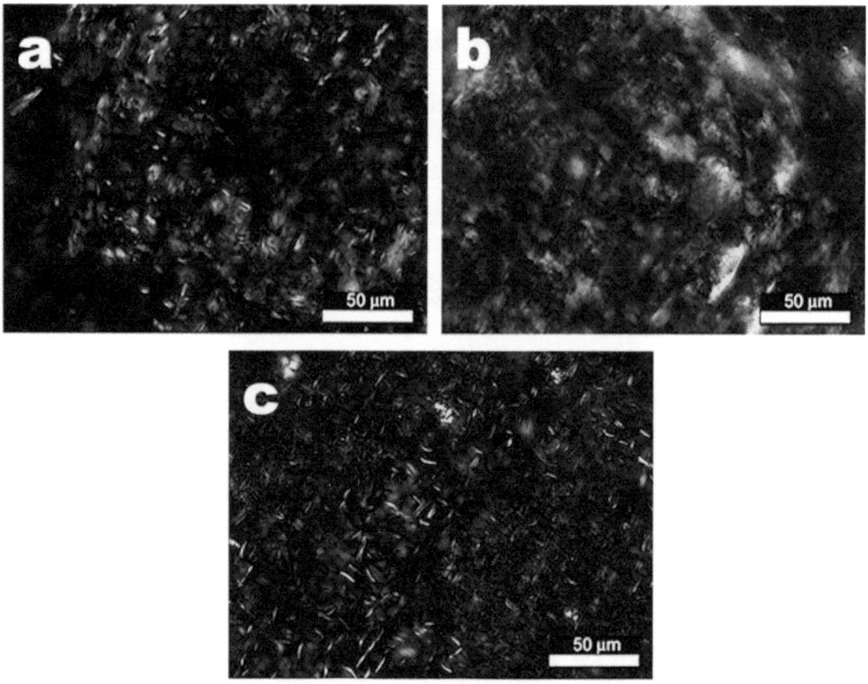

FIGURE 9.2 Microstructure of oleogels formulated with candelilla wax in different types of oils: (a) corn oil; (b) safflower oil; (c) soybean oil (Unpublished data).

In this context, it has been established that to develop molecular self-assembles for engineering an oleogel it is imperative previously to study the molecular composition of each structuring (oil and gelators) and their physicochemical properties (Golodnizky et al., 2021), for example, the degree of polarity and/or solubility capacity (Lan et al., 2015), and the type of interactions that could occur, these last usually are studied by Fourier transform infrared spectroscopy (FTIR) techniques. All the above helps to predict and then obtain the desired macroscopic characteristics of the material and at one point this permits us to select the adequate components and when it is required modify or manipulate the molecular interactions to engineer specific properties.

FIGURE 9.3 Oleogel appearance comparison. From left to right: oleogel, margarine, cream cheese.

For instance, in Figure 9.4a it can be seen the microstructure of a solution of 2% of a mixture of MG with saturated hydrocarbon chain in high oleic safflower oil, at which concentration of MG it is not possible to form a gel (i.e., the solution flows) (Aguilar-Zárate et al., 2019), because there was not enough solid phase formation to support the tridimensional network. Then, we based on the chemical theory to enhance this system and proposed that the MG molecules can interact with phosphatidylcholine (PC) molecule, because both are amphiphilic molecules with polar and nonpolar regions and can form hydrogen bonds between the hydroxyl group of MG and the phosphate group of PC. The above permits us to hypothesize that both molecules could have the ability for self-assembly. So, we reformulated the system using a mixture of 2% of MG and adding only 0.25% of PC with saturated hydrocarbon chains, and the resultant mixed microstructure is shown in Figure 9.2b, in which it can be observed how this small amount of a co-gelator can modify the molecular aggregation of the MG-based oleogel. Additionally, it caused an enhancement in the self-support of the tridimensional network, because of the synergism between both gelators molecules, and it increased the solid content, and all these factors favored the gel formation. Also, it was observed that the addition of PC to the 2% MG oleogel increased the rheological parameters in the system (Figure 9.4d), which was determined using rheological techniques. This property also was improved with the addition of a lower amount of water molecules (i.e., 2.3×10^{-3} moles of water), as can be compared Figure 9.4d. These strategies promoted the self-assembly to reduce the molecular order, that usually is identified by X-ray diffraction techniques, to prevent the formation of certain types of crystals that cause syneresis (i.e., in the MG oleogels the β-polymorphs). This type of procedure to enhance oleogels properties also permits the systems to be more stable, even for more than 1 year (Sawalha et al., 2011, Aguilar-Zárate et al., 2019).

In addition to the use of co-gelators, other important factors can help to manipulate the molecular self-assembly, for example, the application of physical parameters such as sharing and shear rates (Alvarez-Mitre et al., 2012; De La Peña-Gil et al., 2017), ultrasound (da Silva and Danthine, 2021; Lassila et al., 2022), cooling rate or holding temperatures (Pérez-Martínez et al., 2007) during the cooling process, and chemical factors such as the modification of solvent polarity. Specifically, in the next section are presented details of some effects of the gelator type, solvent polarity, and processing parameters in oleogel development and its physicochemical properties.

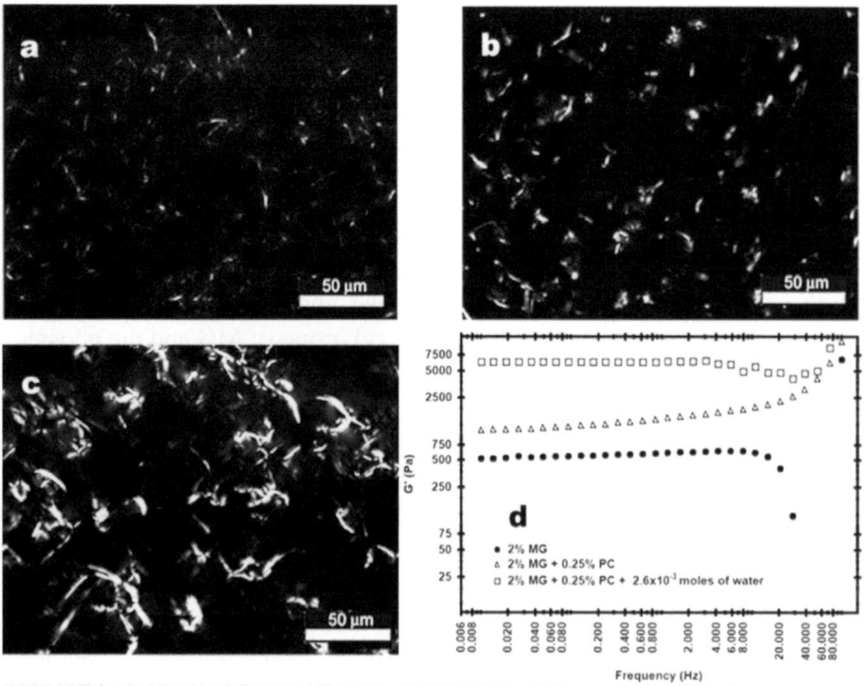

FIGURE 9.4 Appreciation of the variations in microstructure of the systems in safflower oil with (a) 2% of MG, b) mixed system with 2% MG + 0.25% PC, (c) mixed system with 2% MG + 0.25% PC + 2.6 × 10^{-3} moles of water Source: Reprinted with permission from Aguilar-Zárate et al., 2019. © Copyright. Springer. and the variations in (d) the elastic modulus (G′) of each one (unpublished data).

9.2.2 SOLVENT AND GELATOR INFLUENCES IN OLEOGEL ENGINEERING

The chemical properties of the solvent (i.e., vegetable oil) have an important role in the gelator assembly. This is related to the capacity of the solvent to solubilize the gelator molecules (Martínez-Ávila et al., 2019; Rogers et al., 2015). In general, gelators must be relatively insoluble in the oil to promote gelation, usually the interactions should occur between their soluble portion and the oil moieties (Co and Marangoni, 2013). For example, this is the case of PC-based oleogels in which the molecular self-assembly and crystal habit development depend on gelator solubility in the oil, which also is favored by the PC purity and saturation degree (Martínez-Ávila et al., 2019). In vegetable oil, a solvent with a certain amount of polar minor components (Scharfe et al., 2022), the saturated PC developed intertwined crystals that resulted in organogels with a true-gel behavior (G′ of 12.2×10^5–16×10^5 Pa). While, in the oil that is nonpolar (i.e., mineral oil), the saturated PC crystallized in smaller microstructures developing organogels with higher G′ (i.e., G′ of 32×10^5 Pa). In contrast, using the unsaturated PC in both types of oil, the presence of *cis* unsaturations in the fatty acids that compose gelator molecules, favored the formation of inverted micelles in the oleogels developed but had lower G′ (30×10^2–45×10^2 Pa) with a gel-like behavior (Martínez-Ávila et al., 2019).

Another strategy to enhance molecular self-assembly is manipulating the solvent polarity by blending different oils (Wettlaufer et al., 2022) or by the addition of small surface-active molecules (Gravelle et al., 2017; Gravelle and Marangoni, 2018; Giacintucci et al., 2018). This has been the case with soybean oil blended with mineral oil or castor oil and added with oleic acid and oleic alcohol as surface active molecules, which permitted a better gelator molecular assemble and the enhancement of mechanical strength of ethylcellulose oleogels (Blach et al., 2016). In this type of polymeric oleogels the nature of interactions of functional groups involved the formation of covalent bonds between gelator molecules originating the polymeric chains, the conformation of lipid chains because of the solvent-solvent interactions (as they were a blend of solvents), and also the polymer-solvent interactions. This type of interaction had a key role in engineering the rheological properties in the oleogels resulted (Gravelle et al., 2017; Aguilar-Zárate et al., 2019).

On the contrary, edible oleogels are mainly composed of low molecular weight gelators that are metastable, and their crystals self-support the system during long-lasting storage and this crystallization stabilization process can occur in hours, days, or years (Hanabusa and Suzuki, 2014). Here is worth mentioning that in some oleogels and their industrial applications, the crystallization transition is a disadvantage because it causes the collapse of the tridimensional network of the oleogel and oil expulsion consequently (i.e., syneresis). Unlike low molecular weight gelators, polymeric gelators form molecular chains entangled. Polymers and oligomers, which are not individual molecules, never separate from solution as crystals due to the distribution of their molecular weights and the entanglement of their chains. The characteristic of noncrystallization of polymers and oligomers can be used for the molecular design of gelators. For example, a low molecular weight gelator agent (i.e., MG), adheres to a polymer (i.e., ethylcellulose) as "a gelling-promoting compound" (García-Ortega et al., 2023 unpublished results) and this process converts a low molecular weight gelator into a "polymer-based gelator" capable of forming a more stable oleogel. Also, it has been observed that the synergism between different types of gelator molecules permits the formation of more stable oleogels. As has been demonstrated by investigating an oleogeled sunflower oil system with an α-tocopherol/PC/water mixture in an equimolar ratio, in which the molecular self-assembly formed cylindrical micellar aggregates developed a reversible mechanical edible oleogel stable at temperatures above the 35°C (Nikiforidis and Scholten, 2014). In another research work, it was observed that by mixing the phytosterol–ester oryzanol with lecithin in sunflower oil, both gelators may be dispersed at much higher concentrations than they may be individually developing a viscous suspension. But by adding water to these solutions, a solid-like gel was formed and its mechanical properties were close to those desired for a margarine-type spread, while still solubilizing the oryzanol (Matheson et al., 2017). Finally, it is worth mentioning that it is very important to understand the self-assembly process in solvents because this facilitates the incorporation of suitable gelator molecules to form oleogels and also permits to get the adequate material to be incorporated into commercial products (foods, cosmetics, or pharmaceuticals).

9.2.3 RECENT CHALLENGES IN OLEOGELS

Even though the research on oleogels has a long history in nonfood applications such as analytical chemistry, material science, drug delivery (Kumar and Steed, 2014; Jibry et al., 2006; Toro-Vazquez et al., 2007), and cosmetics (Patel, 2017), its focus on food applications is the major topic. Actually, the main challenges in oleogels production are to mimic thermal and textural properties such as hardstock fats, the formulation of food products with oleogels, and the consumer perception of the final material. As previously mentioned, the microstructural properties of the oleogels subsequently affect their physical properties. Also, it is important to indicate that the processing parameters can influence the oleogel formation and the resultant properties (Blake et al., 2014). All the above considerations presented in different research works are important to get a successful oleogelation process and a material with the desired properties. Numerous studies describe the textural properties of oleogels or foods with oleogel, but some of these oleogels developed have sensory limitations, such as undesirable odors, and flavors (Rogers et al., 2014), for example, those developed with waxes (candelilla wax or rice bran wax) (Scharfe et al., 2022; Hwang and Winkler-Moser, 2020). For these reasons, it is important to consider continuing to investigate to enhance such critical characteristics in oleogels.

9.3 FUNCTIONAL PROPERTIES OF OLEOGELS

In response to the increasing demand for healthy food products driven by health-conscious consumers and technological advancements, there is a growing interest in exploring novel solutions in food science. One such advancement that holds promise is the utilization of oleogels, that are considered valuable due to their versatility, ease of processing, and affordability (Martins et al., 2020). Oleogels have gained significant attention in recent years due to their versatile applications in the food industry. Composed of organic solvents and gelling molecules, oleogels form a crystalline tridimensional network, trapping the solvent and transforming liquid fats into solid forms (Manzoor et al., 2022).

Numerous scientific studies have focused on exploring the use of oleogels in food formulation to enhance nutritional properties. These studies have investigated the physicochemical properties, structure, and composition of various oleogelators, emphasizing their suitability for incorporation into edible products (Perța-Crișan et al., 2023). By understanding the physicochemical properties, structure, and composition of oleogelators, researchers can optimize their formulation for specific food applications. Furthermore, investigating the microstructure, rheological properties, oxidative stability, and sensory acceptability of oleogel-based foods are crucial for their successful integration into the market (Patel, 2017; Co and Marangoni, 2012). For the formulation of oleogels, one can use structuring agents that are generally recognized as safe (GRAS) with which manufacturers can develop healthier food options (Pérez-Monterroza et al., 2014). Oleogels too have also shown potential as materials for pharmaceutical and cosmetic applications (Pinto et al., 2021). This section explores the recent advancements and potential of oleogels in various industries, highlighting their advantages such as vegan composition, low gelator, and concentration.

One application of oleogels is that can be use as substitutes of saturated and *trans* fatty acids. Consuming high amounts of saturated and trans fats is associated with adverse health consequences, including detrimental effects on lipoprotein profiles, coronary heart disease, inflammation, oxidative stress, endothelial dysfunction, increased body weight, and metabolic syndrome (Gracia et al., 2017; Marangoni and Garti, 2018; Wu et al., 2017; Micha and Mozaffarian, 2009). Considering the harmful effects of these fatty acids on human health, oleogels are a promising healthy alternative in the food industry (Zoppe et al., 2009). Food-grade oleogels offer a promising solution by reducing or eliminating the use of saturated and *trans* fatty acids and introducing bioactive compounds, while maintaining desirable functional properties of fats, such as mouthfeel, creaminess, and stability (Manzoor et al., 2022). Additionally, oleogels can be customized to have specific melting properties, enabling their application in various food products to control flavor release and prevent oil migration it has even been proven that the incorporation of oleogels into bakery products, such as cookies and cakes, resulting in improved sensory attributes and extended shelf life (Colla et al., 2018; Demirkesen and Mert, 2019).

With the use of oleogels as fats substitutes, it is possible to modulate the fatty acid content, reduce cholesterol levels, lower saturated fat intake, and deliver bioactive compounds with potential benefits for disease prevention (Orsavova et al., 2015). Furthermore, the implementation of oleogels can address specific health conditions such as dysphagia and malnutrition, particularly among vulnerable populations such as the elderly and children (Aguilera and Park, 2016). In addition, liquid vegetable or fish oils can be transformed into solid fats without the inclusion of harmful saturated *trans* fatty acids (Gracia et al., 2017; Micha and Mozaffarian, 2009). This offers an opportunity to improve the fatty acid composition of food products, replacing unhealthy fats with healthier alternatives.

Another area of research focuses on the utilization of natural gums, particularly polysaccharides, as high molecular-weight oleogelators (Davidovich-Pinhas, 2019). Natural gums offer structuring and gelling capabilities, enabling the creation of water/oil gelling emulsions. The structural conformation, internal charge, and polymeric characteristics of natural gums play a crucial role in their gelling behavior (Banaś and Harasym, 2021).

Understanding the molecular interactions within polysaccharide chains allows for the development of sustainable and intelligent strategies in product formulation.

Oleogels also exhibit great potential as materials for pharmaceutical and cosmetic uses. The low concentration of gelators required for gel formation and the composition of oleogels make them attractive in these industries (Balasubramanian et al., 2012). Oleogels can serve as carriers for bioactive compounds, protecting them from oxidation and controlling their release behavior based on molecular size (Pinto et al., 2021; Zhao et al., 2021). In pharmaceutical applications, in situ biodegradable oleogels have shown promise as vehicles for controlled drug delivery, providing sustained release over an extended period (Wang et al., 2014). Oleogels could enhance stability and controlled release of bioactive ingredients, such as vitamins and antioxidants, thus enabling the development of advanced skincare products with improved efficacy (Perţa-Crişan et al., 2023; Sagiri et al., 2015). Some researchers used ultrasonic standing waves to obtain better characteristics of the gels and improve pharmaceutics applications (Lassila et al., 2022).

In the case of industrial applications, oleogels exhibit potential as lubricants, especially under extreme conditions, focused on formulating oleogels with enhanced tribological properties for high-temperature and high-pressure environments. Their study reported significant improvements in wear resistance and load-carrying capacity compared to conventional lubricants (Kumar et al., 2022). The gelling agents employed in the formulation contribute to the formation of a robust network that withstands high temperatures and pressures, ensuring effective lubrication even in demanding industrial settings. Some oleogels that have been studied as those made with high-oleic sunflower vegetable oil and ethylene–vinyl acetate copolymer, maintained their structural integrity and lubricating properties even at elevated temperatures, outperforming traditional lubricants in terms of thermal stability and oxidation resistance. This capability is crucial for industries operating in extreme temperature environments (Martín-Alfonso and Valencia, 2015).

Many researchers have focused on novel formulation strategies to improve the stability and performance of oleogels. Certain types of surfactants, such as ethoxylated fatty alcohols and sodium stearoyl lactylate, have been reported as effective gelling agents (Martín-Alfonso and Valencia, 2015). Polymers, gelatin, agar, polystyrene, polyethylene, hydroxypropyl methylcellulose, and methylcellulose can also be utilized as gelling agents in oleogels (Patel and Dewettinck, 2015; Taylor et al., 2017). Modified cellulose has shown promise as an efficient structuring agent. Their work highlights the superior gelation efficiency compared to traditional gelling agents, leading to enhanced stability and mechanical properties in oleogels (Li et al., 2022).

Others studied the gelation mechanism and the arrangement of structuring agents within the oleogel matrix, aiding in the design of more efficient formulations employing advanced microscopy techniques, such as confocal laser scanning microscopy and atomic force microscopy, to visualize the internal structure of oleogels at the nanoscale level (Davidovich-Pinhas, 2019; Okuro et al., 2020). This type of research has allowed us to know more about these systems and achieve a greater number of applications.

9.4 CONCLUSION

Oleogels have emerged as a promising class of materials with the potential to revolutionize product formulations and contribute to sustainable and innovative solutions in various sectors. This chapter has highlighted molecular self-assembly as a key role in engineering oleogel properties because this process not only facilitates the incorporation of the suitable gelator molecules to form oleogels but also permits to get adequate material to be incorporated into commercial products (foods, cosmetics, or pharmaceuticals). However, despite all the efforts made in more than a decade is important to consider continuing to investigate to achieve the main challenge in oleogel production, that is, to mimic thermal and textural properties such as hardstock fats and explore the full potential of oleogels in industries of food, pharmaceuticals, and cosmetics products. While the applications of oleogels are promising, new challenges arise during the production processes, and they need to be addressed to realize their widespread adoption. Cost considerations, processing challenges, limited applications, and potential sensory changes are critical factors that must be carefully considered during oleogel product formulation.

KEYWORDS

- **oleogelation**
- **molecular self-assembly**
- **functionality**
- **sustainable materials**

REFERENCES

Aguilar-Zárate, M.; De La Peña-Gil, A.; de Álvarez-Mitre, F. M.; Charó-Alonso, M. A.; Toro-Vazquez, J. F. Vegetable and Mineral Oil Organogels Based on Monoglyceride and Lecithin Mixtures. *Food Biophys.* **2019,** *14,* 326–345. https://doi.org/10.1007/s11483-019-09583-1.

Aguilar-Zárate, M.; Macias-Rodriguez, B. A.; Toro-Vazquez, J. F.; Marangoni, A. G. Engineering Rheological Properties of Edible Oleogels with Ethylcellulose and Lecithin. *Carbohydr. Polym.* **2019,** *205,* 98–105. https://doi.org/10.1016/J.CARBPOL.2018.10.032.

Alnoush, W.; Sayed, A.; Alyafei, N.; Sayed, A. Optimization of Contact Angle and Interfacial Tension Measurements for Fluid/Rock Systems at Ambient Conditions. *MethodsX* **2019,** *6,* 1706–1715. https://doi.org/10.1016/j.mex.2019.07.009.

Alvarez-Mitre, F. M.; Morales-Rueda, J. A.; Dibildox-Alvarado, E.; Charó-Alonso, M. A.; Toro-Vazquez, J. F. Shearing as a Variable to Engineer the Rheology of Candelilla Wax Organogels. *Food Res. Int.* **2012,** *49,* 580–587. https://doi.org/10.1016/j.foodres.2012.08.025.

Alvarez-Mitre, F. M.; Morales-Rueda, J. A.; Dibildox-Alvarado, E.; Charó-Alonso, M. A.; Toro-Vazquez, J. F. Shearing as a Variable to Engineer the Rheology of Candelilla Wax Organogels. *Food Res. Int.* **2012,** *49* (1), 580–587. https://doi.org/10.1016/J.FOODRES.2012.08.025.

Alvarez-Mitre, F. M.; Toro-Vazquez, J. F.; Moscosa-Santillán, M. Shear Rate and Cooling Modeling for the Study of Candelilla Wax Organogels Rheological Properties. *J. Food Eng.* **2013,** *119,* 611–618. https://doi.org/10.1016/j.jfoodeng.2013.06.009.

Avendaño-Vásquez, G.; De la Peña-Gil, A.; Charó-Alvarado, M. E.; Charó-Alonso, M. A.; Toro-Vazquez, J. F. Self-Assembly of Symmetrical and Asymmetrical Alkyl Esters in the Neat State and in Oleogels. *Front. Sustain. Food Syst.* **2020,** *4* (September), 1–21. https://doi.org/10.3389/fsufs.2020.00132.

Balasubramanian, R.; Sughir, A. A.; Damodar, G. Oleogel: A Promising Base for Transdermal Formulations. *Asian J. Pharm.* **2012,** *6* (1), 1–9. https://doi.org/10.4103/0973-8398.100118.

Banaś, K.; Harasym, J. Natural Gums as Oleogelators. *Int. J. Mol. Sci.* „**2021,** *22* (23), 12977. https://doi.org/10.3390/IJMS222312977.

Blach, C.; Gravelle, A. J.; Peyronel, F.; Weiss, J.; Barbut, S.; Marangoni, A. G. Revisiting the Crystallization Behavior of Stearyl Alcohol:Stearic Acid (SO:SA) Mixtures in Edible Oil. *RSC Adv.* **2016,** *6* (84), 81151–81163. https://doi.org/10.1039/C6RA15142F.

Blake, A. I.; Co, E. D.; Marangoni, A. G. Structure and Physical Properties of Plant Wax Crystal Networks and Their Relationship to Oil Binding Capacity. *JAOCS, J. Am. Oil Chem. Soc.* **2014,** *91* (6), 885–903. https://doi.org/10.1007/s11746-014-2435-0.

Bot, A.; Den Adel, R.; Roijers, E. C. Fibrils of γ-Oryzanol + β-Sitosterol in Edible Oil Organogels. *JAOCS, J. Am. Oil Chem. Soc.* **2008,** *85* (12), 1127–1134. https://doi.org/10.1007/s11746-008-1298-7.

Co, E. D.; Marangoni, A. G. Organogels: An Alternative Edible Oil-Structuring Method. *JAOCS, J. Am. Oil Chem. Soc.* **2012,** *89* (5), 749–780. https://doi.org/10.1007/s11746-012-2049-3.

Co, E.; Marangoni, A. G. The Formation of a 12-Hydroxystearic Acid/Vegetable Oil Organogel Under Shear and Thermal Fields. *J. Am. Oil Chem. Soc.* **2013,** *90,* 529–544. https://doi.org/10.1007/s11746-012-2196-6.

Colla, K.; Costanzo, A.; Gamlath, S. Fat Replacers in Baked Food Products. *Foods* ,**2018,** *7* (12), 192. https://doi.org/10.3390/FOODS7120192.

da Silva, T. L. T.; Danthine, S. Effect of High-intensity Ultrasound on the Oleogelation and Physical Properties of High Melting Point Monoglycerides and Triglycerides Oleogels. *J. Food Sci.* 2021. https://doi.org/10.1111/1750-3841.15589.

Dassanayake, L. S. K.; Kodali, D. R.; Ueno, S. Formation of Oleogels Based on Edible Lipid Materials. *Curr. Opin. Colloid Interface Sci.* **2011,** *16* (5), 432–439. https://doi.org/10.1016/j.cocis.2011.05.005.

Davidovich-Pinhas, M. Oil Structuring Using Polysaccharides. *Curr. Opin. Food Sci.* **2019,** *27,* 29–35. https://doi.org/10.1016/j.cofs.2019.04.006.

De La Peña-Gil, A.; Álvarez-Mitre, F. M.; González-Chávez, M. M.; Charó-Alonso, M. A.; Toro-Vazquez, J. F. Combined Effect of Shearing and Cooling Rate on the Rheology of Organogels Developed by Selected Gelators. *Food Res. Int.* **2017,** *93,* 52–65. https://doi.org/10.1016/j.foodres.2017.01.010.

Demirkesen, I.; Mert, B. Recent Developments of Oleogel Utilizations in Bakery Products. **2019,** *60* (14), 2460–2479. https://doi.org/10.1080/10408398.2019.1649243.

Giacintucci, V.; Di Mattia, C. D.; Sacchetti, G.; Flamminii, F.; Gravelle, A. J.; Baylis, B.; Dutcher, J. R.; Marangoni, A. G.; Pittia, P. Ethylcellulose Oleogels with Extra Virgin Olive Oil: The Role of Oil Minor Components on Microstructure and Mechanical Strength. *Food Hydrocoll.* **2018,** *84*, 508–514. https://doi.org/10.1016/j.foodhyd.2018.05.030.

Golodnizky, D.; Rosen-Kligvasser, J.; Davidovich-Pinhas, M. The Role of the Polar Head Group and Aliphatic Tail in the Self-Assembly of Low Molecular Weight Molecules in Oil. *Food Struct.* **2021,** *30,* 100240. https://doi.org/10.1016/J.FOOSTR.2021.100240.

Gracia, K. C.; Llanas-Cornejo, D.; Husi, H. CVD and Oxidative Stress. *J. Clin. Med.* ,**2017,** *6* (2), 22. https://doi.org/10.3390/JCM6020022.

Gravelle, A. J.; Davidovich-Pinhas, M.; Barbut, S.; Marangoni, A. G. Influencing the Crystallization Behavior of Binary Mixtures of Stearyl Alcohol and Stearic Acid (SOSA) Using Ethylcellulose. *Food Res. Int.* **2017,** *91,* 1–10. https://doi.org/10.1016/j.foodres.2016.11.024.

Gravelle, A. J.; Davidovich-Pinhas, M.; Zetzl, A. K.; Barbut, S.; Marangoni, A. G. Influence of Solvent Quality on the Mechanical Strength of Ethylcellulose Oleogels. *Carbohydr. Polym.* **2016,** *135,* 169–179. https://doi.org/10.1016/j.carbpol.2015.08.050.

Gravelle, A. J.; Marangoni, A. G. Vegetable Oil Oleogels Structured Using Mixtures of Stearyl Alcohol and Stearic Acid (SO:SA). *Edible Oleogels* **2018,** 193–217. https://doi.org/10.1016/B978-0-12-814270-7.00008-3.

Hanabusa, K.; Suzuki, M. Development of Low-Molecular-Weight Gelators and Polymer-Based Gelators. *Polym. J.* ,**2014,** *46* (11), 776–782. https://doi.org/10.1038/pj.2014.64.

Hwang, H. S.; Winkler-Moser, J. K. Properties of Margarines Prepared from Soybean Oil Oleogels with Mixtures of Candelilla Wax and Beeswax. *J. Food Sci.* **2020,** *85* (10), 3293–3302. https://doi.org/10.1111/1750-3841.15444.

Jibry, N.; Sarwar, T.; Murdan, S. Amphiphilogels as Drug Carriers: Effects of Drug Incorporation on the Gel and on the Active Drug. *J. Pharm. Pharmacol.* **2006,** *58* (2), 187–194. https://doi.org/10.1211/jpp.58.2.0005.

Kumar, D. K.; Steed, J. W. Supramolecular Gel Phase Crystallization: Orthogonal Self-Assembly under Non-Equilibrium Conditions. *Chem. Soc. Rev.* **2014,** *43* (7), 2080–2088. https://doi.org/10.1039/C3CS60224A.

Kumar, R.; Hussainova, I.; Rahmani, R.; Antonov, M. Solid Lubrication at High-Temperatures—A Review. *Materials (Basel).* **2022,** *15* (5), 3–27. https://doi.org/10.3390/ma15051695.

Lan, Y.;, Corradini, M. G.;, Weiss, R. G. Raghvan, S. R.; and Rogers, M. A. To Gel or Not to Gel: Correlating Molecular Gelation with Solvent Parameters. *Chem. Soc. Rev.* **2015,** *44,* 6035–6058. https://doi.org/10.1039/c5cs00136f.

Lassila, P.; Valoppi, F.; Tommiska, O.; Hyvönen, J.; Holmström, A.; Hietala, S.; Salmi, A.; Haeggström, E. Practical Scale Modification of Oleogels by Ultrasonic Standing Waves. *Ultrason. Sonochem.* **2022,** *85,* 1350–4177. https://doi.org/10.1016/j.ultsonch.2022.105970.

Lesellier, E.; Latos, A.; West, C. Ultra High Efficiency/Low Pressure Supercritical Fluid Chromatography (UHE/LP-SFC) for Triglyceride Analysis: Identification, Quantification, and Classification of Vegetable Oils. *Anal. Sci. Adv.* **2021**, *2* (1–2), 33–42. https://doi.org/10.1002/ANSA.202000156.

Li, L.; Liu, G.; Bogojevic, O.; Pedersen, J. N.; Guo, Z. The Influence of Concentration and Temperature on the Formation Of. *Compr. Rev. Food Sci. Food Saf.* **2022,** *21* (3), 2077–2104. https://doi.org/10.1111/1541-4337.12928.

López-Martínez, A.; Morales-Rueda, J. A.; Dibildox-Alvarado, E.; Charó-Alonso, M. A.; Marangoni, A. G.; Toro-Vazquez, J. F. Comparing the Crystallization and Rheological Behavior of Organogels Developed by Pure and Commercial Monoglycerides in Vegetable Oil. *Food Res. Int.* **2014,** *64,* 946–957. https://doi.org/10.1016/j.foodres.2014.08.029.

Manzocco, L.; Calligaris, S.; Da Pieve, S.; Marzona, S.; Nicoli, M. C. Effect of Monoglyceride-Oil-Water Gels on White Bread Properties. *Food Res. Int.* **2012,** *49* (2), 778–782. https://doi.org/10.1016/j.foodres.2012.09.011.

Manzoor, S.; Masoodi, F. A.; Naqash, F.; Rashid, R. Oleogels: Promising Alternatives to Solid Fats for Food Applications. *Food Hydrocoll. Heal.* **2022,** *2,* 100058. https://doi.org/10.1016/J.FHFH.2022.100058.

Marangoni, A. G.; Garti, N. *Edible Oleogels: Structure and Health Implications*, 2nd ed.; Marangoni, A. G., Garti, N., Eds.; Elsevier, AOCS Press and Academic Press: Urbana, IL, 2018.

Martín-Alfonso, J. E.; Valencia, C. Tribological, Rheological, and Microstructural Characterization of Oleogels Based on EVA Copolymer and Vegetables Oils for Lubricant Applications. *Tribol. Int.* **2015,** *90,* 426–434. https://doi.org/10.1016/j.triboint.2015.05.004.

Martínez-Ávila, M.; De la Peña-Gil, A.; Álvarez-Mitre, F. M.; Charó-Alonso, M. A.; Toro-Vazquez, J. F. Self-Assembly of Saturated and Unsaturated Phosphatidylcholine in Mineral and Vegetable Oils. *JAOCS, J. Am. Oil Chem. Soc.* **2019,** *96* (3), 273–289. https://doi.org/10.1002/AOCS.12188.

Martins, A. J.; Vicente, A. A.; Pastrana, L. M.; Cerqueira, M. A. Oleogels for Development of Health-Promoting Food Products. *Food Sci. Hum. Wellness* **2020,** *9* (1), 31–39. https://doi.org/10.1016/J.FSHW.2019.12.001.

Matheson, A. B.; Dalkas, G.; Gromov, A.; Euston, S. R.; Clegg, P. S. The Development of Phytosterol-Lecithin Mixed Micelles and Organogels. **2017**. https://doi.org/10.1039/c7fo01271c.

Micha, R.; Mozaffarian, D. Trans Fatty Acids: Effects on Metabolic Syndrome, Heart Disease and Diabetes. *Nature Reviews Endocrinology*; Nature Publishing Group, **2009**; pp 335–344. https://doi.org/10.1038/nrendo.2009.79.

Miguel Aguilera, J.; June Park, D. Texture-Modified Foods for the Elderly: Status, Technology and Opportunities. *Trends Food Sci. Technol.* **2016,** *57,* 156–164. https://doi.org/10.1016/j.tifs.2016.10.001.

Murashova, N. M.; Prokopova, L. A.; Trofimova, E. S.; Yurtov, E. V. Effects of Oleic Acid and Phospholipids on the Formation of Lecithin Organogel and Microemulsion. *J. Surfactants Deterg.* **2018,** *21* (5), 635–645. https://doi.org/10.1002/jsde.12170.

Nikiforidis, C. V.; Scholten, E. Self-Assemblies of Lecithin and α-Tocopherol as Gelators of Lipid Material. *RSC Adv.* **2014,** *4* (5), 2466–2473. https://doi.org/10.1039/C3RA46584E.

Okuro, P. K.; Martins, A. J.; Vicente, A. A.; Cunha, R. L. Perspective on Oleogelator Mixtures, Structure Design and Behaviour towards Digestibility of Oleogels. *Curr. Opin. Food Sci.* **2020,** *35,* 27–35. https://doi.org/10.1016/J.COFS.2020.01.001.

Orsavova, J.; Misurcova, L.; Vavra Ambrozova, J.; Vicha, R.; Mlcek, J. Fatty Acids Composition of Vegetable Oils and Its Contribution to Dietary Energy Intake and Dependence of Cardiovascular Mortality on Dietary Intake of Fatty Acids. *Int. J. Mol. Sci.* **2015,** *16* (6), 12871–12890. https://doi.org/10.3390/ijms160612871.

Patel, A. R. Oil Structuring: Concepts, Overview and Future Perspectives. In *Food Chemistry, Function and Analysis*; Royal Society of Chemistry, 2017; pp 1–22, Vol. 2018-Janua. https://doi.org/10.1039/9781788010184-00001.

Patel, A. R.; Dewettinck, K. Current Update on the Influence of Minor Lipid Components, Shear and Presence of Interfaces on Fat Crystallization. *Curr. Opin. Food Sci.* **2015,** *3,* 65–70. https://doi.org/10.1016/j.cofs.2015.05.010.

Patel, A. R.; Rajarethinem, P. S.; Gredowska, A.; Turhan, O.; Lesaffer, A.; De Vos, W. H.; Van De Walle, D.; Dewettinck, K. Edible Applications of Shellac Oleogels: Spreads, Chocolate Paste and Cakes. *Food Funct.* **2014,** *5,* 615–822. https://doi.org/10.1039/c4fo00034j.

Pérez-Martínez, D.; Alvarez-Salas, C.; Charó-Alonso, M.; Dibildox-Alvarado, E.; Toro-Vazquez, J. F. The Cooling Rate Effect on the Microstructure and Rheological Properties of Blends of Cocoa Butter with Vegetable Oils. *Food Res. Int.* **2007,** *1* (40), 47–62. https://doi.org/10.1016/J.FOODRES.2006.07.016.

Pérez-Monterroza, E. J.; Márquez-Cardozo, C. J.; Ciro-Velásquez, H. J. Rheological Behavior of Avocado (Persea Americana Mill, Cv. Hass) Oleogels Considering the Combined Effect of Structuring Agents. *LWT - Food Sci. Technol.* **2014,** *59* (2), 673–679. https://doi.org/10.1016/J.LWT.2014.07.020.

Perţa-Crişan, S.; Ursachi, C. Ştefan; Chereji, B. D.; Tolan, I.; Munteanu, F. D. Food-Grade Oleogels: Trends in Analysis, Characterization, and Applicability. *Gels* ,**2023,** *9* (5), 386. https://doi.org/10.3390/GELS9050386.

Pinto, T. C.; Martins, A. J.; Pastrana, L.; Pereira, M. C.; Cerqueira, M. A. Oleogel-Based Systems for the Delivery of Bioactive Compounds in Foods. **2021**. https://doi.org/10.3390/gels7030086.

Rogers, M. A.; Liu, X.; Mallia, V. A.; Weiss, R. G. Dissecting Kinetic Pathways to Formation of the Fibrillar Objects in Molecular Gels Using Synchrotron FT-IR. *CrystEngComm*, **2015,** *17* (42), 8085–8092. https://doi.org/10.1039/C5CE00733J.

Rogers, M. A.; Strober, T.; Bot, A.; Toro-Vazquez, J. F.; Stortz, T.; Marangoni, A. G. Edible Oleogels in Molecular Gastronomy. *Int. J. Gastron. Food Sci.* **2014,** *2* (1), 22–31. https://doi.org/10.1016/j.ijgfs.2014.05.001.

Sagiri, S. S.; Singh, V. K.; Pal, K.; Banerjee, I.; Basak, P. Stearic Acid Based Oleogels: A Study on the Molecular, Thermal and Mechanical Properties. *Mater. Sci. Eng. C*, **2015,** *48,* 688–699. https://doi.org/10.1016/j.msec.2014.12.018.

Sahu, D.; Bharti, D.; Kim, D.; Sarkar, P.; Pal, K. Variations in Microstructural and Physicochemical Properties of Candelilla Wax/Rice Bran Oil–Derived Oleogels

Using Sunflower Lecithin and Soya Lecithin. *Gels* **2021,** *7* (4), 226. https://doi.org/10.3390/GELS7040226.

Sawalha, H.; Venema, P.; Bot, A.; Flöter, E.; van der Linden, E. The Influence of Concentration and Temperature on the Formation of β-Sitosterol and γ-Oryzanol Tubules in Edible Oil Organogels. *Food Biophys.* **2011,** *6* (1), 20–25. https://doi.org/10.1007/s11483-010-9169-9.

Scharfe, M.; Niksch, J.; Flöter, E. Influence of Minor Oil Components on Sunflower, Rice Bran, Candelilla, and Beeswax Oleogels. *Eur. J. Lipid Sci. Technol.* **2022,** *124* (7). https://doi.org/10.1002/EJLT.202100068.

Scharfe, M.; Niksch, J.; Flöter, E. Influence of Minor Oil Components on Sunflower, Rice Bran, Candelilla, and Beeswax Oleogels. *Eur. J. Lipid Sci. Technol.* **2022,** *124* (7), 1–19. https://doi.org/10.1002/ejlt.202100068.

Taylor, M. J.; Tomlins, P.; Sahota, T. S. Thermoresponsive Gels. *Gels*, **2017,** *3* (1). https://doi.org/10.3390/gels3010004.

Toro-Vazquez, J. F.; Morales-Rueda, J. A.; Dibildox-Alvarado, E.; Charó-Alonso, M. A.; Alonzo-Macias, M.; González-Chávez, M. M. Thermal and Textural Properties of Organogels Developed by Candelilla Wax in Safflower Oil. *JAOCS, J. Am. Oil Chem. Soc.* **2007,** *84* (11), 989–1000. https://doi.org/10.1007/s11746-007-1139-0.

Wang, D.; Zhao, J.; Liu, X.; Sun, F.; Zhou, Y.; Teng, L.; Li, Y. Parenteral Thermo-Sensitive Organogel for Schizophrenia Therapy, in Vitro and in Vivo Evaluation. **2014**. https://doi.org/10.1016/j.ejps.2014.04.020.

Wettlaufer, T.; Brykczynski, H.; Flöter, E. Wax-Based Oleogels—Properties in Medium Chain Triglycerides and Canola Oil. *Eur. J. Lipid Sci. Technol.* **2022,** *124* (2), 1–12. https://doi.org/10.1002/EJLT.202100114.

Wu, J. H. Y.; Zheng, M.; Catterall, E.; Downs, S.; Thomas, B.; Veerman, L.; Barendregt, J. J. Contribution of Trans-Fatty Acid Intake to Coronary Heart Disease Burden in Australia: A Modelling Study. *Nutrients* **2017,** *9* (77), 1–9. https://doi.org/10.3390/nu9010077.

Zetzl, A. K.; Gravelle, A. J.; Kurylowicz, M.; Dutcher, J.; Barbut, S.; Marangoni, A. G. Microstructure of Ethylcellulose Oleogels and Its Relationship to Mechanical Properties. *Food Struct.* **2014,** *2,* 27–40. https://doi.org/10.1016/j.foostr.2014.07.002.

Zhao, W.; Wei, Z.; Xue, C. Recent Advances on Food-Grade Oleogels: Fabrication, Application and Research Trends. *Crit. Rev. Food Sci. Nutr.* **2021,** *62* (27), 7659–7676. https://doi.org/10.1080/10408398.2021.1922354.

Zhu, G.; Dordick, J. S. Solvent Effect on Organogel Formation by Low Molecular Weight Molecules. *Chem. Mater.* **2006,** (27), 5988–5995. https://doi.org/10.1021/cm0619297.

Zoppe, J. O.; Peresin, M. S.; Habibi, Y.; Venditti, R. A.; Rojas, O. J. Reinforcing Poly(ε-Caprolactone) Nanofibers with Cellulose Nanocrystals. *ACS Appl. Mater. Interfaces*, **2009,** *1* (9), 1996–2004. https://doi.org/10.1021/AM9003705/ASSET/IMAGES/MEDIUM/AM-2009-003705_0001.GIF.

CHAPTER 10

Production of Fungi for Biological Control in Solid-State Culture Bioreactors: A Sustainable Approach

FERNANDO MÉNDEZ-GONZÁLEZ[1], JOSÉ MIGUEL CASTILLO-MINJAREZ[2], MARÍA ESMERALDA BIBIÁN-LEÓN[3], EBANDRO USCANGA-MORTERA[3], JAVIER BARRIOS-GONZÁLEZ[4], and JOSÉ JUAN BUENROSTRO-FIGUEROA[1*]

[1]*Laboratorio de Biotecnología y Bioingeniería, Centro de Investigación en Alimentación y Desarrollo, Delicias, Chihuahua, Mexico*

[2]*División de Electromecánica Industrial, Universidad Tecnológica de Tecámac, Tecámac, Estado de México, Mexico*

[3]*Posgrado en Botánica, Colegio de Posgraduados-Montecillo, Texcoco, Mexico State, Mexico*

[4]*Departamento de Biotecnología, Universidad Autónoma Metropolitana, Iztapalapa, Mexico City, Mexico*

ABSTRACT

The transition to sustainable agriculture requires strategies that replace the use of harmful chemical substances for human health and the environment. Among them is the production and application of microorganisms with the capacity to agricultural pest populations control. Among these microorganisms are species of filamentous fungi capable of successfully combating the disappearance of insects, weeds, and phytopathogenic fungi. However, the industrial production of high-quality propagules is considered the bottleneck that limits meeting the demand for pesticides. Due to their resistance to environmental conditions and infective capacity, conidia are the preferred

Sustainable Engineering and Agro-Food Processing: Renewable and Nonrenewable Resources.
José Juan Buenrostro-Figueroa, Julio César Tafolla-Arellano, Mónica Chávez-González, Pedro Aguilar-Zárate, A. K. Haghi (Eds.)

propagules to apply in the field. The conidia are produced by solid-state culture in bioreactors with different configurations and technological degrees. Bioreactors with a low level of technology, such as plastic bags, present a simple and low-cost process but with some reproducibility problems. On the contrary, bioreactors with trays, columns, and shakers allow a controlled process, which improves the production, productivity, and quality of the conidia. Therefore, in this chapter, the characteristics of some fungal classifications used in biological control and the different types of bioreactors involved in propagule production will be analyzed.

10.1 INTRODUCTION

One of the milestones left by the United Nations Conference on Environment and Development (UNCED) in Rio de Janeiro in June 1992 was the generation of the global strategic plan for sustainable development called Agenda 21. Within it, actions are suggested to promote agriculture and sustainable rural development. Under these principles, in 2015, the 2030 Agenda was created (Córdova-Albores et al., 2020), where 17 objectives and 169 goals were agreed on in terms of Sustainable Development, which consider the implementation of resilient agricultural practices that increase productivity and food production, promoting the maintenance of ecosystems and strengthen the capacity to adapt to climate change or other natural phenomena (Castillo-Minjarez, 2022).

On the contrary, in Johannesburg, within the framework of the World Summit on Sustainable Development in 2002, human health and the environmental damage caused by chemical exposure were recognized. The foregoing motivated the generation of protocols to regulate agrochemical use and gradually replace them with alternatives of biological origin (Castillo-Minjarez, 2022). For this reason, a great variety of beneficial microorganisms for agriculture have been identified, produced, and applied. Among them are filamentous fungi used for the biological control of pests (Ramírez-Guzmán et al., 2018). These microorganisms present effective treatment of infestations of insects, weeds, and fungi, among other pests; therefore, they have promoted the generation of various commercial products around the world (Castillo-Minjarez, 2022; Deans and Krischik, 2023; Arredondo-Bernal et al., 2020)]. However, the successful application of these biological control agents depends on the large-scale production of propagules, which must have sufficient quality and environmental resistance to fulfill their biological action. Therefore, this manuscript will analyze the general characteristics of these microorganisms and the different processes for producing propagules in solid-state culture bioreactors.

10.2 CHARACTERISTICS OF BIOLOGICAL CONTROL AGENT FUNGI

The fungi used for the biological control of pests are widely distributed in the kingdom Fungi's Phyla and present different properties and mechanisms that allow them to identify and interact with their hosts. Among the different hosts are insects, weeds, and phytopathogenic fungi, which cause losses in the agricultural sector. Therefore, this section addresses the characteristics of filamentous fungi that contribute to the population control of these pests.

10.2.1 FUNGAL INSECTICIDES

The population control of insects by the biological action of filamentous fungi is carried out by organisms called entomopathogenic fungi (EF). EF are heterotrophic organisms with the ability to cause diseases in insects (mycosis) (Inglis et al., 2009), which contributes to the natural reduction of diverse insect populations around the world (Lacey et al., 2015; Tamayo-Mejía et al., 2020). EF are distributed in the kingdom Fungi in six phyla (Basidiomycota, Blastocladiomycota, Chytridiomycota, Kickxellomycota, Entomophthormycota y Ascomycota) (Sharma et al., 2020). Among them, the phylum Ascomycota has stood out for its wide application spectrum since it includes genera of fungi such as *Aschersonia*, *Hypocrella*, *Regiocrella*, *Metarhizium*, *Cordyceps*, *Beuveria*, *Akanthomyces*, *Penicillum*, and *Paecilomyces* which are capable of infecting 13 different orders of insects distributed in various ecosystems (Santos et al., 2022). EF have certain application advantages compared to entomopathogenic bacteria since their chitin-rich cell wall maintains stability and resistance to adverse environmental factors (Jaronski and Jackson, 2012; Wraight et al., 2001), their conidia have hydrophobic properties that allow them to adhere to the host, so it is not Ingestion necessary to start the infectious cycle (Jeffs and Khachatourians, 1997; Kamp and Bidochka, 2002), and the dissemination of spores produced on the corpses of the hosts allows them to persist in infestation areas (Rangel et al., 2006; Sun et al., 2002).

The EF attack mechanism on insect hosts consists of two phases: the infective and the reproductive. In the infective phase, the spore contacts the host, which can be by adhesion to the cuticle, ingestion, or aspiration (Santos et al., 2022). Subsequently, the fungus produces the germ tube and penetrating hyphae (Holder and Keyhani, 2005). During cuticle penetration, the fungus uses an enzymatic system that includes proteases, chitinases, lipases, and lipoxygenases (Sabbour, 2002) to degrade the structural compounds (Montesinos-Matías et al., 2011). If the infection is successful, the fungus proliferates in the hemolymph, colonizing the insect's tissues and producing toxins to attack the immune system (Inglis et al., 2009). Subsequently, the infection period ends with the host's death (Holder and Keyhani, 2005) and the spread of conidia produced on the insect's corpse.

Most EF-based commercial products use its spores as an active ingredient. EF spores are produced in liquid (blastospores) and solid (conidia) culture processes at different scales (small, medium, and large scale) (Méndez-González et al., 2022). Between both processes, the solid-state culture presents advantages in the production, quality, and resistance of the spores (Méndez-González et al., 2022; Sala et al., 2019).

10.2.2 FUNGAL HERBICIDES

Biological control of weeds is carried out by phytopathogenic organisms and/or phytotoxins (Bailey, 2014). Most bioherbicides are of microbial origin, among them, mycoherbicides are the most abundant on the market (Triolet et al., 2020). A large number of fungal herbicide species are distributed in the genera *Phoma*, *Fusarium*, *Sclerotinia*, *Phytophthora*, *Alternaria*, *Condrostereum*, *Puccinia*, and *Colletotrichum* (Triolet et al., 2020; Roberts et al., 2022). These phytopathogenic fungi can identify their host and trigger an infectious process that involves the biosynthesis of enzymes, growth regulators, effector proteins, and toxins, which allow the degradation of cellular components, combat the immune system, and colonization of intracellular spaces in the plant (Peng et al., 2021). Regularly, this infectious process begins when the spores contact with the host and adhere to it through mucilage secretion (Hamer et al., 1979). After that, the spores germinate, and the polarized growth of the hyphae begins on the surface of the plant. During this stage, the fungus will perceive physical (such as hydrophobicity) and chemical stimuli (such as the presence of cutin and plant waxes) that will induce the formation of the appressorium to begin the process of invasion of plant tissues (Doehlemann et al., 2017; Wilson and Talbot, 2009). Once the intracellular space is colonized, the interaction between the host and the host depends on the type of microorganism, which can be necrotrophic, biotrophic, or hemibiotrophic (Horbach et al., 2011). Weeds infected by mycoherbicides present symptoms such as loss of seed viability, morphological defects, and nutrient uptake disorders, which will eventually lead to death (Radhakrishnan et al., 2018) and population control.

Among the commercial products formulated based on these microorganisms are Collego™/LockDown™ (Bowers, 1986), BioMal® (Mortensen, 1988), Lubao 1, and Lubao 2 (Nandhini et al., 2019), which use *Colletotrichum gloeosporioides* as the active principle and Di-Bak™ presents a formulation based on *Macrophomina phaseolina*, *Lasiodiplodia pseudotheobromae*, and *Neoscytalidium novaehollandiae*. In the case of Di-Bak™, the commercial presentation consists of capsules with propagules of the three species, which are produced by solid-state culture using millet as a substrate (Galea, 1909).

10.2.3 ANTIFUNGALS

The control of fungal pests is carried out biologically by antagonistic mechanisms (parasitism, antibiosis or competition for nutrients, or space) (Mukhopadhyay and Kumar, 2020). Fungal genera such as *Trichoderma*, *Epicoccum*, and *Akanthomyces* (Ons et al., 2020; Zewdie et al., 2021) contain antagonistic species to phytopathogenic fungi. Among them, the genus *Trichoderma* has the large number of effective species for the control of phytopathogens, which includes *Trichoderma harzianum, Trichoderma koningii*, *Trichoderma viride*, *Trichoderma atroviride*, *Trichoderma pseudokoningii, Trichoderma longibrachiatum*, *Trichoderma hamatum*, *Trichoderma polysporum*, *T. asperellum*, and *Trichoderma reesei* (Amerio et al., 2020; Mukhopadhyay and Kumar, 2020; Monte, 2001). Regularly, the mechanism of action of these microorganisms consists of the penetration of the host hypha and/or the segregation of enzymes that degrade the fungal cell wall. Furthermore, some species can produce antibiotics that limit the growth of the phytopathogen and its ability to compete for nutrients (Monte, 2001). Some commercial products that use propagules of these fungi as active ingredients include bacterial agents (such as *Mesorhizobium cicero*, *Bacillus subtilis*, and *Pseudomonas flourescens*, among others) that enhance their biological activity (Jambhulkar et al., 2018; Dubey et al., 2015).

10.3 PROPAGULES PRODUCTION IN SOLID-STATE CULTURE BIOREACTORS

The propagules production of the biological control agents described in the previous section is carried out by solid-state culture, involving simple and technified processes. Among the simple ones are the use of plastic containers (Grzywacz et al., 2014) and bags (Jaronski, 2023). These processes are popular since they can be carried out regardless of the region's technological level and have low equipment costs (Méndez-González et al., 2022). On the contrary, the technified processes involve bioreactors (static and agitated) with monitoring and control systems, which improve production yields and product quality (Méndez-González et al., 2020) at the expense of a greater amount of initial investment and cost maintenance. Each available process has different characteristics, which in this section are analyzed.

10.3.1 BAG BIOREACTORS

Bag bioreactors (BB) used to produce fungal conidia are generally fabricated with high-density polypropylene (Méndez-González et al., 2020) with the capacity for 200 and 1000 g of substrate (Jaronski, 2023; Méndez-González et al., 2018). The manufacturing material contains micropores that allow gas exchange by diffusivity (Jenkins and Goettel, 2011); however, the low rate of CO_2 removal can cause its accumulation within the system (CO_2 >5% v/v) (Arias Barreto et al., 2017). In some cases, the bags are partially closed, or plugs with greater porosity than the bag are used to facilitate gas exchange.

The BB culture process described by Cherry et al. (Cherry et al., 1999) was designed for the *Metarhizium* genus fungi; however, it can be used to produce propagules of other control agents (Table 10.1). The process presents advantages such as low initial investment and can be implemented regardless of the level of technological advancement in the region (Swanson, 1997) since it requires little equipment and consumables are available locally (Cherry et al., 1999). However, in terms of process, it presents variations in productivity and quality of the conidia (Jenkins and Grzywacz, 2000), which are regularly caused by variations in the process conditions (temperature, moisture content, nutrients, light, pH, gas concentrations, among others) (Luz and Fargues, 1998; Arthurs and Thomas, 2001). Due to the bioreactor characteristics is difficult to implement monitoring and control instruments for process conditions (Méndez-González et al., 2018). Eventually, the lack of control of these parameters affects the production cost and marketing Price (Cherry et al., 1999). Moreover, the small bioreactor capacity (~1 kg) makes it necessary to process a considerable number of units per batch, which causes high labor and area requirements for incubation (Méndez-González et al., 2018; Méndez-González et al., 2018).

TABLE 10.1 Conidia Production of Biological Control Agents in Bag Bioreactors.

Microorganism	Substrate	Culture time (days)	Conidia production (conidia × 109/gdm)	Reference
Metarhizium anisopliae	Rice grains	13–18	0.9–1.0	(Cherry et al., 1999)
	Rice grains	15	0.8–1.0	(Alcantara-Vargas et al., 2020)
	Parboiled rice	20	2.3–2.6	(Barra-Bucarei et al., 2016)
	Rice grains	9	0.88	(Méndez-González et al., 2018)
Metarhizium robertsii	Rice grains	7	0.3–0.9	(Méndez-González et al., 2018)
	Rice grains	10.29	0.9–1.3	(Méndez-González et al., 2020)
	Rice grains	28	0.03 - 0.04	(Mathulwe et al., 2022)
	Parboiled rice	10	1.5	(Ferreira et al., 2021)
Akanthomyces lecanii		15	0.3–1.5	(Machado et al., 2010)
	Pearl millet	9	1.0	(Chaparro et al., 2021)
L. pseudotheobromae	Rice bran	14	1.8–3.1	(Adetunji and Oloke, 2013)
B. bassiana	Cooked rice	15	8.6	(Posada-Flórez, 2008)
	Rice grains and sorghum	15	29	(Dhar and Kaur, 2011)
Penicillium frequentans	Peat:vermiculite and nutrients	10–30	0.8–1.1	(De Cal et al., 2002)
Cordyceps fumosoreosea	Parboiled rice	10	3.57	(Rojas et al., 2023)
Trichoderma asperellum	Jatropha cake	7	8.5	(Rayhane et al., 2019)

dm: dry mass

10.3.2 TRAY BIOREACTORS

Tray bioreactors (TB) are a thin layer of substrate spread over a rectangular surface (Méndez-González et al., 2018; Lonsane et al., 1985). In its classic form (trays without perforation), the transport of CO_2 and O_2 between the bed and the headspace is carried out by diffusion and natural convection (Méndez-González et al., 2020; Raghava Rao et al., 1993). On the contrary, heat removal is carried out by conduction and evaporation (Figueroa-Montero et al., 2011; Diniz da Silva et al., 2022). The gas and heat transport rate in the bioreactor limits the maximum height of the culture bed (da Cunha et al., 2019). In cultures of *Beauveria bassiana* and *M. anisopliae*, a temperature gradient of up to 4°C has been observed with increasing bed height from 2 to 8 cm and from 2 to 6 cm, respectively. The above causes conidia production to decrease by around 63 and 76% for *B. bassiana* (Xie et al., 2012) and *M. anisopliae* (da Cunha et al., 2019), respectively. Therefore, to avoid overheating and the anaerobic zones formation in the culture bed, its height for the conidia production in TB is between 2 and 4 cm (Méndez-González et al., 2020; Xie et al., 2012; Ye et al., 2006). However, in the industrial production of *B. bassiana* conidia, beds with a height of 10 cm have been used (Alves and Pereira, 1989).

Traditionally, these bioreactors are incubated in temperature-regulated areas (similar to bag bioreactors), where the trays can be arranged on racks, one above the other a few centimeters apart (Lonsane et al., 1985). However, in some systems, TBs are incubated in closed chambers, in which instruments to measure temperature, moisture, temperature, and gaseous atmosphere composition can be implemented (Figueroa-Montero et al., 2011; da Cunha et al., 2019; Ye et al., 2006). In addition, the incubation chambers can be supplied with forced aeration, which improves the efficiency of the O_2, CO_2, and heat transport (Ye et al., 2006). Additionally, the base of the TB can be perforated, which allows airflow to pass through the bed, improving the O_2 supply and the removal of heat and CO_2 (Lonsane et al., 1985).

Increasing the TB dimensions is difficult due to the risk of overheating, anaerobic zones generation, and compromising the culture asepsis (Pandey, 1991) for this reason, TB scale-up is carried out by increasing the number of trays inside the incubation chamber, decreasing the space between them (Brand et al., 2010). The TB characteristics allow a controlled process; however, production yields (Table 10.2), intensive labor (Pandey, 1991), and area for the process (Lonsane et al., 1985) are similar to those required for the BB.

10.3.3 PACKED COLUMN BIOREACTORS

Packed column bioreactors (PCBs) are cylindrical systems (Méndez-González et al., 2018) supplied with forced aeration (Mitchell et al., 1993). Forced aeration promotes gas exchange (O_2 supply and CO_2 removal) and heat removal by convective mechanisms (Méndez-González et al., 2020; Lopes-Perez et al., 2018). Additionally, during the process, it is possible to modify the aeration moisture content as a strategy to control the temperature and/or moisture content in the packed bed (Lonsane et al., 1985). The

TABLE 10.2 Conidia Production of Biological Control Agents in Tray Bioreactors.

Microorganism	Substrate	Culture time (days)	Conidia production (conidia × 109/ gdm)	Reference
B. bassiana	Rice grains	8	2.5–2.9	(Xie et al., 2012)
	Rice grains	7	2.0–2.8	(Ye et al., 2006)
		10	3.72	(Alves and Pereira, 1989)
M. anisopliae	Rice grains	14	0.76–1.0	(Méndez-González et al., 2018)
				(da Cunha et al., 2019)
M. robertsii	Rice grains	7	0.7–1.0	(Méndez-González et al., 2018)
	Rice grains	12.49	0.7–1.1	(Méndez-González et al., 2020)
Thichoderma virens	Rice grains	4.5	10–30	(Jou and Lo, 2011)

dm: dry mass

PCB's design facilitates the use of instrumentation for the monitoring and control of the process conditions (temperature, humidity, pH, CO_2, and O_2 concentration, among others) (da Cunha et al., 2020; Henrique et al., 2022). The conidia from diverse biological control agents are produced on PCBs (Table 10.3). Additionally, the forced aeration in PCBs allows reached highest production and productivity of conidia than those obtained in the tray bioreactor (TB) and bag bioreactor (BB) (Méndez-González et al., 2020; Méndez-González et al., 2018). The above has motivated studies aimed at the propagules production in PCBs at a larger scale. However, the PCB´s dimensional increase gradually decreases the packed bed porosity (Benenati and Brosilow, 1962), generating overheating and anaerobic zones (Méndez-González et al., 2022), which hinders microorganism growth and conidia production (da Cunha et al., 2020; Méndez-González et al., 2022; Van Breukelen et al., 2011). Therefore, to improve the bed structure, the use of low-density supports and substrates and the addition of texturizers have been evaluated. The low-density packing materials reduce shrinkage and agglomeration of the culture bed (Van Breukelen et al., 2011; De Vrije et al., 2001) but decrease the volumetric productivity of the bioreactor (Méndez-González et al., 2022; Jones et al., 2004). On the contrary, the texturizers addition, such as sugarcane bagasse (da Cunha et al., 2020) and spherical-style packing (Méndez-González et al., 2022), allow PCBs scale up without reducing the conidia production. The above demonstrates that the texturizers addition contributes to solving the problems associated with the solid matrix structure, which could allow operating PCBs on an industrial scale.

TABLE 10.3 Conidia Production of Biological Control Agents in Packed Column Bioreactors.

Microorganism	Substrate	Culture time (d)	Conidia production (conidia × 109/ gdm)	Reference
M. anisopliae	Rice grains	10	1.7–2.2	(da Cunha et al., 2020)
	Rice grains and sugarcane bagasse	10	3.2–3.7	(da Cunha et al., 2020)
	Rice grains	7	6	(Arzumanov et al., 2005)
	Rice bran	9	9.7	(Dorta and Arcas, 1998)
	Rice grains	7	1.5	(Méndez-González et al., 2018)
	Hemp, glucose, and yeast extract	12.5	8	(Van Breukelen et al., 2011)
	Wheat bran	12.5	5.6	(Van Breukelen et al., 2011)
	Rice bran	12.5	3.4	(Van Breukelen et al., 2011)
	Wheat grains	12.5	0.9	(Van Breukelen et al., 2011)
M. robertsii	Rice grains	8	1.3–1.8	(Méndez-González et al., 2020)
	Rice grains and spherical style packings	8.3	1.0–1.5	(Méndez-González et al., 2022)
B. bassiana	Potato residues and sugarcane bagasse	10	9.8	(Santa et al., 2004)
	Potato Peel	7–8	1.3	(Sala et al., 2021)
	Beer draff	8	2.5	(Sala et al., 2023)
Coniothyrium minitans	Hemp, glucose, and yeast extract	18	12.8	(De Vrije et al., 2001)
	Oat grains	14	2.6–5.1	(Jones et al., 2004)
Tichoderma longibranchiatum	Sugarcane bagasse and wheat bran (principally)	4	1.31	(De la Cruz-Quiroz et al., 2017)
T. harzianum	Sugarcane bagasse and wheat bran (principally)	4	1.08	(De la Cruz-Quiroz et al., 2017)
	Potato peel	5–6	6.4	(Sala et al., 2021)
	Beer draff	5–6	7.5	(Sala et al., 2021)
Trichoderma yunnanese	Sugarcane bagasse and wheat bran (principally)	4	1.2	(De la Cruz-Quiroz et al., 2017)
T. asperellum	Sugarcane bagasse and wheat bran (principally)	4	1.23–1.30	(De la Cruz-Quiroz et al., 2017)
	Jatropha cake	7	8.5	(Rayhane et al., 2019)

dm: dry mass

10.3.4 AGITATED BIOREACTORS

Agitated bioreactors (ABs) are presented with a rotating drum or internal mixing configuration (De Vrije et al., 2001). In both, the stirring can be continuous or intermittent and contributes to heat transfer and O_2 and CO_2 transport (Brand et al., 2010). Similar to the PCBs in the AB, can supply forced aeration (Martínez-Ramírez et al., 2021) which improves the oxygen availability and the removal of heat and CO_2. ABs allow controlled processes at an industrial scale (up to 25 t) under low aseptic conditions (Durand and Chereau, 1988; Xue et al., 1992). Rotary drum bioreactors are cylindrical and rotate on their horizontal axis (Mitchell et al., 1993). The smooth movement of the drum causes low mechanical damage to the microorganism or the substrate; however, it can present substrate agglomerations, which cause heterogeneity in the bed (Brand et al., 2010). Internally mixed bioreactors are cylindrical (horizontal and vertical) and have a blade system that mechanically mixes the substrate (Durand, 2003). The movement of the blades can cause mechanical damage to the mycelium and the substrate, but it contributes to the uniformity of the bed (Lonsane et al., 1992; Durand, 2003). The mechanical damage produced to the mycelium has limited the use of the internal mixed bioreactors for fungal conidia production (Méndez-González et al., 2022). However, aerated packed columns with mechanical stirring and a capacity of 50 L (Durand, 2003) have been used successfully for the conidia production of *C. minitans* and *Paecelomyces lilacinus*, reaching a production yield of up to 12.8×10^9 conidia/gdm (dm, dry mass) at 13 days (De Vrije et al., 2001). It is possible that, in this type of bioreactor, the correct adjustment of the intensity and period of agitation is the key to the successful propagule production from filamentous fungi used in biological control.

10.4 CONCLUIDING REMARKS

Diverse filamentous fungi species are important biological control agents, which due to their high efficiency, can be applied instead of chemical pesticides. The propagules of these fungi are produced in solid-state culture bioreactors. In these bioreactors, processes with different levels of technology make it possible to produce these biopesticides around the world. In most fungal agents, conidia production and productivity increase in PCBs. However, the conditions for scale-up in this bioreactor type still are not determined. Among the strategies implemented for scale-up is the texturizers addition whose effect on the solid matrix structure could be the key to operating packed columns at industrial scale. This could allow satisfying the growing demand for filamentous fungi used in biological pest control.

KEYWORDS

- **biopesticides**
- **conidia production**
- **sustainable agriculture**
- **bioprocess**
- **biorreactors.**

REFERENCES

Adetunji, C. O.; Oloke, J. K. Effect of Wild and Mutant Strain of *Lasiodiplodia pseudotheobromae* Mass Produced on Rice Bran as a Potential Bioherbicide Agents for Weeds under Solid State Fermentation **2013,** *1* (02), 18–23. https://doi.org/10.7324/JABB.2013.1204.

Alcantara-Vargas, E.; Espitia-López, J.; Garza-López, P. M.; Angel-Cuapio, A. Producción y Calidad de Conidios de Cepas de Entomopatógenos Del Género *Metarhizium anisopliae*, Aislados En Zonas Agrícolas Del Estado de México. *Rev. Mex. Biodivers* **2020,** *91* (July), 912912. https://doi.org/10.22201/ib.20078706e.2020.91.2912.

Alves, S.; Pereira, R. Production of *Metarhizium anisoplae* (Metsch.) Sorok. and *Beauveria bassiana* (Bals.) Vuill. in Plastic Trays. *Ecossistema* **1989,** *14* (May), 188–192.

Amerio, N. S.; Castrillo, M. L.; Bich, G. A.; Zapata, P. D.; Villalba, L. L. *Trichoderma* En La Argentina: Estado Del Arte. *Ecología Austral* **2020,** *30* (1), 113–124. https://doi.org/10.25260/EA.20.30.1.0.945.

Arias Barreto, A.; Abalone, R.; Gastón, A.; Ochandio, D.; Cardoso, L.; Bartosik, R. Validation of a Heat, Moisture and Gas Concentration Transfer Model for Soybean (Glycine Max) Grains Stored in Plastic Bags (Silo Bags). *Biosyst. Eng.* **2017,** *158,* 23–37. https://doi.org/10.1016/j.biosystemseng.2017.03.009.

Arredondo-Bernal, H. C.; González-Cabrera, J. El Comercio Del Control Biológico En México. In *Fundamento y Práctica del Control Biológico de plagas y Enfermedades*; Arredondo-Bernal, H. C., Tamayo-Mejía, F., Rodriguez-del-Bosque, L. Á., Eds.; bba: México, 2020; pp 599–629.

Arthurs, S.; Thomas, M. B. Effects of Temperature and Relative Humidity on Sporulation of *Metarhizium anisopliae* Var. Acridum in Mycosed Cadavers of *Schistocerca gregaria*. *J. Invertebr. Pathol.* **2001,** *78* (2), 59–65. https://doi.org/10.1006/jipa.2001.5050.

Arzumanov, T.; Jenkins, N.; Roussos, S. Effect of Aeration and Substrate Moisture Content on Sporulation of *Metarhizium anisopliae* Var. Acridum. *Process Biochem.* **2005,** *40* (3–4), 1037–1042. https://doi.org/10.1016/j.procbio.2004.03.013.

Bailey, K. L. The Bioherbicide Approach to Weed Control Using Plant Pathogens. In *Integrated Pest Management*; Abrol, D., Ed.; Elsevier, 2014; pp 245–266. https://doi.org/10.1016/B978-0-12-398529-3.00014-2.

Barra-Bucarei, L.; Vergara, P.; Cortes, A. Conditions to Optimize Mass Production of *Metarhizium anisopliae* (Metschn.) Sorokin 1883 in Different Substrates. *Chil. J. Agric. Res.* **2016,** *76* (4), 448–454. https://doi.org/10.4067/S0718-58392016000400008.

Benenati, R. F.; Brosilow, C. B. Void Fraction Distribution in Beds of Spheres. *AIChE J.* **1962,** *8* (3), 359–361. https://doi.org/10.1002/aic.690080319.

Bowers, R. C. Commercialization of CollegoTM –An Industrialist's View. *Weed Sci.* **1986,** *34* (S1), 24–25. https://doi.org/10.1017/S0043174500068326.

Brand, D.; Soccol, C. R.; Sabu, A.; Roussos, S. Production of Fungal Biological Control Agents through Solid State Fermentation: A Case Study on *Paecelomyces lilacinus* Againts Root-Knot Nematodes. *Micología Aplicada Internacional* **2010,** *22* (1), 31–48.

Castillo-Minjarez, J. M. Contexto de Los Bioplaguicidas Comerciales de Base Fúngica En México. *Mex. J. Technol. Eng.* **2022,** *1* (1), 15–27.

Chaparro, M. L.; Sanabria, P. J.; Jiménez, A. M.; Gómez, M. I.; Bautista, E. J.; Mesa, L. A Circular Economy Approach for Producing a Fungal-Based Biopesticide Employing Pearl Millet as a Substrate and Its Economic Evaluation. *Bioresour. Technol. Rep.* **2021,** *16,* 100869. https://doi.org/10.1016/j.biteb.2021.100869.

Cherry, A. J.; Jenkins, N. E.; Heviefo, G.; Bateman, R.; Lomer, C. J. Operational and Economic Analysis of a West African Pilot-Scale Production Plant for Aerial Conidia of *Metarhizium* spp. for Use as a Mycoinsecticide against Locusts and Grasshoppers. *Biocontrol. Sci. Technol.* **1999,** *9* (1), 35–51. https://doi.org/10.1080/09583159929893.

Córdova-Albores, L. C.; Zelaya-Molina, L. X.; Ávila-Alistac, N.; Valenzuela-Ruíz, V.; Cortés-Martínez, N. E.; Parra-Cota, F. I.; Burgos-Canul, Y. Y.; Chávez-Díaz, I. F.; Fajardo-Franco, M. L.; De los Santos-Villalobos, S. Omics Sciences Potential on Bioprospecting of Biological Control Microbial Agents: The Case of the Mexican Agro-Biotechnology. *Rev. Mex. Fitopatol. Mexican J. Phytopathol.* **2020,** *39* (1), 239–246. https://doi.org/10.18781/R.MEX.FIT.2009-3.

da Cunha, L. P.; Casciatori, F. P.; de Cenço Lopes, I.; Thoméo, J. C. Production of Conidia of the Entomopathogenic Fungus *Metarhizium anisopliae* ICB 425 in a Tray Bioreactor. *Bioprocess Biosyst. Eng.* **2019,** *42* (11), 1757–1768. https://doi.org/10.1007/s00449-019-02172-z.

da Cunha, L. P.; Casciatori, F. P.; Vicente, I. V.; Garcia, R. L.; Thoméo, J. C. *Metarhizium anisopliae* Conidia Production in Packed-Bed Bioreactor Using Rice as Substrate in Successive Cultivations. *Process Biochem.* **2020,** *97* (July), 104–111. https://doi.org/10.1016/j.procbio.2020.07.002.

Dalla Santa, H. S.; Sousa, N. J.; Brand, D.; Dalla Santa, O. R.; Pandey, A.; Sobotka, M.; Páca, J.; Soccol, C. R. Conidia Production of *Beauveria* sp. by Solid-State Fermentation for Biocontrol of *Ilex paraguariensis* Caterpillars. *Folia Microbiol (Praha)* **2004,** *49* (4), 418–422. https://doi.org/10.1007/BF02931603.

De Cal, A.; Larena, I.; Guijarro, B.; Melgarejo, P. Mass Production of Conidia of *Penicillium frequentans*, a Biocontrol Agent Against Brown Rot of Stone Fruits. *Biocontrol. Sci. Technol.* **2002,** *12* (6), 715–725. https://doi.org/10.1080/0958315021000039897.

De la Cruz-Quiroz, R.; Robledo-Padilla, F.; Aguilar, C. N.; Roussos, S. Forced Aeration Influence on the Production of Spores by *Trichoderma* Strains. *Waste Biomass Valori.* **2017,** *8* (7), 2263–2270. https://doi.org/10.1007/s12649-017-0045-4.

De Vrije, T.; Antoine, N.; Buitelaar, R. M.; Bruckner, S.; Dissevelt, M.; Durand, A.; Gerlagh, M.; Jones, E. E.; Lüth, P.; Oostra, J.; Ravensberg, W. J.; Renaud, R.; Rinzema, A.; Weber, F. J.; Whipps, J. M. The Fungal Biocontrol Agent *Coniothyrium minitans*: Production by Solid-State Fermentation, Application and Marketing. *Appl. Microbiol. Biotechnol.* **2001,** *56* (1–2), 58–68. https://doi.org/10.1007/s002530100678.

Deans, C.; Krischik, V. The Current State and Future Potential of Microbial Control of Scarab Pests. *Applied Sciences* **2023,** *13* (2), 766. https://doi.org/10.3390/app13020766.

Dhar, P.; Kaur, G. Response Surface Methodology for Optimizing Process Parameters for the Mass Production of *Beauveria bassiana* Conidiospores. *Afr. J. Microbiol. Res.* **2011,** *5* (17). https://doi.org/10.5897/AJMR10.088.

Diniz da Silva, M. P.; Dutra, R. S.; Casciatori, F. P.; Grajales, L. M. A Two-Phase Model for Simulation of Water Transfer during Lipase Production by Solid-State Cultivation in a Tray Bioreactor Using Babassu Residues as Substrate. *Chem. Eng. Process.* **2022,** *177,* 108981. https://doi.org/10.1016/j.cep.2022.108981.

Doehlemann, G.; Ökmen, B.; Zhu, W.; Sharon, A. Plant Pathogenic Fungi. *Microbiol. Spectr.* **2017,** *5* (1). https://doi.org/10.1128/microbiolspec.FUNK-0023-2016.

Dorta, B.; Arcas, J. Sporulation of *Metarhizium anisopliae* in Solid-State Fermentation with Forced Aeration. *Enzyme Microb. Technol.* **1998,** *23* (7–8), 501–505. https://doi.org/10.1016/S0141-0229(98)00079-9.

Dorta, B.; Ertola, R. J.; Arcas, J. Characterization of Growth and Sporulation of *Metarhizium anisopliae* in Solid-Substrate Fermentation. *Enzyme Microb. Technol.* **1996,** *19* (6), 434–439. https://doi.org/10.1016/S0141-0229(96)00017-8.

Dubey, S. C.; Singh, V.; Priyanka, K.; Upadhyay, B. K.; Singh, B. Combined Application of Fungal and Bacterial Bio-Agents, Together with Fungicide and *Mesorhizobium* for Integrated Management of *Fusarium wilt* of Chickpea. *BioControl* **2015,** *60* (3), 413–424. https://doi.org/10.1007/s10526-015-9653-8.

Durand, A. Bioreactor Designs for Solid State Fermentation. *Biochem. Eng. J.* **2003,** *13* (2–3), 113–125.

Durand, A.; Chereau, D. A New Pilot Reactor for Solid-State Fermentation: Application to the Protein Enrichment of Sugar Beet Pulp. *Biotechnol. Bioeng.* **1988,** *31* (5), 476–486. https://doi.org/10.1002/bit.260310513.

Ferreira, J. M.; Pinto, S. M. N.; Soares, F. E. F. *Metarhizium robertsii* Protease and Conidia Production, Response to Heat Stress and Virulence against *Aedes aegypti* Larvae. *AMB Express* **2021,** *11* (1), 166. https://doi.org/10.1186/s13568-021-01326-1.

Figueroa-Montero, A.; Esparza-Isunza, T.; Saucedo-Castañeda, G.; Huerta-Ochoa, S.; Gutiérrez-Rojas, M.; Favela-Torres, E. Improvement of Heat Removal in Solid-State Fermentation Tray Bioreactors by Forced Air Convection. *J. Chem. Technol. Biotechnol.* **2011,** *86* (10), 1321–1331. https://doi.org/10.1002/jctb.2637.

Galea, V. J. Use of Stem Implanted Bioherbicide Capsules to Manage an Infestation of *Parkinsonia aculeata* in Northern Australia. *Plants* **2021,** *10* (9), 1909. https://doi.org/10.3390/plants10091909.

Grzywacz, D.; Moore, D.; Rabindra, R. J. Mass Production of Entomopathogens in Less Industrialized Countries. In *Mass Production of Beneficial Organisms*; Morales-Ramos,

J., Rojas, G., Shapiro-Ilan, D., Eds.; Elsevier, 2014; pp 519–561. https://doi.org/10.1016/B978-0-12-391453-8.00015-7.

Hamer, J. E.; Howard, R. J.; Chumley, F. G.; Valent, B. A Mechanism for Surface Attachment in Spores of a Plant Pathogenic Fungus. *Science (1979)* **1988,** *239* (4837), 288–290. https://doi.org/10.1126/science.239.4837.288.

Henrique, J. P.; Casciatori, F. P.; Thoméo, J. C. Automatic System for Monitoring Gaseous Concentration in a Packed-Bed Solid-State Cultivation Bioreactor. *Chem. Eng. Sci.* **2022,** *259,* 117793. https://doi.org/10.1016/j.ces.2022.117793.

Holder, D. J.; Keyhani, N. O. Adhesion of the Entomopathogenic Fungus *Beauveria* (Cordyceps) *bassiana* to Substrata. *Appl. Environ. Microbiol.* **2005,** *71* (9), 5260–5266. https://doi.org/10.1128/AEM.71.9.5260-5266.2005.

Horbach, R.; Navarro-Quesada, A. R.; Knogge, W.; Deising, H. B. When and How to Kill a Plant Cell: Infection Strategies of Plant Pathogenic Fungi. *J. Plant Physiol.* **2011,** *168* (1), 51–62. https://doi.org/10.1016/j.jplph.2010.06.014.

Inglis, G. D.; Goettel, M. S.; Butt, T. M.; Strasser, H. Use of Hyphomycetous Fungi for Managing Insect Pests. In *Fungi as Biocontrol Agents: Progress, Problems and Potential*; Butt, T. M., Jackson, C., Magan, N., Eds.; CABI: Wallingford, 2009; pp 23–69. https://doi.org/10.1079/9780851993560.0023.

Jambhulkar, P. P.; Sharma, P.; Manokaran, R.; Lakshman, D. K.; Rokadia, P.; Jambhulkar, N. Assessing Synergism of Combined Applications of *Trichoderma harzianum and Pseudomonas fluorescens* to Control Blast and Bacterial Leaf Blight of Rice. *Eur. J. Plant Pathol.* **2018,** *152* (3), 747–757. https://doi.org/10.1007/s10658-018-1519-3.

Jaronski, S. T. Mass Production of Entomopathogenic Fungi—State of the Art. In *Mass Production of Beneficial Organisms*; Morales-Ramos, J., Rojas, G., Shapiro-Ilan, Eds.; Elsevier, 2023; pp 317–357. https://doi.org/10.1016/B978-0-12-822106-8.00017-8.

Jaronski, S. T.; Jackson, M. A. Mass Production of Entomopathogenic Hypocreales. In *Manual of Techniques in Invertebrate Pathology*, Elsevier, 2012; pp 255–284. https://doi.org/10.1016/B978-0-12-386899-2.00008-7.

Jeffs, L. B.; Khachatourians, G. G. Estimation of Spore Hydrophobicity for Members of the Genera *Beauveria, Metarhizium*, and *Tolypocladium* by Salt-Mediated Aggregation and Sedimentation. *Can. J. Microbiol.* **1997,** *43* (1), 23–28. https://doi.org/10.1139/m97-004.

Jenkins, N. E.; Goettel, M. S. Methods for Mass-Production of Microbial Control Agents of Grasshoppers and Locusts. *Mem. Ent. Soc. Can.* **2011,** *129* (S171), 37–48. https://doi.org/10.4039/entm129171037-1.

Jenkins, N. E.; Grzywacz, D. Quality Control of Fungal and Viral Biocontrol Agents - Assurance of Product Performance. *Biocontrol. Sci. Technol.* **2000,** *10* (6), 753–777. https://doi.org/10.1080/09583150020011717.

Jones, E. E.; Weber, F. J.; Oostra, J.; Rinzema, A.; Mead, A.; Whipps, J. M. Conidial Quality of the Biocontrol Agent *Coniothyrium minitans* Produced by Solid-State Cultivation in a Packed-Bed Reactor. *Enzyme Microb. Technol.* **2004,** *34* (2), 196–207. https://doi.org/10.1016/j.enzmictec.2003.10.002.

Jou, R.-Y.; Lo, C.-T. Heat and Mass Transfer Measurements for Tray-Fermented Fungal Products. *Int. J. Thermophys* **2011,** *32* (1–2), 523–536. https://doi.org/10.1007/s10765-011-0934-x.

Kamp, A. M.; Bidochka, M. J. Conidium Production by Insect Pathogenic Fungi on Commercially Available Agars. *Lett. Appl. Microbiol.* **2002,** *35* (1), 74–77. https://doi.org/10.1046/j.1472-765X.2002.01128.x.

Lacey, L. A.; Grzywacz, D.; Shapiro-Ilan, D. I.; Frutos, R.; Brownbridge, M.; Goettel, M. S. Insect Pathogens as Biological Control Agents: Back to the Future. *J. Invertebr. Pathol.* **2015,** *132,* 1–41. https://doi.org/10.1016/j.jip.2015.07.009.

Lonsane, B. K.; Ghildyal, N. P.; Budiatman, S.; Ramakrishna, S. V. Engineering Aspects of Solid State Fermentation. *Enzyme Microb. Technol.* **1985,** *7* (6), 258–265. https://doi.org/10.1016/0141-0229(85)90083-3.

Lonsane, B. K.; Saucedo-Castañeda, G.; Raimbault, M.; Roussos, S.; Viniegra-González, G.; Ghildyal, N. P.; Ramakrishna, M.; Krishnaiah, M. M. Scale-up Strategies for Solid State Fermentation Systems. *Process Biochem.* **1992,** *27* (5), 259–273.

Lopes-Perez, C.; Casciatori, F. P.; Thoméo, J. C. Strategies for Scaling-up Packed-Bed Bioreactors for Solid-State Fermentation: The Case of Cellulolytic Enzymes Production by a Thermophilic Fungus. *Chem. Eng. J.* **2019,** *361* (December 2018), 1142–1151. https://doi.org/10.1016/j.cej.2018.12.169.

Luz, C.; Fargues, J. Factors Affecting Conidial Production of *Beauveria bassiana* from Fungus-Killed Cadavers of *Rhodnius prolixus*. *J. Invertebr. Pathol.* **1998,** *72* (2), 97–103. https://doi.org/10.1006/jipa.1998.4774.

Machado, A. C. R.; Monteiro, A. C.; Almeida, A. M. B. de; Martins, M. I. E. G. Production Technology for Entomopathogenic Fungus Using a Biphasic Culture System. *Pesqui Agropecu Bras* **2010,** *45* (10), 1157–1163. https://doi.org/10.1590/S0100-204X2010001000015.

Martínez-Ramírez, C.; Esquivel-Cote, R.; Ferrera-Cerrato, R.; Martínez-Ruiz, J. A.; Rodríguez-Serrano, G.; Saucedo-Castañeda, G. Solid-State Culture of *Azospirillum brasilense*: A Reliable Technology for Biofertilizer Production from Laboratory to Pilot Scale. *Bioprocess Biosyst. Eng.* **2021,** *44* (7), 1525–1538. https://doi.org/10.1007/s00449-021-02537-3.

Mathulwe, L. L.; Malan, A. P.; Stokwe, N. F. Mass Production of Entomopathogenic Fungi, *Metarhizium robertsii* and *Metarhizium pinghaens*, for Commercial Application Against Insect Pests. *J. Vis. Exp.* **2022,** No. 181. https://doi.org/10.3791/63246.

Mattedi, A.; Sabbi, E.; Farda, B.; Djebaili, R.; Mitra, D.; Ercole, C.; Cacchio, P.; Del Gallo, M.; Pellegrini, M. Solid-State Fermentation: Applications and Future Perspectives for Biostimulant and Biopesticides Production. *Microorganisms* **2023,** *11* (6), 1408. https://doi.org/10.3390/microorganisms11061408.

Méndez-González, F.; Castillo-Minjarez, J. M.; Loera, O.; Favela-Torres, E. Current Developments in the Resistance, Quality, and Production of Entomopathogenic Fungi. *World J. Microbiol. Biotechnol.* **2022,** *38* (7), 115. https://doi.org/10.1007/s11274-022-03301-9.

Méndez-González, F.; Chávez-Escalante, G.; Loera, O.; Favela-Torres, E. What Does Respirometric Analysis Tell Us about *Metarhizium robertsii*. *J. Appl. Biotechnol. Bioeng.* **2022,** *9* (4), 94–96. https://doi.org/10.15406/jabb.2022.09.00292.

Méndez-González, F.; Figueroa-Montero, A.; Saucedo-Castañeda, G.; Loera, O.; Favela-Torres, E. Addition of Spherical-style Packing Improves the Production of Conidia by *Metarhizium robertsii* in Packed Column Bioreactors. *J. Chem. Technol. Biotechnol.* **2022,** *97* (6), 1517–1525. https://doi.org/10.1002/jctb.6993.

Méndez-González, F.; Loera, O.; Favela-Torres, E. Conidia Production of *Metarhizium anisopliae* in Bags and Packed Column Bioreactors. *Curr. Biotechnol.* **2018,** *7* (1), 65–69. https://doi.org/10.2174/2211550105666160926123350.

Méndez-González, F.; Loera, O.; Saucedo-Castañeda, G.; Favela-Torres, E. Forced Aeration Promotes High Production and Productivity of Infective Conidia from *Metarhizium robertsii* in Solid-State Fermentation. *Biochem. Eng. J.* **2020,** *156,* 107492. https://doi.org/10.1016/j.bej.2020.107492.

Méndez-González, F.; Loera-Corral, O.; Saucedo-Castañeda, G.; Favela-Torres, E. Bioreactors for the Production of Biological Control Agents Produced by Solid-State Fermentation. In *Current Developments in Biotechnology and Bioengineering*; Pandey, A., Larroche, C., Soccol, C. R., Eds.; Elsevier, 2018; pp 109–121. https://doi.org/10.1016/B978-0-444-63990-5.00007-4.

Mitchell, D. A.; Lonsane, B. K.; Reanud, R.; Almanza, S.; Maratray, J.; Desgranes, C.; Crooke, P. S.; Hong, K.; Tanner, R. D.; Malaney, G. C. General Principles of Bioreactor Design and Operation for SSC. In *Solid State Cultivation*; Doelle, H., Mitchell, D. A., Rolz, C. E., Eds.; Elsevier Applied Science: London, 1993; pp 115–136.

Monte, E. Understanding *Trichoderma*: Between Biotechnology and Microbial Ecology. *International Microbiology* **2001,** No. 4, 1–4.

Montesinos-Matías, R.; Viniegra-González, G.; Alatorre-Rosas, R.; Loera, O. Relationship between Virulence and Enzymatic Profiles in the Cuticle of Tenebrio Molitor by 2-Deoxy-d-Glucose-Resistant Mutants of *Beauveria bassiana* (Bals.) Vuill. *World J. Microbiol. Biotechnol.* **2011,** *27* (9), 2095–2102. https://doi.org/10.1007/s11274-011-0672-z.

Mortensen, K. The Potential of an Endemic Fungus, *Colletotrichum gloeosporioides*, for Biological Control of Round-Leaved Mallow (*Malva pusilla*) and Velvetleaf (*Abutilon theophrasti*). *Weed Sci.* **1988,** *36* (4), 473–478. https://doi.org/10.1017/S0043174500075226.

Mukhopadhyay, R.; Kumar, D. Trichoderma: A Beneficial Antifungal Agent and Insights into Its Mechanism of Biocontrol Potential. *Egypt J. Biol. Pest. Control* **2020,** *30* (1), 133. https://doi.org/10.1186/s41938-020-00333-x.

Nandhini, C.; Ganesh, P.; Yoganathan, K.; Kumar, D. Efficacy of *Colletotrichum gloeosporioides*, Potential Fungi for Bio Control of *Echinochloa* Crus-Galli (*Barnyard rass*). *J. Drug Deliv. Ther.* **2019,** *9* (6-s), 72–75. https://doi.org/10.22270/jddt.v9i6-s.3751.

Ons, L.; Bylemans, D.; Thevissen, K.; Cammue, B. P. A. Combining Biocontrol Agents with Chemical Fungicides for Integrated Plant Fungal Disease Control. *Microorganisms* **2020,** *8* (12), 1930. https://doi.org/10.3390/microorganisms8121930.

Pandey, A. Aspects of Fermenter Design for Solid-State Fermentations. *Process Biochem.* **1991,** *26* (6), 355–361. https://doi.org/10.1016/0032-9592(91)85026-K.

Peng, Y.; Li, S. J.; Yan, J.; Tang, Y.; Cheng, J. P.; Gao, A. J.; Yao, X.; Ruan, J. J.; Xu, B. L. Research Progress on Phytopathogenic Fungi and Their Role as Biocontrol Agents. *Front. Microbiol.* **2021,** 12. https://doi.org/10.3389/fmicb.2021.670135.

Posada-Flórez, F. J. Production of *Beauveria bassiana* Fungal Spores on Rice to Control the Coffee Berry Borer, *Hypothenemus hampei*, in Colombia. *J. Insect Sci.* **2008,** *8* (41), 1–13. https://doi.org/10.1673/031.008.4101.

Radhakrishnan, R.; Alqarawi, A. A.; Abd_Allah, E. F. Bioherbicides: Current Knowledge on Weed Control Mechanism. *Ecotoxicol. Environ. Saf.* **2018,** *158,* 131–138. https://doi.org/10.1016/j.ecoenv.2018.04.018.

Raghava Rao, K. S. M. S.; Gowthaman, M. K.; Ghildyal, N. P.; Karanth, N. G. A Mathematical Model for Solid State Fermentation in Tray Bioreactors. *Bioprocess Eng.* **1993,** *8* (5–6), 255–262. https://doi.org/10.1007/BF00369838.

Ramírez-Guzmán, N. K.; Torres-León, C.; Martínez-Terrazas, E.; De la Cruz-Quiroz, R.; Flores-Gallegos, A. C.; Rodríguez-Herrera, R.; Aguilar, C. N. Biocontrol as an Efficient Tool for Food Control and Biosecurity. In *Food Safety and Preservation*; Grumezescu,

A. M., Holban, A. M., Eds.; Academic Press, 2018; pp 167–193. https://doi.org/10.1016/B978-0-12-814956-0.00007-X.

Rangel, D. E. N.; Anderson, A. J.; Roberts, D. W. Growth of *Metarhizium anisopliae* on Non-Preferred Carbon Sources Yields Conidia with Increased UV-B Tolerance. *J. Invertebr. Pathol.* **2006,** *93* (2), 127–134. https://doi.org/10.1016/j.jip.2006.05.011.

Rayhane, H.; Josiane, M.; Gregoria, M.; Yiannis, K.; Nathalie, D.; Ahmed, M.; Sevastianos, R. From Flasks to Single Used Bioreactor: Scale-up of Solid State Fermentation Process for Metabolites and Conidia Production by *Trichoderma asperellum. J. Environ. Manage.* **2019,** *252,* 109496. https://doi.org/10.1016/j.jenvman.2019.109496.

Roberts, J.; Florentine, S.; Fernando, W. G. D.; Tennakoon, K. U. Achievements, Developments and Future Challenges in the Field of Bioherbicides for Weed Control: A Global Review. *Plants* **2022,** *11* (17), 2242. https://doi.org/10.3390/plants11172242.

Rojas, V. M. A.; Iwanicki, N. S.; D'Alessandro, C. P.; Fatoretto, M. B.; Demétrio, C. G. B.; Delalibera, I. Characterization of Brazilian *Cordyceps fumosorosea* Isolates: Conidial Production, Tolerance to Ultraviolet-B Radiation, and Elevated Temperature. *J. Invertebr. Pathol.* **2023,** *197,* 107888. https://doi.org/10.1016/j.jip.2023.107888.

Sabbour, M. M. The Role of Chemical Additives in Enhancing the Efficacy of *Beauveria bassiana* and *Metarhizium anisopliae* Against the Potato Tuber Moth *Phthorimaea operculella* (Zeller) (Lepidoptera:Gelechiidae). *Pak. J. Biol. Sci.* **2002,** *5* (11), 1155–1159. https://doi.org/10.3923/pjbs.2002.1155.1159.

Sala, A.; Barrena, R.; Artola, A.; Sánchez, A. Current Developments in the Production of Fungal Biological Control Agents by Solid-State Fermentation Using Organic Solid Waste. *Crit. Rev. Environ. Sci. Technol.* **2019,** *49* (8), 655–694. https://doi.org/10.1080/10643389.2018.1557497.

Sala, A.; Barrena, R.; Meyling, N. V.; Artola, A. Conidia Production of the Entomopathogenic Fungus *Beauveria bassiana* Using Packed-Bed Bioreactor: Effect of Substrate Biodegradability on Conidia Virulence. *J. Environ. Manage.* **2023,** *341,* 118059. https://doi.org/10.1016/j.jenvman.2023.118059.

Sala, A.; Vittone, S.; Barrena, R.; Sánchez, A.; Artola, A. Scanning Agro-Industrial Wastes as Substrates for Fungal Biopesticide Production: Use of *Beauveria bassiana* and *Trichoderma harzianum* in Solid-State Fermentation. *J. Environ. Manage.* **2021,** *295,* 113113. https://doi.org/10.1016/j.jenvman.2021.113113.

Santos, A. C. da S.; Lopes, R. da S.; de Oliveira, L. G.; Diniz, A. G.; Shakeel, M.; Lima, E. Á. de L. A.; da Costa, A. F.; Lima, V. L. de M. Entomopathogenic Fungi: Current Status and Prospects. In *New and Future Development in Biopesticide Research: Biotechnological Exploration*; De Mandal, S., Ramkumar, G., Karthi, S., Jin, F., Eds.; Springer Nature Singapore: Singapore, 2022; pp 55–91. https://doi.org/10.1007/978-981-16-3989-0_2.

Sharma, A.; Srivastava, A.; Shukla, A. K.; Srivastava, K.; Srivastava, A. K.; Saxena, A. K. Entomopathogenic Fungi: A Potential Source for Biological Control of Insect Pests. In *Phytobiomes: Current Insights and Future Vistas*; Solanki, M., Kashyap, P., Kumari, B., Eds.; Springer Singapore: Singapore, 2020; pp 225–250. https://doi.org/10.1007/978-981-15-3151-4_9.

Sun, J.; Fuxa, J. R.; Henderson, G. Sporulation of *Metarhizium anisopliae* and *Beauveria bassiana* on *Coptotermes formosanus* in Vitro. *J. Invertebr. Pathol.* **2002,** *81* (2), 78–85. https://doi.org/10.1016/S0022-2011(02)00152-0.

Swanson, D. Economic Feasibility of Two Technologies for Production of Mycopesticide in Magadascar. *Mem. Ent. Soc. Can.* **1997,** *129* (S171), 101–113. https://doi.org/10.4039/entm129171101-1.

Tamayo-Mejía, F.; Alatorre-Rosas, R.; Delgado-Fernández, S.; Ocampo-Hernández, J. A. Principios de Aplicación de Entomopatógenos. In *Fundamento y Práctica Del Control Biológico de Plagas y Enfermedades*; Arredondo-Bernal, H. C., Tamayo-Mejía, F., Rodríguez-del-Bosque, L. A., Eds.; Colegio de Posgraduados: Texcoco, 2020; pp 405–426.

Triolet, M.; Guillemin, J.; Andre, O.; Steinberg, C. Fungal-based Bioherbicides for Weed Control: A Myth or a Reality? *Weed Res.* **2020,** *60* (1), 60–77. https://doi.org/10.1111/wre.12389.

Van Breukelen, F. R.; Haemers, S.; Wijffels, R. H.; Rinzema, A. Bioreactor and Substrate Selection for Solid-State Cultivation of the Malaria Mosquito Control Agent *Metarhizium anisopliae*. *Process Biochem.* **2011,** *46* (3), 751–757. https://doi.org/10.1016/j.procbio.2010.11.023.

Wilson, R. A.; Talbot, N. J. Under Pressure: Investigating the Biology of Plant Infection by *Magnaporthe oryzae*. *Nat. Rev. Microbiol.* **2009,** *7* (3), 185–195. https://doi.org/10.1038/nrmicro2032.

Wraight, S. P.; Jackson, M. A.; Kokc, S. L. de. Production, Stabilization and Formulation of Fungal Biocontrol Agents. In *Fungi as Biocontrol Agents: Progress, Problems and Potential*; CABI: Wallingford, 2001; pp 253–287. https://doi.org/10.1079/9780851993560.0253.

Xie, L.; Chen, H. M.; Yang, J. Bin. Conidia Production by *Beauveria bassiana* on Rice in Solid-State Fermentation Using Tray Bioreactor. *Adv. Mat. Res.* **2012,** *610–613,* 3478–3482. https://doi.org/10.4028/www.scientific.net/AMR.610-613.3478.

Xue, M.; Liu, D.; Zhang, H.; Qi, H.; Lei, Z. A Pilot Process of Solid State Fermentation from Sugar Beet Pulp for the Production of Microbial Protein. *J. Ferment Bioeng.* **1992,** *73,* 203–205.

Ye, S. D.; Ying, S. H.; Chen, C.; Feng, M. G. New Solid-State Fermentation Chamber for Bulk Production of Aerial Conidia of Fungal Biocontrol Agents on Rice. *Biotechnol. Lett.* **2006,** *28* (11), 799–804. https://doi.org/10.1007/s10529-006-9004-z.

Zewdie, B.; Tack, A. J. M.; Ayalew, B.; Adugna, G.; Nemomissa, S.; Hylander, K. Temporal Dynamics and Biocontrol Potential of a Hyperparasite on Coffee Leaf Rust across a Landscape in Arabica Coffee's Native Range. *Agric. Ecosyst. Environ.* **2021,** *311,* 107297. https://doi.org/10.1016/j.agee.2021.107297.

CHAPTER 11

Utilization of Genome Edition to Design Tropical Plants with Increased Resistance to Biotic Stress for Agricultural Sustainability

JOSÉ ALBERTO VALENZUELA-AVILÉS[1],
EDUARDO ANTONIO TRILLO HERNÁNDEZ[2,3],
MIGUEL ANGEL HERNÁNDEZ OÑATE[4], and
MARTÍN ERNESTO TIZNADO HERNÁNDEZ[5*]

[1]*Estudiante de Doctorado. Coordinación de Tecnología de Alimentos de Origen Vegetal, Centro de Investigación en Alimentación y Desarrollo A.C. Sonora. Código Postal. México*

[2]*Unidad de Tecnología de Alimentos-Secretaría de Investigación y Posgrado, Universidad Autónoma de Nayarit, Ciudad de la Cultura S/N, Colonia Centro, Tepic, Nayarit, Mexico*

[3]*Estancias Posdoctorales-Consejo Nacional de Ciencia y Tecnología, Coordinación de Apoyos a Becarios e Investigadores, Dirección de Posgrado, Ciudad de México*

[4]*CONACYT-Coordinación de Tecnología de Alimentos de Origen Vegetal, Centro de Investigación en Alimentación y Desarrollo, Colonia La Victoria. Hermosillo, Sonora. México*

[5]*Coordinación de Tecnología de Alimentos de Origen Vegetal, Centro de Investigación en Alimentación y Desarrollo A.C. Hermosillo, Sonora. México*

Sustainable Engineering and Agro-Food Processing: Renewable and Nonrenewable Resources.
José Juan Buenrostro-Figueroa, Julio César Tafolla-Arellano, Mónica Chávez-González, Pedro Aguilar-Zárate, A. K. Haghi (Eds.)

ABSTRACT

Biotic stress causes important losses in agriculture yield and quality of fruits during postharvest. The utilization of chemicals to control biotic stresses during agricultural production is not sustainable. This is because chemical management is expensive, environmentally not sustainable, induces resistant strains, and the residues carry out adverse effects on human health. For instance, the induction of resistant strains demands the utilization of higher amounts of chemicals to reduce the losses due to the attack of different organisms and renders the chemical product ineffective. This in turn increases the contamination of soils in such a way that they cannot be used to grow plants for food production any longer. On the contrary, there is an increased demand for food due to population growth. In this sense, it is clear the need to design plants with higher resistance to the different biotic stress to reduce the use of chemicals and increases the amount of agricultural products. The ability to edit the genome allows the possibility to overcome the inherently complex use of traditional genetic engineering techniques in tropical plants to create resistant plants. The experimental approach of genome editing represents a powerful and relatively easy tool to face the impact of biotic stress on tropical fruit production. In this chapter, we will describe the different experimental approaches developed to carry out the plant's genome edition. Also, it will describe the different efforts carried out to improve plant resistance by using as an approach the genome edition with the goal of increasing the resistance of tropical plants to biotic stress.

11.1 INTRODUCTION

The continuous growth of the human population demands more production from agriculture. In this context, biotic stresses reduce the total amount of agricultural products at harvest and reduce the quality characteristics during postharvest. Consequently, there is a large amount of pre- and postharvest losses (Anuar et al., 2023). Traditionally, it had been utilized different chemical compounds to reduce the effects of deleterious organisms' attacks. However, the utilization of agrochemicals is not sustainable because of the induction of resistance strains, contamination of the environment, and negative effects on human health due to the presence of residues in agricultural products (Iriti and Vitalini, 2020; Rocha et al., 2015). By other side, the develop of plants with a higher yield using a mendelian approach is length, expensive and unprecise. Further, in this context, the utilization of DNA recombinant technology is a very promising option. In this regard, the first experimental approaches developed to create transgenic plants appears to be insufficient to create plants with better phenotypical characteristics of pathogen resistance. Fortunately, the continued development of the technology has brought better protocols for DNA sequencing and bioinformatic analysis (Kang et al., 2019). These protocols have led to the development of a huge amount of data which includes the DNA sequence of millions of genes from different species. Besides that, today it is possible to carry out the annotation of genes and in this way, it is possible to know the physiological and biochemical functions of many genes (Dmitriev et al., 2022). Also, the knowledge about the gene expression in biochemical pathways, organelles, tissues, stages of development, and most importantly, in response to both biotic and abiotic stress is available, at least in silico. With all this knowledge, the recent development of several genetic editing protocols allows the design of specific changes in genes playing a role in different physiological phenomena to induce specific phenotypic characteristics (Ceasar and Ignacimuthu, 2023). In this regard, the modern gene editing protocols allow the elimination of DNA regions, the exchange of specific DNA regions, the induction of point mutations to introduce changes in the primary structure of the protein as well as the alteration of the epigenome by the introduction of methyl groups of elimination of the histones acetyl groups to induce expression changes in target genes (Pulecio et al., 2017). With all this knowledge and experimental protocols, today it is possible to induce changes in genes to develop specific phenotypes such as the resistance to biotic stresses (Wang et al., 2022; Rojas-Vasquez and Gatica-Arias, 2020). In this review, it will be described the different genetic editing protocols as well as different experimental approaches using these protocols to induce a phenotype of resistance in tropical plants.

11.2 DESCRIPTION OF THE DIFFERENT GENOME EDITION TECHNIQUES

Previously, methods for generating crop varieties with favorable characteristics or improved yield for agriculture were based on the deliberate selection of crops and crosses with better performance, based on Mendelian genetics. Subsequently, in the 90s, transgenic technology significantly improved crop characteristics by integrating recombinant genes from other organisms, thereby obtaining transgenic phenotypes that had advantages compared to wild crops. However, transgenesis technology exhibits a few disadvantages such as random insertions, gene silencing, variable gene expression, and side effects that are not present in genetic editing, not to mention the social rejection of transgenic organisms (Shah et al., 2018). The transgenesis interchange for exogenous donors is a procedure that has been reported for decades. This techniques are based in the phenomena of DNA repair via homologous recombination (Cermak et al., 2011) which process gave rise to the creation of the first genomic editing techniques (Fernández et al., 2017).

Since genetic editing emerged, the way of studying biology underwent an exponential revolutionary change (Qi, 2017). Genetic editing consists of making targeted modifications to an organism's DNA which makes it possible to remove, add or modify nucleotides at specific locations. This technology allows any change in the genome sequence of an organism, generating specific modifications such as substitutions, deletions, or single-nucleotide insertions, as well as precise and specific insertions or deletions of entire genes (Sovová et al., 2017).

The nucleases used for genetic editing are designed to recognize specific DNA sequences and create a double-stranded break (DSB). Some examples of these nucleases that have been effectively used to edit the plant genome are zinc finger nucleases (ZFNs), transcription activator-like effector nucleases (TALEN), and clustered regularly interspaced short palindromic repeats (CRISPR)-associated protein 9 (CRISPR/Cas9), being the latter the most recent and versatile tool for genetic editing. CRISPR/Cas9 has been successfully implemented in several plant species mainly because it is simpler to utilize as compared with ZFN and TALEN (Zhang and Qi, 2019).

11.2.1 ZINC FINGER NUCLEASES

The first method used for genome editing was the ZFNs. First described in 1996, these were the result of a fusion between zinc finger proteins and the type-II endonuclease domain of Fok I to generate tailor-made sequence-specific endonucleases (Kim et al., 1996). The endonuclease domain consists of a nonspecific Fok I located at the N terminal end. To correctly and accurately create a DSB, the Fok I domain needs to dimerize, which happens when the two ZFNs bind to opposite strands of the same target site (Mitra et al., 2023).

The zinc finger domain, located at the C terminal end, is capable of binding to specific sites in the DNA. It consists of one or several zinc finger motifs in tandem, each formed by approximately 30 amino acids creating a ββα structure (Sufyan et al., 2023). The name "zinc finger" comes from a zinc ion that interacts between the conserved residues Cys_2-His_2 in the α-helix, which chelates the zinc ion stabilizing the whole structure. The zinc finger binds to the DNA, especially in G-rich sites, inserting the α-helix into the DNA's major groove. Further, each zinc finger binds to a 3-base region in the target DNA (Urnov et al., 2010). The specificity is given by the amino acids in the α helix's positions –1 through +6. By changing these amino acids, it is possible to create several sequence specific zinc fingers (Durai et al., 2005).

In the technique of genome editing mediated by ZFN, some disadvantages of the technique were reported for the construct assembly, implementation, and phenotype analysis (2011).

Some challenges commonly found are poor integration frequency at specific genome sites which demands the utilization of a selectable marker which in most cases is an antibiotic resistance gene. This experimental approach had been utilized to determinate the success of the transgenesis in mammalian cells (Cohen-Tannoudji et al., 1998; Porteus and Carroll, 2005).

One of the main drawbacks of this technique is that the zinc finger may allow mispairing in its recognition of the nucleotide triplet. Further, it is necessary to engineer each finger domain and the nuclease that they are going to be fused into the construct. This requires a high level of skill, and it is also costly and time consuming (Randhawa and Sengar, 2021).

11.2.2 TRANSCRIPTIONAL ACTIVATOR-LIKE EFFECTOR NUCLEASES

TALEN was reported in 2009 as an established method (Boch et al., 2009). This biotechnology tool was created with the fusion of transcription activator like-effector (TALE) and the restriction endonuclease FokI. TALEN was the first protocol that could be built relatively easily with the goal of targeting any specific gene in the genome with high accuracy and high precision (Becker and Boch, 2021).

The nucleases utilized in TALEN were isolated from pathogenic bacteria *Xanthomona*. This technique consists of introducing the construct in the plant cell, where they locate the nucleus and bind the target promoter to induce gene expression (Szurek et al., 2002). The functional domains of TALE are three, two functional nuclear localization signals joined by an acidic activation domain (Van den Ackerveken et al., 1996). are flanking the C-terminal domain which interacts with the plant transcription factor IIA. Further, on the opposite N-terminal, is the bacterial secretion signal and a nonspecific DNA-binding activity that is required for the general affinity of the protein with the DNA (Yuan et al., 2016). To create a TALEN construct, it is needed 152 amino acids in the N-terminal end and between 28 and 63 amino acids in the C-terminal end which are fused to FokI nuclease (Mussolino et al., 2011). Further, it has been shown that C-terminal longer than 63 amino acids reduce the efficiency of TALEN (Mussolino et al., 2011). Finally, TALEN as ZFN worked in pair because both employee the FokI nuclease dimer to break DNA within the spacer between two TALEN binding sites. The design of the TALEN construct has been well reviewed (Becker and Boch, 2021).

TALEN technology are efficient in the induction of targeted alteration in a wide kind of models that were previously difficult to genetically manipulate such as fruit flies, roundworms, zebrafish, frogs, mice, and pigs (Joung and Sander, 2013). Until 2013, TALEN was used in genetic therapy diseases to the alteration of human genes through the induction of nonhomologous end joining into the gene-coding sequences related to the diseases (Reyon et al., 2012). The ability to induce the mutation in several models should contribute to the development of treatment for human diseases as reported in pigs. They created a progeny of pigs with hypercholesterolemia, through the silencing of low-density lipoprotein (Carlson et al., 2012).

11.2.3 CLUSTERED REGULARLY INTERSPACED SHORT PALINDROMIC REPEATS CRISPR/CAS SYSTEM

CRISPR and CRISPR-associated genes are part of an adaptive immune system in prokaryotes against phage infection and plasmid transfer in nature, first discovered in the the *E. coli* genome in 1987. However, it was not until 2007 that this phenomenon was found to be an adaptive and heritable immune system.

There are three types of CRISPR/Cas systems, distinguished by the sequence and structure of the Cas proteins. The type II system, which is the most straightforward, is used for genetic editing. In this system, immunity is acquired after exposure to exogenous DNA. The exogenous DNA is processed by Cas, generating short fragments of the invading DNA called spacers, which are then integrated into the CRISPR arrays in the host chromosome. Each CRISPR array contains a spacer separated by repeated sequences called repeats. The choice of protospacers relies, in part, on the identification of protospacer adjacent motifs (PAM) in the viral genetic material (Jiang and Doudna, 2015).

In this way, the host maintains a genetic record that allows it to act against the same invader in future infections. During subsequent encounters with the invading DNA, the CRISPR arrays, are transcribed into a precursor CRISPR RNA (pre-crRNA) of about 40 nucleotides in length. This pre-crRNA combines with the trans-activating CRISPR RNA (tracrRNA), which is a noncoding RNA. They are then processed by RNase II to obtain mature crRNAs that contain a spacer and a partial repeat sequence. Finally, this processed crRNA guides the Cas9 protein to the complementary DNA sequence, causing invader DNA cleavage (Jinek et al., 2012).

The CRISPR/Cas system used as a genetic editing tool is a type II CRISPR/Cas9 derived from *Streptococcus pyogenes*. In this system, a chimeric RNA called guide RNA (gRNA) is used, which consists of the fusion of tracrRNA and crRNA. The crRNA is designed to have an arbitrary sequence of 20 nucleotides complementary to the target site. This reduces the system to two components: the Cas9 protein and the gRNA. It has also been shown that gRNAs are more efficient than the two components separately (crRNA and tracrRNA) in plants. Cas9 recognition is dependent on the presence of the PAM followed by the complementary sequence to the gRNA. 5'-NGG-'3 is the PAM sequence specific for the Cas9 from *S. pyogenes* (Sovová et al., 2017).

The Cas9 endonuclease produces a site-specific DSB of three nucleotides toward the 5' end of the PAM, generating blunt ends. The specificity of Cas9 is determined by the 17–20 nucleotide sequence present in the gRNA and the PAM. However, imperfect alignments can result in cuts at nontarget positions called off-targets, so gRNA sequences must be carefully chosen to reduce the probability of binding to an undesired sequence. The design of standard-length gRNAs may be limited because they are usually transcribed by RNA polymerase III under the control of a U3 or U6 promoter, which requires guanine and adenosine residues, respectively, for optimal transcription (Hryhorowicz et al., 2017).

Genome editing driven by CRISPR has been in constant improvement. The second generation of variables CRISPR nucleases system has emerged as catalytical-inactivated CAS9 (dCAS9), which includes a CAS nuclease with a mutation in the catalytic domain called RuvC creating a nonfunctional nuclease (Hilton et al., 2015). Further, this nuclease is unable to break the DNA but it can be able to target any locus in the genome and interfere with the transcription process, bringing to the world a new system of reversible gene silencing guided by RNA guide with the same precision as conventional CRISPR (Qi et al., 2013). The difference in architecture between dCAS9 and CRISPR-CAS9 is over the nuclease complex because nonfunctional nuclease just requires a crRNA not the complete scaffold Trans Active RNA and crRNA. Further, the PAM of dCAS9 is rich-T which significantly increases the accuracy scope in the genome (Pan et al., 2021). Thereby CRISPR/dCas9 has been used for genetic screening, overexpression of proteins of interest, and exploitation of a varied array of transcriptional activators and repressors in regulating the expression of target genes (Moradpour and Abdulah, 2020). Besides, there is a CRISP construct with a deaminase enzyme included which carries out the change of the base cytosine to thymine and guanine to adenine by deamination inducing point mutation in the target gene which can lead to a specific change in one amino acid for instance (Shimatani et al., 1917). Besides, it has been designed CRISP constructs including DNA methyl transferases catalytic domain and histone acetyltransferase to induce specific epigenetic changes; in this way, controlling gene expression (Qi et al., 2023).

Since its discovery as a genetic editing tool, CRISPR/Cas9 has become the preferred tool for crop improvement, modifying traits that will enhance yields as well as product quality (Wang and Doudna, 2023) and tolerance to biotic (Tyagi et al., 2021) or abiotic stress (Bhat et al., 2020). Some successfully edited crops using CRISPR/Cas9 include rice, wheat, corn, tomato, soybean, citrus fruits, potato, cotton, alfalfa, watermelon, grape, cassava, lettuce, cocoa, carrot, banana, and barley, among others (Khan et al., 2018).

11.3 APPLICATION OF GENOME EDITION IN TROPICAL CROPS TO IMPROVE BIOTIC STRESS RESISTANCE

In this section, it will be described experiments from the year 2018 to date. Readers interested in experiments carried out before that year can consult several interesting reviews (Ricroch et al., 2017; Romero and Gatica-Arias, 2019).

To increase the resistance of *Theobroma cacao* to the pathogen *Phytophthora tropicalis* infection, two single guides RNA were designed against exons two and three of the nonexpressor of pathogenesis-related 3 (*TcNPR3*) gene. This gene has been shown in *T. cacao* to act as a repressor of the transcription factor nonexpressor of PR1 which plays a role in the response to pathogens (Shi et al., 2013). The right activity of the two single guides was demonstrated in vitro before carrying out the transient transformation by Agrobacterium infection. It was found that the mutation of the gene *TcNPR3* by gene edition induced a reduction in the area size of fungi infection in cocoa leaves. Further, it was recorded a reduction in the ratio of fungi DNA to cacao DNA. Altogether, it was concluded that the edition of gene *TcNPR3* successfully increased the fungi resistance of *T. cacao* plants (Fister et al., 2018). The plant of cassava (*Manihot esculenta*) (Ricroch et al., 2017) can be infected by the cassava brown streak virus with the consequence of yield reduction (Hillocks et al., 2001). It has been reported that a virus protein interacts with the protein host eukaryotic initiation factor 4E (elF4E). Based on this, an experiment was carried out to edit the *nCBP-1* and *nCBP-2* isoforms of *elF4E* gene. The two RNA guides were designed to edit the first exon–intron junction in the case of both genes. The interaction of *nCBP-1* and *nCBP-2* proteins with the virus protein was demonstrated by an experiment using yeast two hybrid system. Transgenic plants of the cassava cultivar 60444 were carried out by Agrobacterium infection and several lines harboring the edited genes *nCBP-1* and *nCBP-2* separate and together were created. It was reported that the aerial symptom development of the plants harboring the two genes edited was reduced in severity as compared with the wild type or the plants harboring each of the genes edited. Furthermore, root necrosis in the plants harboring the two genes edited of nCBP-2 was lower as compared with the wild-type plants and plants harboring the nCBP-1 edited gene. Further, the amount of virus in the plants with the two genes edited was between 43 and 45% lower as compared with the wild type (Gomez et al., 2019). This experiment clearly shows the promising application of gene edition to increase cassava biotic stress resistance.

An experiment was carried out to alter the genome of the African cassava mosaic virus to confer resistance to cassava plants. With this objective, a single guide RNA was designed to edit the AC_2 y AC_3 virus genes, and transgenic cassava plants were created by Agrobacterium infection. The construct failed to induce resistance in the cassava plant due to the development of mutations by single nucleotide insertions in the virus which confers the phenotype of resistance from the CRISPR-associated endonuclease degradation (Mehta et al., 2019). This experiment suggests the importance of using the gene edition technology with caution.

To induce resistance to citrus canker by *Xanthomonas citri* spp. citri bacterial infection, it was carried out the edition of the *CsLOB1* transcription factor in Duncan grapefruit (*Citrus x paradisi*), which is a highly susceptible variety. It was designed a guide RNA able to interact with the two alleles of the gene. The transformation was carried out in epicotyl tissues by Agrobacterium-mediated infection. The transformed tissues were grafted on "Carrizo" rootstock [(*Citrus sinensis* (L.) *Osbeck x Poncirus trifoliata* (Khan et al., 2018) Raf.]. Out of the different tissues grafted, six were obtained and utilized for further studies. The resistance of the six plants was tested using a solution with a concentration of 5×10^8 CFU ml^{-1}. Two of the plants did not develop the typical symptoms of the disease whereas two of them showed a milder symptom as compared with the wild type (Jia et al., 2017).

Virus-infecting plants need to use plant cell proteins to play a role in the translation of RNA. Based on this, an experiment to develop cucumber (*Cucumis sativus* L.) plants resistant to infection of the *Potyviridae* family of viruses was done by editing the eukaryotic translation initiation factor 4E (eIF4E). Two constructs including RNA guides designed based on the sequence of the first and third gene exons were created. Transgenic cucumber plants were created by Agrobacterium infection of cut cotyledon explants. Generation T3 without harboring the transferred DNA was utilized to analyze the resistance to virus attack. Inoculation of CVYV virus with flies in homozygous plants showed complete resistance, whereas heterozygous and wild-type plants were highly susceptible. Further, no virus DNA was found in the homozygous line by reverse transcription polymerase chain reaction. In contrast, only in some plants depending upon the line, it was observed complete resistance to the ZYMV virus infection. Further, weak resistance was observed against the PRSV-W (Israeli isolate) virus in almost all the plants. On the other side, plants did not show any resistance against the CMV and CGMMV viruses (Chandrasekaran et al., 2016).

The fungus *Magnaporthe oryzae* which is the cause of the rice blast is considered one of the most devastating biotic stresses of rice (*Oryza sativa*) (Dean et al., 2012). To increase the blast resistance in rice, the transcription factor *OsERF922* was knocked down by using CRISPR-Cas9 in the variety Kuiku131. The guide RNA was designed including the ATG of the gene and the construct was introduced by Agrobacterium infection strain EHA105. Transgenic plants were segregated until the T_2 generation to eliminate the T-DNA insertion. Six plants with both insertions and different numbers of base deletions were tested for fungi resistance. The development of lesions by the infection was statistically significant smaller as compared with the wild type. Besides of that it was recorded no changes in different agronomic characteristics between the mutant lines and the wild type. This last characteristic is very important because sometimes the genetic transformation induces negative changes in the plants (Wang et al., 2016).

An experiment was carried out to understand the physiological role of the exocyst SEC3 subunit of rice, which is part of the protein complex functioning in the exocytosis phenomena. Although it was not the main objective, the mutation by gene edition of the *OsSEC3A* gene induced a resistance to the *M. oryzae* fungi infection. Two single guides RNA were designed to edit the third and tenth exons of the gene. Several insertions and deletions were recorded in the two exons. In the plants harboring the mutated *OsSEC3A* gene, it was found a higher expression of the pathogenesis-related proteins *PR1a*, *NPR1*, and *PBZ1* as well as genes induced by salicylic acid. Furthermore, it was recorded a fewer number of lesions and a slight reduction in lesion size although this last was not statistically significant. Besides the aforementioned, several agronomic traits were affected by the mutation such as plant height, weight of a thousand grains, length of panicle, and fertility which suggest that this strategy to increase the biotic resistance is not adequate (Ma et al., 2018).

The virus tomato yellow leaf curl is the causal agent of large tomato fruit losses which can reach up to 100%. Further, it has been reported that this virus is difficult to control by using different approaches (Moriones and Navas-Castillo, 2000). Resistance against the infection by tomato yellow leaf curl virus was induced in *Nicotiana benthamiana* and *Solanum lycopersicum* by the design of single guide RNAs to edit the sequences of the coat protein and the replicase. The effectivity of the single guides RNA was tested in *N. benthamiana* before its utilization in tomato plants. The creation of tomato transgenic plants was carried out by Agrobacterium infection of cotyledons from the tomato variety "Moneymaker." Six independent transgenic plants harboring the construct to target either the coat protein gene or the replicase gene were created and seeds from the T_2 generation were obtained. In the case of both groups of transgenic plants, it was recorded the presence of deletions and insertions. It was found by rolling circle amplification assays that the transgenic plants accumulate lower quantities of virus DNA. Further, it was recorded that this characteristic was stable through different generations by infecting T_3 generation plants. However, analysis in *N. benthamiana* plants showed that the replicase gene mutated at the target sequence but still the gene was able to code for a functional protein. In this way, the virus was able to continue active and at the same time to overcome the inhibition by the CRISPR-Cas construct (Tashkandi et al., 2018).

The infection of tomato plants by the parasitic plant *Phelipanche aegyptiaca* cause great losses in production by reducing the leaf chlorophyll content and photosynthetic rate and in this way the plant growth (Mauromicale et al.,

2008). The edition of the tomato gene *MAX1* (Solyc08g062950) was carried out by designing a single guide RNA to edit the third exon of the gene. Several transgenic plants were created by Agrobacterium infection and T1 generation tomato plants were obtained. This allowed us to observe that the mutation remained stable through a sexual generation. Analysis of the DNA region showed the presence of 9 bp deletions in all of them. Further, these plants showed phenotypic changes s the MAX1 tomato mutant plant. Also, by using Sanger DNA sequencing, no mutations in other DNA regions were found. The tomato transgenic plants were cultivated with *P. aegyptiaca* seeds to observe the level of resistance. It was found that plants heterozygous and homozygous for the mutation showed a low number of germinated parasite tubercles and shoots, which suggests that the approach was successful (Bari et al., 2021).

11.4 CONCLUSIONS AND GENERAL PERSPECTIVES

The utilization of the recently developed genome editing techniques will allow the design of plants with higher resistance to the different biotic stresses both before and after harvest. This will reduce the negative impact on the environment and human health due to the utilization of agrochemicals. Further, this in turn will lead to both the reduction in postharvest agricultural products losses and increase in agricultural production. As a consequence, in the future, these technologies will have a positive impact on food safety, and they will assure food security. Finally, there is the hope that the long-lasting discussion about the negative effects of the traditional techniques of transgenic plant creation and the consequent rejection of the consumer will not continue any longer.

KEYWORDS

- **agriculture production**
- **tropical plants**
- **genome edition**
- **biotic stress resistance**
- **sustainable**

REFERENCES

Anuar, M. S. K.; Hashim, A. M.; Ho, C. L.; Wong, M. Y.; Sundram, S.; Saidi, N. B.; Yusof, M. T. Synergism: Biocontrol Agents and Biostimulants in Reducing Abiotic and Biotic Stresses in Crop. *World J. Microbiol. Biotechnol.* **2023,** *39* (5). DOI: 10.1007/s11274-023-03579-3.

Bari, V. K.; Nassar, J. A.; Aly, R. CRISPR/Cas9 Mediated Mutagenesis of MORE AXILLARY GROWTH 1 in Tomato Confers Resistance to Root Parasitic Weed *Phelipanche aegyptiaca*. *Sci. Rep.* **2021,** *11* (1), 3905.

Becker, S.; Boch, J. TALE and TALEN Genome Editing Technologies. *Gene Genome Edit.* **2021,** *2,* 100007.

Bhat, M. A.; Bhat, M. A.; Kumar, V.; Wani, I. A.; Bashir, H.; Shah, A. A.; Rahman, S.; Jan, A. T. The Era of Editing Plant Genomes Using CRISPR/Cas: A Critical Appraisal. *J. Biotechnol.* **2020,** *324,* 34–60.

Boch, J.; Scholze, H.; Schornack, S.; Landgraf, A.; Hahn, S.; Kay, S.; Lahaye, T.; Nickstadt, A.; Bonas, U. Breaking the Code of DNA Binding Specificity of TAL-type III Effectors. *Science* **2009,** *326* (5959), 1509–1512.

Carlson, D. F.; Tan, W.; Lillico, S. G.; Stverakova, D.; Proudfoot, C.; Christian, M.; Voytas, D. F.; Long, C. R.; Whitelaw, C. B. A.; Fahrenkrug, S. C. Efficient TALEN-Mediated Gene Knockout in Livestock. *Proc. Natl. Acad. Sci.* **2012,** *109* (43), 17382–17387.

Ceasar, S. A.; Ignacimuthu, S. CRISPR/Cas Genome Editing in Plants: Dawn of *Agrobacterium* Transformation for Recalcitrant and Transgene-free Plants for Future Crop Breeding. *Plant Physiol. Biochem.* **2023,** *196,* 724–730. DOI: 10.1016/j.plaphy.2023.02.030.

Cermak, T.; Doyle, E. L.; Christian, M.; Wang, L.; Zhang, Y.; Schmidt, C.; Baller, J. A.; Somia, N. V.; Bogdanove, A. J.; Voytas, D. F. Efficient Design and Assembly of Custom TALEN and Other TAL Effector-based Constructs for DNA Targeting. *Nucleic Acids Res.* **2011,** *39* (12), e82-e82.

Chandrasekaran, J.; Brumin, M.; Wolf, D.; Leibman, D.; Klap, C.; Pearlsman, M.; Sherman, A.; Arazi, T.; Gal-On, A. Development of Broad Virus Resistance in Non-transgenic Cucumber Using CRISPR/Cas9 Technology. *Mol. Plant Pathol.* **2016,** *17* (7), 1140–1153.

Cohen-Tannoudji, M.; Robine, S.; Choulika, A.; Pinto, D.; El Marjou, F.; Babinet, C.; Louvard, D.; Jaisser, F. I-Sce I-induced Gene Replacement at a Natural Locus in Embryonic Stem Cells. *Mol. Cell. Biol.* **1998,** *18* (3), 1444–1448.

da Rocha, M. P.; Dourado, P. L. R.; Rodrigues, M. D.; Raposo, J. L.; Grisolia, A. B.; de Oliveira, K. M. P. The Influence of Industrial and Agricultural Waste on Water Quality in the Agua Boa Stream (Dourados, Mato Grosso do Sul, Brazil). *Environ. Monit. Assess.* **2015,** *187* (7). DOI: 10.1007/s10661-015-4475-9.

Dean, R.; Van Kan, J. A.; Pretorius, Z. A.; Hammond-Kosack, K. E.; Di Pietro, A.; Spanu, P. D.; Rudd, J. J.; Dickman, M.; Kahmann, R.; Ellis, J. The Top 10 Fungal Pathogens in Molecular Plant Pathology. *Mol. Plant Pathol.* **2012,** *13* (4), 414–430.

Dmitriev, A. A.; Pushkova, E. N.; Melnikova, N. V. Plant Genome Sequencing: Modern Technologies and Novel Opportunities for Breeding. *Mol. Biol.* **2022,** *56* (4), 495–507. DOI: 10.1134/s0026893322040045.

Durai, S.; Mani, M.; Kandavelou, K.; Wu, J.; Porteus, M. H.; Chandrasegaran, S. Zinc Finger Nucleases: Custom-Designed Molecular Scissors for Genome Engineering of Plant and Mammalian Cells. *Nucleic Acids Res.* **2005,** *33* (18), 5978–5990.

Fernández, A.; Josa, S.; Montoliu, L. A history of Genome Editing in Mammals. *Mamm. Genome* **2017,** *28* (7–8), 237–246.

Fister, A. S.; Landherr, L.; Maximova, S. N.; Guiltinan, M. J. Transient Expression of CRISPR/Cas9 Machinery Targeting TcNPR3 Enhances Defense Response in *Theobroma cacao*. *Front. Plant Sci.* **2018,** *9,* 268.

Gomez, M. A.; Lin, Z. D.; Moll, T.; Chauhan, R. D.; Hayden, L.; Renninger, K.; Beyene, G.; Taylor, N. J.; Carrington, J. C.; Staskawicz, B. J. Simultaneous CRISPR/Cas9-Mediated Editing of Cassava eIF 4E Isoforms nCBP-1 and nCBP-2 Reduces Cassava Brown Streak Disease Symptom Severity and Incidence. *Plant Biotechnol. J.* **2019,** *17* (2), 421–434.

Hillocks, R.; Raya, M.; Mtunda, K.; Kiozia, H. Effects of Brown Streak Virus Disease on Yield and Quality of Cassava in Tanzania. *J. Phytopathol.* **2001,** *149* (7-8), 389–394.

Hilton, I. B.; D'Ippolito, A. M.; Vockley, C. M.; Thakore, P. I.; Crawford, G. E.; Reddy, T. E.; Gersbach, C. A. Epigenome Editing by a CRISPR-Cas9-based Acetyltransferase Activates Genes from Promoters and Enhancers. *Nat. Biotechnol.* **2015,** *33* (5), 510–517. DOI: 10.1038/nbt.3199.

Hryhorowicz, M.; Lipiński, D.; Zeyland, J.; Słomski, R. CRISPR/Cas9 Immune System as a Tool for Genome Engineering. *Arch. Immunol. Ther. Exp.* **2017,** *65,* 233–240.

Iriti, M.; Vitalini, S. Sustainable Crop Protection, Global Climate Change, Food Security and Safety-Plant Immunity at the Crossroads. *Vaccines* **2020,** *8* (1). DOI: 10.3390/vaccines8010042.

Jia, H.; Zhang, Y.; Orbović, V.; Xu, J.; White, F. F.; Jones, J. B.; Wang, N. Genome Editing of the Disease Susceptibility Gene Cs LOB 1 in Citrus Confers Resistance to Citrus Canker. *Plant Biotechnol. J.* **2017,** *15* (7), 817–823.

Jiang, F.; Doudna, J. A. The Structural Biology of CRISPR-Cas Systems. *Curr. Opin. Struct. Biol.* **2015,** *30,* 100–111.

Jinek, M.; Chylinski, K.; Fonfara, I.; Hauer, M.; Doudna, J. A.; Charpentier, E. A Programmable Dual-RNA–guided DNA Endonuclease in Adaptive Bacterial Immunity. *Science* **2012,** *337* (6096), 816–821.

Joung, J. K.; Sander, J. D. TALENs: A Widely Applicable Technology for Targeted Genome Editing. *Nat. Rev. Mol. Cell Biol.* **2013,** *14* (1), 49–55.

Kang, Y. N.; Kang, C. S.; Kim, C. History of Nucleotide Sequencing Technologies: Advances in Exploring Nucleotide Sequences from Mendel to the 21st Century. *Hortic. Sci. Technol.* **2019,** *37* (5), 549–558. DOI: 10.7235/hort.20190055.

Khan, M. H. U.; Khan, S. U.; Muhammad, A.; Hu, L. M.; Yang, Y.; Fan, C. C. Induced Mutation and Epigenetics Modification in Plants for Crop Improvement by Targeting CRISPR/Cas9 Technology. *J. Cell. Physiol.* **2018,** *233* (6), 4578–4594. DOI: 10.1002/jcp.26299.

Kim, Y.-G.; Cha, J.; Chandrasegaran, S. Hybrid Restriction Enzymes: Zinc Finger Fusions to Fok I Cleavage Domain. *Proc. Natl. Acad. Sci.* **1996,** *93* (3), 1156–1160.

Ma, J.; Chen, J.; Wang, M.; Ren, Y.; Wang, S.; Lei, C.; Cheng, Z.; Sodmergen. Disruption of OsSEC3A Increases the Content of Salicylic Acid and Induces Plant Defense Responses in Rice. *J. Exp. Bot.* **2018,** *69* (5), 1051–1064. DOI: 10.1093/jxb/erx458 From NLM.

Mauromicale, G.; Monaco, A. L.; Longo, A. M. G. Effect of Branched Broomrape (*Orobanche ramosa*) Infection on the Growth and Photosynthesis of Tomato. *Weed Sci.* **2008,** *56* (4), 574–581. DOI: 10.1614/WS-07-147.1 From Cambridge University Press Cambridge Core.

Mehta, D.; Stürchler, A.; Anjanappa, R. B.; Zaidi, S. S.-E.-A.; Hirsch-Hoffmann, M.; Gruissem, W.; Vanderschuren, H. Linking CRISPR-Cas9 Interference in Cassava to the Evolution of Editing-Resistant Geminiviruses. *Genome Biol.* **2019,** *20* (1), 1–10.

Method of the Year 2011. *Nat. Methods* **2012,** *9* (1), 1–1. DOI: 10.1038/nmeth.1852.

Mitra, S.; Anand, U.; Ghorai, M.; Kant, N.; Kumar, M.; Radha; Jha, N. K.; Swamy, M. K.; Proćków, J.; de la Lastra, J. M. P. Genome Editing Technologies, Mechanisms and Improved Production of Therapeutic Phytochemicals: Opportunities and Prospects. *Biotechnol. Bioeng.* **2023,** *120* (1), 82–94.

Moradpour, M.; Abdulah, S. N. A. CRISPR/dCas9 Platforms in Plants: Strategies and Applications Beyond Genome Editing. *Plant Biotechnol. J.* **2020,** *18* (1), 32–44. DOI: 10.1111/PBI.13232.

Moriones, E.; Navas-Castillo, J. Tomato Yellow Leaf Curl Virus, An Emerging Virus Complex Causing Epidemics Worldwide. *Virus Res.* **2000,** *71* (1–2), 123–134.

Mussolino, C.; Morbitzer, R.; Lütge, F.; Dannemann, N.; Lahaye, T.; Cathomen, T. A Novel TALE Nuclease Scaffold Enables High Genome Editing Activity in Combination with Low Toxicity. *Nucleic Acids Res.* **2011,** *39* (21), 9283–9293.

Pan, C.; Sretenovic, S.; Qi, Y. CRISPR/dCas-Mediated Transcriptional and Epigenetic Regulation in Plants. *Curr. Opin. Plant Biol.* **2021,** *60,* 101980–101980. DOI: 10.1016/J.PBI.2020.101980.

Porteus, M. H.; Carroll, D. Gene Targeting Using Zinc Finger Nucleases. *Nat. Biotechnol.* **2005,** *23* (8), 967–973. DOI: 10.1038/nbt1125.

Pulecio, J.; Verma, N.; Mejia-Ramirez, E.; Huangfu, D.; Raya, A. CRISPR/Cas9-Based Engineering of the Epigenome. *Cell Stem Cell* **2017,** *21* (4), 431–447. DOI: 10.1016/j.stem.2017.09.006.

Qi, L. S.; Larson, M. H.; Gilbert, L. A.; Doudna, J. A.; Weissman, J. S.; Arkin, A. P.; Lim, W. A. Repurposing CRISPR as an RNA-γuided Platform for Sequence-specific Control of Gene Expression. *Cell* **2013,** *152* (5), 1173–1183. DOI: 10.1016/j.cell.2013.02.022.

Qi, Q.; Hu, B.; Jiang, W.; Wang, Y.; Yan, J.; Ma, F.; Guan, Q.; Xu, J. Advances in Plant Epigenome Editing Research and Its Application in Plants. *Int. J. Mol. Sci.* **2023,** *24* (4), 3442.

Qi, Y. Genome Editing is Revolutionizing Biology. *Cell Biosci.* **2017,** *7* (1), 1–1. DOI: 10.1186/S13578-017-0162-6/METRICS.

Randhawa, S.; Sengar, S. The Evolution and History of Gene Editing Technologies. *Prog. Mol. Biol. Transl. Sci.* **2021,** *178,* 1–62.

Reyon, D.; Tsai, S. Q.; Khayter, C.; Foden, J. A.; Sander, J. D.; Joung, J. K. FLASH Assembly of TALENs for High-Throughput Genome Editing. *Nat. Biotechnol.* **2012,** *30* (5), 460–465.

Ricroch, A.; Clairand, P.; Harwood, W. Use of CRISPR Systems in Plant Genome Editing: Toward New Opportunities in Agriculture. *Emerg. Topics Life Sci.* **2017,** *1* (2), 169–182. DOI: 10.1042/etls20170085 (acccessed 4/19/2023).

Rojas-Vasquez, R.; Gatica-Arias, A. Use of Genome Editing Technologies for Genetic Improvement of Crops of Tropical Origin. *Plant Cell Tissue Organ Cult. (PCTOC)* **2020,** *140,* 215–244.

Romero, F. M.; Gatica-Arias, A. CRISPR/Cas9: Development and Application in Rice Breeding. *Rice Sci.* **2019,** *26* (5), 265–281. DOI: 10.1016/j.rsci.2019.08.001.

Shah, T.; Andleeb, T.; Lateef, S.; Noor, M. A. Genome Editing in Plants: Advancing Crop Transformation and Overview of Tools. *Plant Physiol. Biochem.* **2018,** *131,* 12–21.

Shi, Z.; Zhang, Y.; Maximova, S. N.; Guiltinan, M. J. TcNPR3 from Theobroma Cacao Functions as a Repressor of the Pathogen Defense Response. *BMC Plant Biol.* **2013,** *13,* 1–12.

Shimatani, Z.; Ariizumi, T.; Fujikura, U.; Kondo, A.; Ezura, H.; Nishida, K. Targeted Base Editing with CRISPR-Deaminase in Tomato. In *Plant Genome Editing with Crispr Systems:*

Methods and Protocols, Qi, Y., Ed.; Methods in Molecular Biology, 2019; Vol. 1917, pp 297–307.

Sovová, T.; Kerins, G.; Demnerová, K.; Ovesná, J. Genome Editing with Engineered Nucleases in Economically Important Animals and Plants: State of the Art in the Research Pipeline. *Curr. Issues Mol. Biol.* **2017,** *21* (1), 41–62.

Sufyan, M.; Daraz, U.; Hyder, S.; Zulfiqar, U.; Iqbal, R.; Eldin, S. M.; Rafiq, F.; Mahmood, N.; Shahzad, K.; Uzair, M. An Overview of Genome Engineering in Plants, Including Its Scope, Technologies, Progress and Grand Challenges. *Funct. Integr. Genomics* **2023,** *23* (2), 119.

Szurek, B.; Rossier, O.; Hause, G.; Bonas, U. Type III-dependent Translocation of the Xanthomonas AvrBs3 Protein into the Plant Cell. *Mol. Microbiol.* **2002,** *46* (1), 13–23.

Tashkandi, M.; Ali, Z.; Aljedaani, F.; Shami, A.; Mahfouz, M. M. Engineering Resistance Against Tomato Yellow Leaf Curl Virus Via the CRISPR/Cas9 System in Tomato. *Plant Signal. Behav.* **2018,** *13* (10), e1525996.

Tyagi, S.; Kumar, R.; Kumar, V.; Won, S. Y.; Shukla, P. Engineering Disease Resistant Plants Through CRISPR-Cas9 Technology. *GM Crops & Food* **2021,** *12* (1), 125–144.

Urnov, F. D.; Rebar, E. J.; Holmes, M. C.; Zhang, H. S.; Gregory, P. D. Genome Editing with Engineered Zinc Finger Nucleases. *Nat. Rev. Genet.* **2010,** *11* (9), 636–646.

Van den Ackerveken, G.; Marois, E.; Bonas, U. Recognition of the Bacterial Avirulence Protein AvrBs3 Occurs Inside the Host Plant Cell. *Cell* **1996,** *87* (7), 1307–1316.

Wang, F.; Wang, C.; Liu, P.; Lei, C.; Hao, W.; Gao, Y.; Liu, Y.-G.; Zhao, K. Enhanced Rice Blast Resistance by CRISPR/Cas9-targeted Mutagenesis of the ERF Transcription Factor Gene OsERF922. *PloS One* **2016,** *11* (4), e0154027.

Wang, J. Y.; Doudna, J. A. CRISPR Technology: A Decade of Genome Editing is Only the Beginning. *Science* **2023,** *379* (6629), eadd8643.

Wang, Y.; Zafar, N.; Ali, Q.; Manghwar, H.; Wang, G.; Yu, L.; Ding, X.; Ding, F.; Hong, N.; Wang, G. CRISPR/Cas Genome Editing Technologies for Plant Improvement against Biotic and Abiotic Stresses: Advances, Limitations, and Future Perspectives. *Cells* **2022,** *11* (23), 3928.

Zhang, Y.; Qi, Y. CRISPR Enables Directed Evolution in Plants. *Genome Biol.* **2019,** *20* (1), 1–4.

CHAPTER 12

Sustainable Use of Chihuahua Desert Endemic Plants: An Overview of Traditional Uses, Phytochemistry and Biological Properties

J. OMAR DÍAZ-RIVAS[1], O. MIRIAM RUTIAGA-QUIÑONES[2], M. HUMBERTO CHAIREZ-RAMIREZ[3], and JULIO CÉSAR TAFOLLA-ARELLANO[4*]

[1]*Laboratorio de química analítica, Instituto de Investigación y Transferencia de Tecnología de Zonas Áridas. Parras de la Fuente, Coahuila, México*

[2]*Departamento de Ings. Química-Bioquímica, Laboratorio de tecnología postcosecha. Tecnológico Nacional de México Instituto Tecnológico de Durango, Durango, Dgo, México*

[3]*Unidad de investigación Biomédica de Cáncer, Instituto de Investigaciones Biomédicas, Universidad Nacional Autónoma de México & Instituto Nacional de Cancerología, Secretaria de Salud, Tlalpan, CDMX, México*

[4]*Laboratorio de Biotecnología y Biología Molecular. Departamento de Ciencias Básicas. Universidad Autónoma Agraria Antonio Narro. Saltillo, Coahuila, México*

ABSTRACT

Currently, the search for natural alternatives that improve the quality of life has increased, and factors such as poor diet, stress, anxiety, or

Sustainable Engineering and Agro-Food Processing: Renewable and Nonrenewable Resources.
José Juan Buenrostro-Figueroa, Julio César Tafolla-Arellano, Mónica Chávez-González, Pedro Aguilar-Zárate, A. K. Haghi (Eds.)

self-medication have been a fundamental part of the development of health problems in the population. Recently, the study and development of functional or nutraceutical foods have focused on the use of some plants that have been recognized for their biological potential in multiple applications such as biological control, food, materials development, or in the health sector. These herbal sources have been used by our ancestors in traditional ways, such as the generation of infusions, topicals, ointments or as spices in the preparation of food. However, the consumption of this type of matrices has been lost over the years, considering them in many cases as weeds or unwanted products by producers. The importance of the study of these sources is due to the phytochemicals, compounds from the secondary metabolism of plants, which are produced as a defense system against environmental conditions or predators to guarantee their survival, however, this class of compounds has been shown to have important biological activities. The most important are the called phenolic compounds, among which stand out some phenolic acids, flavonoids, terpenes, ellagitannins, among others. One way in which these compounds are synthesized is naturally, through biotic or abiotic factors, or they can be controlled using some hormones. Some studies of these compounds have focused on pest control, obtaining some phytopharmaceuticals, animal and human nutrition, and the evaluation of their antioxidant, anti-inflammatory, and antihistamine capacity. Thus, with the production and sustainable management of these ancestral matrices, plus the development of these herbal products, they can be useful candidates aimed at the needs that the populations have.

12.1 IMPORTANCE OF ANCESTRAL KNOWLEDGE

In the UNESCO Universal Declaration on Cultural Diversity of 2001, it is established that traditional and ancestral knowledge is a heritage whose value is not limited solely to the original communities, but that such knowledge constitutes an important resource for all humanity, as it enriches mutual knowledge through dialogue, and allow the preservation of the wide spectrum of cultural diversity existing in each territory.

All traditional knowledge is all about the knowledge accumulated by popular cultures and transmitted orally (Lizcano and Fernández, 2014). In this sense, Pedrero (2011) argues that current research should recognize and highlight ancestral knowledge, as in the case of food preparation, where producers preserve ancestral knowledge; this allows for obtaining healthy, natural foods with flavors that manage to differentiate it (McClung et al., 2014). In the traditional processes to prepare food, emulsions, infusions, among others, local knowledge is implicit, which is part of its natural and cultural heritage, reason why it is created, developed, and transmitted in community settings (García, 2007).

The first peoples, generators of the enormous knowledge that plants represent for the prevention and cure of many diseases, have been shared over time with those people who respect the origins, methods, and traditional and natural forms of their use. They have done it for many years, it is part of their culture, it is part of the process of keeping knowledge and traditions current, and it is part of their ecological and social resilience. Sharing the "secrets" leads to a process that must be understood, valued, and adhered to high ethical values (Rosado-May, 2014). According to a precise meaning, the so-called ancestral knowledge has the following specific characteristics (Estermann, 2006): (1) Knowledge built in the community vital praxis; not by identifiable individual authors in the case of claiming intellectual property and/or copyright; (2) This knowledge must be created by the wisest people, that is, those who have the greatest amount of information acquired naturally in those areas where it is generated; (3) Saved and transmitted from generation to generation orally; not in writing, which is intended to guarantee the veracity and repeatability of the sources

The topic of medicinal plants brings together all the characteristics of Mexico and the world (Rosado-May, 2014). The great development that the Mayas, the Aztecs, and other cultures had at the time, cannot be explained without considering the role of the deep knowledge they had of their natural resources, especially medicinal plants. People without food, education, and health cannot reach the very high levels that the great pre-Hispanic cultures had in their time.

Ethnobotany is a mixed discipline between natural sciences and social sciences that studies the relationship between plants and man (Balick and Cox, 1996). Human beings, from their origin and in all existing cultures, have always resorted to vegetables to cover their most basic needs: to feed themselves, to cure the different ailments most common in each region, to dress, to provide shelter and heat, or to process food, among others. Each culture has obtained the necessary resources in one way or another and has used them according to its needs, in such a way that cultural evolution has gone hand in hand with evolution in the knowledge of the natural environment and understanding of its different components and operation (Towle, 2017).

The loss of ancestral knowledge regarding the use of plants by different cultures is a reality that has already been demonstrated worldwide (Benz, et al., 2000); Ramírez (2007); Aswani et al. (2018). The causes of this loss, although they vary depending on the place, are closely related to the rapid environmental, socioeconomic, and cultural changes that currently occur in different cultures and societies due to globalization, and that somehow imply a distance from nature, both in our day-to-day life and at an educational level. In this scenario, the loss of biodiversity, the progressive modernization, and cultural homogenization are presented as key pieces in the decline of ancestral knowledge and uses related to the use of plants, and in this same way, the inclusion of these aspects within the scope formal and nonformal education (Reyes-García et al., 2009).

12.2 IMPORTANCE OF THE CHIHUAHUAN DESERT PLANTS

In Mexico, the Chihuahuan desert provides a great diversity of plants that produce compounds with important biological activity. It is considered the largest desert in North America (Cloudsley, 1977) and the second most diverse in the world. It originates in the highlands of Mexico between the Sierra Madre Occidental and the Sierra Madre Oriental, continues north to southern Arizona, New Mexico, and Texas, and includes parts of the Mexican states of Chihuahua, Coahuila, Durango, Nuevo León, San Luis Potosi, Tamaulipas, and Zacatecas. Its extension in Mexico occupies around one-sixth of the Mexican territorial surface (Henrickson and Johnston, 2007). Also, it encompasses one of the most biologically rich arid regions on earth; its floristic and endemic richness enhances the biological diversity of our country (Balleza and Villaseñor, 2011). The annual temperature range is 5.5–35°C and annual rainfall ranges from 175 to 400 mm (Henrickson and Johnston, 2007). The Chihuahuan Desert stands out as one of the most biologically diverse ecological regions worldwide, not only due to its overall biodiversity but also due to its exceptional richness of endemic species. The high degree of endemism is the result of the effects of isolation, physiography, dynamic changes in climate during the last 10,000 years, and the colonization of habitats by specialist elements. Numerous organisms, which are in some categories of risk, are housed here from species in special protection to those in danger of extinction (SEMARNAT, 2010).

Due to the richness of these plants, vegetables, and herbs used in traditional medicine have gained wide acceptance as the main sources of phytochemical discovery. For this reason, information on the antioxidant properties of natural products is becoming relevant in the fields of nutrition and nutraceutical product development. The phytochemicals present in these plants associated with their bioactivity are included in several plant processes, where from germination to maturity biochemical strategies are established that allow them to defend themselves against pathogens and predators. These strategies involve the synthesis of compounds capable of promoting and inhibiting environmental aggression through abiotic and biotic stress. Among these most notable plants are the lechuguilla (*Agave lechuguilla*), candelilla (*Euphorbia antisyphilitica*), mesquite (*Prosopis* spp.), gobernadora (*Larrea tridentata*), oregano (*Lippia graveolens*), and some cactus as well as succulents. Most of these plants produce some saponins, terpenes, steroids, carbohydrates, and phenolic compounds (Tucuch-Pérez et al., 2020; Rábago et al., 2020; García-Pérez et al., 2012).

12.3 SECONDARY METABOLISM AND PRODUCTION OF PHYTOCHEMICALS

In all organisms, metabolism is the set of biochemical reactions that take place in the cells to synthesize complex compounds from simpler ones or degrade complex substances and obtain simpler ones (Kornberg, 2019). In this way, plants have a secondary metabolism in addition to the primary metabolism that enables them to produce several compounds of diverse chemical nature. The products of this metabolism, called secondary metabolites, are involved in plants' protection against biotic and abiotic stresses. These metabolites are differentially distributed among plant groups, displaying important biological properties, and performing sustainable ecological functions. They are characterized by different uses and potential applications as medicines, insecticides, herbicides, perfumes, or dyes (Pagare et al., 2015).

Polyphenols are metabolites produced by all plants which constitute a family of varied substances from simple to complex structures derived from the metabolism of shikimic acid and polyacetate. To date, many polyphenolic compounds have been characterized in plants and grouped in various classes. These classes exist many variations around the basic chemical skeleton essentially concerning the degrees of oxidation, hydroxylation, methylation, glycosylation, and the possible connections to other molecules (Bruneton, 1993).

These molecules have been investigated in their mechanisms of action on their possible effects on biological models discovering one of the most important effects around these compounds which is to inhibit and delay the oxidation process by free radicals (Munin and Edwards, 2011). This effect is possible due to two fundamental properties: Interaction with proteins or with ions, and radical scavenging activity. Polyphenols can act using different modes of action depending on the approach studied (Leopoldini et al., 2011).

These effects have increased the interest in phytochemicals due to their multiple applications in the health, industrial, and food or agricultural sectors (Akrem et al., 2007). From the great variety of secondary metabolites, polyphenols, including phenolic acids, flavonoids, stilbenes, coumarins, lignans and tannins (Shahidi and Ambigaipalam, 2015), stand out due to their structural diversity and essential for diverse functions in plants, such as defense, growth, and development, which are influenced by environmental factors and plant-herbivorous interactions (Cheynier et al., 2015).

Polyphenols occur in different forms in plants, including aglycones, esters, glycosides, etc. These different forms show variable suitability for different extraction conditions and vary in their susceptibility to degradation (Ross et al., 2009). These compounds are thought to provide health benefits through several mechanisms, including free-radical scavenging; protection and regeneration of other dietary antioxidants; and chelating of prooxidant metal ions (Magalhães et al., 2009; Huang et al., 2007). The solubility of phenolics depends on their chemical nature, which may vary from simple to polymerized. There is a possibility of interaction of phenolics with other plant components such as carbohydrates and proteins that may lead to the formation of complexes that may be quite insoluble.

12.4 BIOACTIVITIES OF NATURAL COMPOUNDS FROM ENDEMIC PLANTS OF CHIHUAHUA'S DESERT

Since ancient times, plants have been an important hub for the development of civilization, through the development of agriculture for food, their use as a source of cosmetic products, as well as their importance in the pharmaceutical field. In Mexico, agriculture represents an important bright spot for the economy, only in 2022 the exports of agro-food and agro-industry sectors surpass 50 and 28 billion dollars, respectively (Secretaría de Agricultura y Desarrollo Rural - México, 2023). Apart from the economic importance of plants in Mexico, there are many uses for plants including but not limited to cosmetic and medicinal use, construction and textile materials, fuel sources, etc. The most common semi-desertic plants include species such as *Flourensia cernua*, *Eucalyptus camaldulensis*, *E. antisyphilitica* Zucc, *L. tridentata*, *L. graveolens*, *Opuntia*, *Agave*, and *Prosopis* (Palomo-Ligas et al., 2022), and these species might be of importance for all of the aforementioned industries. Humankind continues its exponential population growth, and people start to be concerned about synthetic-derived products for food and health care. People have increased their interest in lifestyle habits improvement, especially there is a growing interest in natural-derived products, particularly, those derived from phytochemicals that have great acceptance. These factors may play a crucial role in the sustainable use of the plant biodiversity of the Chihuahua semi-desert which might result in important advances in food and drug research, thus contributing to regional economic development.

Mexico is considered the 5th most megadiverse country, this diversity is reflected in its nearly 50,000 plant species, many of which are primarily used as food. Chihuahua's semi-desert is rich in a diversity of plants that have been used as food since prehispanic times (Arellano-Perusquía et al., 2019). Since prehispanic times, certain endemic plants from the Chihuahua's semi-desert have been used as food sources. However, besides their primary use, some of these endemic plants may be used for the formulation of functional foods or nutraceuticals, because of their rich content in phytochemicals. Nowadays, there is an increasing interest to study the impact of food and diet on human health. It is well accepted that diets that include regular consumption of plant-derived foods, as well as grains and fish are considered healthy diets and have a generally positive impact on well-being. On the contrary, rich diets on ultraprocessed and fast food, as well as carbonated beverages negatively impact health status. There is a big list of compounds responsible for the beneficial effects of food. Nutrients such as ω-3 or nonnutritional compounds such as phytochemicals are widely associated with those effects. Healthy foods, for example, ω-3-rich foods mediate the regulation of the intestinal microbiome (Fu et al., 2021), on the contrary, the negative effects of ultraprocessed foods have been associated with inflammatory processes, endothelial dysfunction, and insulin resistance (Segal and Opie, 2015).

Despite relevant advances in pharmaceutical fields, there is an important number (70–95%) of people in developing countries that still use phytochemical-rich ancestry matrices as primary therapies for disease management and control (Heredia et al., 2022). Chronic degenerative diseases (CDDs) are responsible for approximately 41 million deaths annually, which represent 74% of total deaths worldwide. The principal leading causes of mortality and morbidity worldwide related to CDDs include cardiovascular diseases, cancers, chronic respiratory diseases, and diabetes (World Health Organization, 2020).

Plants represent the most important source of natural bioactive compounds with more than 391,000 species worldwide (Díaz et al., 2022). Particularly, Mexico has 3500 plant species used in traditional medicine (Govea-Salas et al., 2017). Each with a unique phytochemical composition, plants are a potential source of possible therapeutical compounds for the treatment of many health disorders. Plants are so important as sources of drugs that approximately 60% of current drugs are natural products or derivatives of these (e.g. aspirin, digoxin, paclitaxel, and morphine), and only 5–15% of plants have been studied under scientific rigor (Mathur and Hoskins, 2017). As promising sources of drugs, phytochemicals have been extensively studied as novel molecules for CDDs. Mexican semi-desertic plants are composed principally of phenolic acids, flavonoids, and ellagitannins which have been related to several bioactivities including prevention and treatment of aging diseases, cancers, heart diseases, diabetes, anti-inflammatory effects, etc.

However, we must clarify that food and medicines are two different entities, while medicines are designed to prevent, treat, or cure a certain disease; food has the main purpose of fulfilling the nutritional requirements of the human being. However, in recent years, there has been a growing interest in the consumption of functional foods, this type of food is like conventional ones, but their consumption produces health benefits in addition to their nutritional properties. This effect is regularly associated with phytochemicals, which are secondary metabolites of low molecular weight that regulate the plant-environment interactions, and they are usually present in low concentrations. It has been estimated that there are more than 100,000 phytochemicals and a classic occidental diet involves the consumption of 1.5 g per day. The main groups of phytochemicals are polyphenols, carotenoids, phytosterols, terpenes, phytoestrogens, glucosinolates, etc. (Leitzmann, 2016), and their constant intake has been associated with a wide range of health benefits. Because of the wide range of health benefits associated with the consumption of phytochemicals, the term "nutraceutical" was coined linking nutrition and pharmaceutical. A nutraceutical is a substance that is a food or part of a food that provides therapeutic effects or a health benefit, and those effects can prevent or treat a certain disease (DeFelice, 1995).

12.4.1 *OPUNTIA FICUS*

In this sense, different endemic plants from Chihuahua's deserts have been used for both food and medicinal uses. Prickly pear cactus and its fruit (tuna) are widely used in Mexican cuisine. Given its resistance to dry climates and the versatility of the products that can be obtained from its use, the nopal is recognized as a plant capable of growing in lands where other crops do not prosper, due to its unique characteristics that provide resilience to the limiting conditions generated by prolonged droughts and even the phenomenon of desertification.

The nopal can also be used as a vegetable, prickly pear, grana cochineal (dye), to produce cosmetics and medicines, as fodder for cattle, and to produce electricity and biogas. In addition, it can generate sustainable development poles, both economic, social, and environmental. *O. ficus-indica* and its fruit are highly valued in Mexican food and are consumed by most of the population not only in Chihuahua's desert but also all over the country. Apart from the culinary use of this cactus, it is rich in many compounds such as vitamins C and E, carotenoids, amino acids, and antioxidant compounds (phenols, flavonoids, betaxanthin, and betacyanin) (El-Mostafa et al., 2014). *O. ficus-indica* has been related to hypoglycemic control (Nuñez et al., 2013), and it has documented antidiabetic properties (López-Romero et al., 2014).

12.4.2 *LIPPIA GRAVEOLENS* (OREGANO)

The Mexican oregano, known as *L. graveolens* grows in dry or semi-dry climates, in soils of limestone origin, so the species has the potential for domestication, in addition to the aforementioned. On the contrary, it stands out for having great economic potential and significant international demand, thanks to its uses in the pharmaceutical and cosmetic industry. In

Mexico, oregano production exceeds 80 tons. In this way, all its advantages and utilities make it one of the most widely used products in the world. *L. graveolens* has been widely used as an additive to many foods, as well as in traditional medicine as a therapy for multiple pathologies due to its phytochemical richness. All these benefits offered by this plant are thanks to its multiple chemical compounds, among which are caffeic, chlorogenic, rosmarinic, apigenol, luteolol, quercetin and some tannins (Soto-Dominguez et al., 2012). Oregano essential oils are used as active ingredients in the soft drink, cosmetic, soap and perfume industries. They are used, for example, to prepare antirheumatic ointments, ointments for dermatitis, and as a disinfectant and healing agent. In small industries, oregano is an ingredient that is added to sauces, dressings, olives, pickles, seasoning pastes, cheeses, pre-cooked soups, packaged beans, chili pickles, and powdered moles. The mixture of oregano with thyme is incorporated into an infinite number of canned foods, sauces, preserves, pizzas, cooked vegetables, salads, and stews, among others. The essential oil contains terpenic hydrocarbons, alcohols, oxides, phenols, ethers, ketones, and esters, recognizing carvacrol, thymol, p-cymene, eugenol, cymene, pinene, and linalool as the ones that appear in the highest percentage (Ortega-Nieblas et al., 2011).

12.4.3 *EUPHORBIA ANTISYPHILITICA ZUCC (CANDELILLA)*

The candelilla plant (*E. antisyphilitica* Zucc) is one of the few plants in the world that contains a high number of hydrocarbons, which facilitates the extraction of natural wax. The discovery and use of this plant for medicinal purposes are attributed to the pre-Hispanic natives of northern Mexico, who boiled the plant's rods in clay pots to extract the wax and placed it on the string of their bows, thus They protected them from temperature changes. Many inhabitants of the desert territories of Mexico depend on the collection of the candelilla plant and its use. The traditional candelilla production system covers approximately 3 million hectares and is the total or partial livelihood of numerous peasants in the states of Coahuila, Zacatecas, Durango, and Chihuahua. Other uses include anticaking agent (release agent), bulking agent, emulsifier, and coating agent. Chemically, among the most reported compounds in candelilla are some organic acids and some lipids that are related to the presence of candelilla wax as mentioned above. In addition, it has been reported that the candelilla plant has many high-quality polyphenols with potential applications in different industries, acting as antimicrobial and antioxidant agents, and providing other technological advantages. Among the compounds already reported we find ellagic acid, gallic acid, lignin, and some tannins (Navarro-Guajardo et al., 2018).

12.4.4 *LARREA TRIDENTATA (GOBERNADORA)*

Larrea tridentata, also known as chaparral, gobernadora or creosote bush, is a shrubby plant that dominates the desert areas of the Mexican states of San Luis Potosí, Coahuila, Chihuahua, Durango, Sonora, Zacatecas, Baja California Norte and Sur, and in the Southwestern United States in Arizona, California, Nevada, Texas, and New Mexico. This plant is a source of natural products with 50% of the leaves as extractable matter said to contain aglycone flavonoids, lignans, some phenolic acids, glycosylated flavonoids, sapogenins, essential oils, halogen alkaloids, terpenoids, mevalonic acid waxes. shikimic. and fatty acid pathways (Abou-Gazar et al., 2004). All the applications that this plant has are attributed to its antioxidant capacity. Due to its varied chemical composition, it has been used as an antiseptic, ointment against burns, acne, psoriasis, hemorrhoids, healing and allergic problems, and various antibiotic, antifungal and antiviral properties. It is also said to have analgesic and anti-inflammatory properties, acts against arthritis, rheumatism and for the treatment of respiratory diseases such as colds, coughs, flu, bronchitis, and tuberculosis (Barragán et al., 1994; Sheikh et al., 1997).

12.4.5 *PROSOPIS SPP. (MESQUITE)*

Another edible-plant from the Chihuahua's semi-desert that is commonly used is Mesquite (*Prosopis* spp.). The fruit from this tree is used as food for both humans and farm animals. *Prosopis* spp. flour has been used for fiber enrichment of bread products (Bigne et al., 2016) and as a functional ingredient in gluten-free products with an important increase in organoleptic properties (Korus et al., 2022). Agave is an important plant found in desertic and semi-desertic regions in Mexico, and it depends on the species for its use, being one of the main uses for beverage production (tequila and mezcal). Chihuahua's semi-desert have species such as *A. lechuguilla* whose flowers are used as food in the Region of Mapimi Durango; agave is commonly used as material for fibers (Reyes-Agüero et al., 2000) and might be an alternative for biofuel production (Carmona and Morales-Martínez, 2017). These reported functional activities results promising for the sustainable use of endemic plants from this semi-desertic region.

12.4.5.1 NUTRACEUTICALS EFFECT

There are several well-documented nutraceuticals with beneficial health effects, semi-desertic plants from Chihuahua are composed of a broad spectrum of bioactive compounds. *Opuntia ficus-indica* is rich in mucilage, vitamin C, gallic acid, 6-isorhamnetin 3-O-robinobioside, isorhamnetin glycosides, feruloyl derivatives, tannins, and sinapoyl diglucoside, kaempferol, quercetin, isoquercetin and ferulic acid (El-Mostafa et al., 2014). *Prosopis* spp. are rich in gallic acid, catechin, gallocatechin, epicatechin gallate, rutin, and luteolin, and have an excellent antioxidant capacity, inhibition of low-density lipoproteins (LDL) oxidation, and cardioprotective potential (Gallegos-Infante et al., 2013; García-Andrade et al., 2013).

A great number of studies describe the bioactivities of these compounds. Quercetin whose antioxidant, anti-inflammatory, and anticancer effects that are well-accepted (Hashemzaei et al., 2017; Xu et al., 2019; Yuan et al., 2020). Other nutraceuticals offering a multi-target approach are gallic and ferulic acids with antibacterial, antioxidant, and anticancer effects reported (Gao et al., 2019; Gao et al., 2018; Ibitoye and Ajiboye, 2019). Catechins have antiviral (Song et al., 2005), antibacterial (Escandón et al., 2016) as well as anticancer effects (Chen et al., 2020). Other types of nutraceuticals with very well-accepted health benefits are prebiotics found in *A. lechuguilla* like inulin, which improves the absorption of vitamin D and E in children with celiac disease (Drabińska et al., 2018; Jiménez-Muñóz et al., 2016). As mentioned above nutraceuticals are effective strategies to treat a wide range of diseases like obesity related to insulin resistance (luteolin) (Zhang et al., 2016), diabetes (kaemferol) (Al-Numair et al., 2015), cancer (Jiang et al., 2017), etc. All those beneficial effects of nutraceuticals highlight the relevance of the multi-target approach for treating complex diseases.

12.5 CONCLUSION

Here we can conclude that the endemic semi-desertic plants of Chihuahua have relevant roles not only as a part of the ecosystem but also as important sources of food, and medicinal materials which may promote the economic development of neighboring communities enhancing quality of life. The endemic plant species of the Mexican semi-desert are a potential source of bioactive ingredients for the food industry; however, further research is still required to apply these products, and thus meet the growing demand of consumers for healthy, natural foods with nutraceutical properties.

KEYWORDS

- **phytochemicals**
- **bioactivities**
- **secondary metabolism**
- **phenolics**
- **sustainable**
- **valorization**

REFERENCES

Abou-Gazar, H.; Bedir, E.; Takamatsu, S.; Ferreira, D.; Khan, I. A. Antioxidant Lignans from *Larrea tridentata. Phytochemistry* **2004**, *65* (17), 2499–2505. https://doi.org/10.1016/j.phytochem.2004.07.009.

Al-Numair, K. S.; Chandramohan, G.; Veeramani, C.; Alsaif, M. A. Ameliorative Effect of Kaempferol, a Flavonoid, on Oxidative Stress in Streptozotocin-Induced Diabetic Rats. *Redox Report* **2015,** *20* (5), 198–209. https://doi.org/10.1179/1351000214Y.0000000117

Arellano-Perusquía, A.; Bañuelos-Hernández, B.; Urquieta-Ramírez, L.; Albarrán-Tamayo, F.; Aguilar-Ruiz, C. A.; González-Márquez, M. A.; Junquera-Martínez, S.; Cruz-Hernández, A. Mexican Indigenous Species with Agroecological Uses. In *Endemic Species*; Cano Carmona, E., Maria Musarella, C., Cano Ortiz, A., Eds.; 1st ed. InTechOpen, 2019; pp 1–10, Vol. 1. https://doi.org/10.5772/intechopen.83400

Aswani, S.; Lemahieu, A.; Sauer, W. H. Global Trends of Local Ecological Knowledge and Future Implications. *PLoS ONE* **2018,** *13* (4). https://doi.org/10.1371/journal.pone.0195440

Balick, M. J. Y.; Cox, P. A. *Plants, People, and Culture: The Science of Ethnobotany*; Scientific American Library: New York, 1996.

Balleza, J. J.; Villaseñor, J. L..Contribución Del Estado de Zacatecas (México) a la Conservación de la Riqueza Florística del Desierto Chihuahuense. *Acta Bot. Mex.* **2011,** *94,* 61–89.

Barragán, S.; Alvarez, G.; Zavalza, A. In *In Vitro Evaluation of the Antifungic Activity of Larrea tridentata Against Human Pathogenic Fungi*, Proceedings of the First Mexican Congress of Ethnobiology, Summaries, Cuernavaca, México, 1994; p 17.

Benz, B. F.; Cevallos, J.; Santana, F.; Rosales, J. Y.; Graf, S. Losing Knowledge about Plant use in the Sierra de Manantlan Biosphere Reserve, Mexico. *Econ. Bot.* **2000,** *54,* 183–191. https://doi.org/10.1007/BF02907821

Bigne, F.; Puppo, M. C.; Ferrero, C. Fibre Enrichment of Wheat Flour with Mesquite (*Prosopis* spp.): Effect on Breadmaking Performance and Staling. *LWT* **2016,** *65,* 1008–1016. https://doi.org/10.1016/j.lwt.2015.09.028

Bruneton, J. *Pharmacognosie: Phytochimie, Plantes Medicinales*; Tec& Doc Lavoisier: Paris, France (No. 581.634 B7) 1993.

Carmona, J. E.; Morales-Martínez, T. Chemical, Structural and Functional Properties of lechuguilla (*A. lechuguilla* Torr.). *Rev. Mex. Cien. Forestales* **2017,** *8* (42), 1–16.

Chen, B. H.; Hsieh, C. H.; Tsai, S. Y.; Wang, C. Y.; Wang, C. C. Anticancer Effects of Epigallocatechin-3-gallate Nanoemulsion on Lung Cancer Cells Through the Activation of AMP-Activated Protein Kinase Signaling Pathway. *Sci. Rep.* **2020,** *10* (1), 1–11. https://doi.org/10.1038/s41598-020-62136-2

Cheynier, V.; Tomas-Barberan, F. A.; Yoshida, K. Polyphenols: From Plants to a Variety of Food and Nonfood Uses. *J. Agric. Food Chem.* **2015,** *63,* 7589–7594. https://doi.org/10.1021/acs.jafc.5b01173

Cloudsley, J. L. *Man and the Biology of Arid Zones*; University Park Press: Baltimore, USA, 1977; pp 255.

DeFelice, S. L. The Nutraceutical Revolution: Its Impact on Food Industry R&D. *Trends Food Sci. Technol.* **1995,** *6* (2), 59–61. https://doi.org/10.1016/S0924-2244(00)88944-X

Díaz, S.; Kattge, J.; Cornelissen, J. H. C.; Wright, I. J.; Lavorel, S.; Dray, S.; Reu, B.; Kleyer, M.; Wirth, C.; Prentice, I. C.; Garnier, E.; Bönisch, G.; Westoby, M.; Poorter, H.; Reich, P. B.; Moles, A. T.; Dickie, J.; Zanne, A. E.; Chave, J.; Zotz, G. The Global Spectrum of Plant form and Function: Enhanced Species-Level Trait Dataset. *Sci. Data* **2022,** *9* (1), 1–18. https://doi.org/10.1038/s41597-022-01774-9

Drabińska, N.; Krupa-Kozak, U.; Abramowicz, P.; Jarocka-Cyrta, E. Beneficial Effect of Oligofructose-Enriched Inulin on Vitamin d and e Status in Children With Celiac Disease on a Long-Term Gluten-Free Diet: A Preliminary Randomized, Placebo-Controlled Nutritional Intervention Study. *Nutrients* **2018,** *10* (11), 1–9. https://doi.org/10.3390/nu10111768

El Akrem, H.; Manaf, A.; Marielle, B.; Moktar, H. The Effects of Solvents and Extraction Method of the Phenolic Contents and Biological Activities In Vitro of Tunisian *Quercus coccifera* L. and *Juniperus phoenicea* L. Fruit Extracts. *Food Chem.* **2007,** *105,* 1126–1134. https://doi.org/10.1016/j.foodchem.2007.02.010

El-Mostafa, K.; El Kharrassi, Y.; Badreddine, A.; Andreoletti, P.; Vamecq, J.; El Kebbaj, M. S.; Latruffe, N.; Lizard, G.; Nasser, B.; Cherkaoui-Malki, M. Nopal Cactus (*Opuntia ficus-indica*) as a Source of Bioactive Compounds for Nutrition, Health and Disease. *Molecules* **2014,** *19* (9), 14879–14901. https://doi.org/10.3390/molecules190914879

Escandón, R. A.; del Campo, M.; López-Solis, R.; Obreque-Slier, E.; Toledo, H. Antibacterial Effect of kaempferol and (−)-epicatechin on *Helicobacter pylori. Eur. Food Res. Technol.* **2016,** *242* (9), 1495–1502. https://doi.org/10.1007/s00217-016-2650-z

Fu, Y.; Wang, Y.; Gao, H.; Li, D.; Jiang, R.; Ge, L.; Tong, C.; Xu, K. Associations Among Dietary Omega-3 Polyunsaturated Fatty Acids, the Gut Microbiota, and Intestinal Immunity. *Mediat. Inflamm.* **2021,** 2021. https://doi.org/10.1155/2021/8879227

Gallegos-Infante, J. A.; Rocha-Guzman, N. E.; Gonzalez-Laredo, R. F.; Garcia-Casas, M. A. Efecto Del Procesamiento Térmico Sobre La Capacidad Antioxidante de Pinole a Base de Vainas de Mezquite (*Prosopis laevigata*). *CYTA - J. Food* **2013,** *11* (2), 162–170. https://doi.org/10.1080/19476337.2012.712057

Gao, J., Hu, J.; Hu, D.; Yang, X. A Role of Gallic Acid in Oxidative Damage Diseases: A Comprehensive Review. *Nat. Prod. Commun.* **2019,** *14* (8). https://doi.org/10.1177/1934578X19874174

Gao, J., Yu, H.; Guo, W.; Kong, Y.; Gu, L.; Li, Q.; Yang, S.; Zhang, Y.; Wang, Y. The Anticancer Effects of Ferulic Acid is Associated With Induction of Cell Cycle Arrest and Autophagy in Cervical Cancer Cells. *Cancer Cell Int.* **2018,** *18* (1), 1–9. https://doi.org/10.1186/s12935-018-0595-y

García, M. *Conocimiento Tradicional De Los Pueblos Indígenas de México y Recursos Genéticos*; CDI: México, 2007.

García-Andrade, M.; González-Laredo, R. F.; Rocha-Guzmán, N. E.; Gallegos-Infante, J. A.; Rosales-Castro, M.; Medina-Torres, L. Mesquite Leaves (*Prosopis laevigata*), A Natural Resource with Antioxidant Capacity and Cardioprotection Potential. *Ind. Crops Prod.* **2013,** *44,* 336–342. https://doi.org/10.1016/j.indcrop.2012.11.030

García-Pérez, E.; Fernando Francisco, C. Á.; Gutiérrez-Uribe, J. A.; García-Lara, S. Revisión de la Producción, Composición Fitoquímica Y Propiedades Nutracéuticas Del Orégano Mexicano. *Rev. Mexicana Cienc. Agríc.* **2012,** *3,* 339–353

Govea-Salas, M.; Morlett-Chávez, J.; Rodriguez-Herrera, R.; Ascacio-Valdés, J. Some Mexican Plants Used in Traditional Medicine. *In Aromatic and Medicinal Plants - Back to Nature*; 2017. https://doi.org/10.5772/66637

Hashemzaei, M.; Far, A. D.; Yari, A.; Heravi, R. E.; Tabrizian, K.; Taghdisi, S. M.; Sadegh, S. E.; Tsarouhas, K.; Kouretas, D.; Tzanakakis, G.; Nikitovic, D.; Anisimov, N. Y.; Spandidos, D. A.; Tsatsakis, A. M.; Rezaee, R. Anticancer and Apoptosis-Inducing Effects of Quercetin In Vitro and In Vivo. *Oncol. Rep.* **2017,** *38* (2), 819–828. https://doi.org/10.3892/or.2017.5766

Henrickson, J.; M. C. Johnston. *A Flora of the Chihuahuan Desert Region*; Published by J. Henrickson: Los Angeles, USA, 2007; p 1695.

Heredia, D.; Green, I.; Klaasen, J.; Rahiman, F. Importance and Relevance of Phytochemicals Present in *Galenia africana. Scientifica* **2022,** *2022,* 1–12. https://doi.org/10.1155/2022/5793436

Huang, Z.; Wang, B.; Eaves, D. H.; Shikany, J. M.; Pace, R. D. Phenolic Compound Profile of Selected Vegetables Frequently Consumed by African Americans in the Southeast US. *Food Chem.* **2007,** *103,* 1395–1402. https://doi.org/10.1016/j.foodchem.2006.10.077

Ibitoye, O. B.; Ajiboye, T. O. Ferulic Acid Potentiates the Antibacterial Activity of Quinolone-Based Antibiotics Against *Acinetobacter baumannii. Microb. Pathog.* **2019,** *126,* 393–398. https://doi.org/10.1016/j.micpath.2018.11.033

Jiang, L.; Wang, W.; He, Q.; Wu, Y.; Lu, Z.; Sun, J.; Liu, Z.; Shao, Y.; Wang, A. Oleic Acid Induces Apoptosis and Autophagy in the Treatment of Tongue Squamous Cell Carcinomas. *Sci. Rep.* **2017,** *7* (1), 1–11. https://doi.org/10.1038/s41598-017-11842-5

Jiménez-Muñóz, E.; Prieto-García, F.; Prieto-Méndez, J.; Acevedo-Sandoval, O. A.; Rodríguez-Laguna, R. Caracterización Fisicoquímica de Cuatro Especies de Agaves Con Potencialidad en la Obtención de Pulpa de Celulosa Para Elaboración de Papel. *DYNA (Colombia)* **2016,** *83* (197), 233–243. https://doi.org/10.15446/dyna.v83n197.52243

Kornberg, H. Metabolism, Encyclopaedia Britannica Online. Encyclopaedia Britannica, Inc, 2019.

Korus, J.; Witczak, M.; Korus, A.; Juszczak, L. Mesquite (*Prosopis* L.) as a Functional Ingredient in Gluten-Free Dough and Bread. LWT **2022,** *168* (April). https://doi.org/10.1016/j.lwt.2022.113957

Leitzmann, C. Characteristics and Health Benefits of Phytochemicals. *Forschende Komplementarmedizin* **2016,** *23* (2), 69–74. https://doi.org/10.1159/000444063

Leopoldini, M.; Russo, N.; Toscano, M. The Molecular Basis of Working Mechanism of Natural Polyphenolic Antioxidants. *Food Chem.* **2011,** *125* (2), 288–306. https://doi.org/10.1016/j.foodchem.2010.08.012

Lizcano, E.; Fernández, J. Saberes Locales y Conocimientos CientíΘcos: El Caso de los "Lagos Asesinos" de Camerún. *Intersticios: Revista Sociol. Pensamiento Crítico* **2014,** *8* (2), 123–138

López-Romero, P.; Pichardo-Ontiveros, E.; Avila-Nava, A.; Vázquez-Manjarrez, N.; Tovar, A. R.; Pedraza-Chaverri, J.; Torres, N. The Effect of Nopal (*Opuntia ficus* Indica) on Postprandial Blood Glucose, Incretins, and Antioxidant Activity in Mexican Patients with Type 2 Diabetes after Consumption of Two Different Composition Breakfasts. *J. Acad. Nutrit. Dietetics* **2014,** *114* (11), 1811–1818. https://doi.org/10.1016/j.jand.2014.06.352

Magalhães, A. S.; Silva, B. M.; Pereira, J. A.; Andrade, P. B.; Valentão, P.; Carvalho, M. Protective Effect of Quince (*Cydonia oblonga* Miller) Fruit Against Oxidative Hemolysis of Human Erythrocytes. *Food Chem. Toxicol.* **2009,** *47,* 1372–1377. https://doi.org/10.1016/j.fct.2009.03.017

Mathur, S.; Hoskins, C. Drug development: Lessons from Nature. *Biomed. Rep.* **2017,** *6* (6), 612–614. https://doi.org/10.3892/br.2017.909

McClung, E.; Martínez, D.; Ibarra, E.; Adriano, C. Los Orígenes Prehispánicos de Una Tradición Alimentaria En La Cuenca de México. *An. Antropol.* **2014,** *48* (1), 97–121.

Munin, A.; Edwards-Lévy, F. Encapsulation of Natural Polyphenolic Compounds; a Review. *Pharmaceutics* **2011,** *3,* 793–829: https://doi.org/10.3390/pharmaceutics3040793.

National Park Service - US. Chihuahuan Desert Ecoregion. Chihuahuan Desert Inventory & Monitoring Network, 2022. https://www.nps.gov/im/chdn/ecoregion.htm

Navarro-Guajardo, N.; García-Carrillo, E. M.; Espinoza-González, C.; Téllez-Zablah, R.; Dávila-Hernández, F.; Romero-García, J.; Ledezma-Pérez, A. Candelilla Wax as Natural Slow-Release Matrix for Fertilizers Encapsulated by Spray Chilling. *J. Renew. Mater.* **2018,** *6* (3), 226.

Nuñez, M.; Paredes, O.; Reynoso, R. Functional and Hypoglycemic Properties of Nopal Cladodes (*O. ficus*-indica) at Different Maturity Stages Using in Vitro and in Vivo Tests. *J. Agric. Food Chem.* **2013,** *61,* 10981–10986.

Ortega-Nieblas, M. M.; Robles-Burgueño, M. R.; Acedo-Félix, E.; González-León, A.; Morales-Trejo, A.; Vázquez-Moreno, L. Chemical Composition and Antimicrobial Activity of Oregano (*Lippia palmeri* S.WATS) ESSENTIAL OIL. *Rev. Fitotec. Mex.* **2011,** *34* (1), 11. https://doi.org/10.35196/rfm.2011.1.11.

Pagare, S.; Bhatia, M.; Tripathi, N.; Pagare, S.; Bansal, Y. K. Secondary Metabolites of Plants and Their Role: Overview. *Curr. Trends Biotechnol. Pharm.* **2015,** *9,* 293–304.

Palomo-Ligas, L.; Flores, S. D. N.; García-Ortiz, J. D.; Pérez-Juárez, C. M.; Flores-Gallegos, A. C.; Rodríguez-Herrera, R.; Cruz-Requena, M. Molecular Effects of Bioactive Compounds from Semi-Desert Plants and Their Uses as Potential Ingredient in Food Products. In *Molecular Mechanisms of Functional Food,* 1st ed.; Campos-Vega, R.; Oomah, B. D., Eds.; John Wiley and Sons Inc, 2022. https://doi.org/10.1002/9781119804055.ch6

Pedrero, E. Cultura CientíΘca y Pseudociencia. *Rev. Cuba. Fís.* **2011,** *28* (1), 77–80.

Rábago-Panduro, L. M.; Martín-Belloso, O.; Welti-Chanes, J.; Morales-de la Peña M. Changes in Bioactive Compounds Content and Antioxidant Capacity of Pecan Nuts [*Carya illinoinensis* (Wangenh. K. Koch)] During Storage. *Rev. Mex. Ing. Quím.* **2020,** *19,* 1439–1452.

Ramírez, C. R. Ethnobotany and the Loss of Traditional Knowledge in the 21st Century. *Ethnobot. Res. Appl.* **2007,** *5*, 245–247. Recuperado de shorturl.at/huF15

Reyes-Agüero, J. A.; Aguirre-Rivera, J. R.; Peña-Valdivia, C. B. Biología y Aprovechamiento de *Agave lechuguilla* Torrey. *Botan. Sci.* **2000,** *88* (67), 75–88. https://doi.org/10.17129/botsci.1626

Reyes-García, V.; Broesch, J.; Calvet-Mir, L.; Fuentes-Peláez, N.; McDade, T. W.; Parsa, S.; Traner, S.; Huanca, T.; Leonard, W. R.; Martínez-Rodríguez, M. R. Cultural Transmission of Ethnobotanical Knowledge and Skills: An Empirical Analysis from an Amerindian Society. *Evol. Hum. Behav.* **2009,** *30*, 274–285. https://doi.org/10.1016/j.evolhumbehav.2009.02.001

Rosado-May, F. J. Las plantas medicinales: síntesis de saberes ancestrales y recursos genéticos resultado de coevolución cultura-naturaleza. In Serralta Peraza, L.E.S., Sierra Aguilar, F., Dzul Batun M., Franco Monsreal, J., Eds.; *Herbolaria Maya, Patrimonio Biocultural para el Mundo. Universidad Intercultural Maya de Quintana Roo y Secretaría de Educación y Cultura de Quintana Roo*; Chetumal, Q. Roo: México, 2014; pp 14–16. ISBN: 978-607-96524-3-2.

Ross, K. A.; Beta, T.; Arntfield, S. D. A comparative study on the Phenolic Acids Identified and Quantified in Dry Beans using HPLC as Affected by Different Extraction and Hydrolysis Methods. *Food. Chem.* **2009,** *113,* 336–344. https://doi.org/10.1016/j.foodchem.2008.07.064

Secretaría de Agricultura y Desarrollo Rural - México. Rompe México Barrera de los 50 Mil Millones de Dólares en Exportaciones Agroalimentarias. Prensa Gobierno de México 2023. https://www.gob.mx/agricultura/prensa/rompe-mexico-barrera-de-los-50-mil-millones-de-dolares-en-exportaciones-agroalimentarias

Segal, L.; Opie, R. S. A nutrition Strategy to Reduce the Burden of Diet Related Disease: Access to Dietician Services Must Complement Population Health Approaches. *Front. Pharmacol.* **2015,** *6* (Aug), 1–7. https://doi.org/10.3389/fphar.2015.00160

SEMARNAT. Norma Oficial Mexicana NOM-059- SEMARNAT-2010. Protección Ambiental. Especies Nativas de México de Flora y Fauna Silvestres-Categorías de Riesgo y Especificaciones Para Su Inclusión, Exclusión o Cambio-Lista de Especies en Riesgo. Diario Oficial de la Federación. *Secretaria de Medio Ambiente y Recursos Naturales (SEMARNAT).* México, D.F.: México. 2010; p 77.

Shahidi, F.; Ambigaipalan, P. Phenolics and Polyphenolics in Foods, Beverages and Spices: Antioxidant Activity and Health Effects-A Review. *J. Funct. Food* **2015,** *18,* 820–897. https://doi.org/10.1016/j.jff.2015.06.018

Sheikh, N. M.; Philen, R. M.; Love, L. A. Chaparral-Associated Hepatotoxicity. *Arch. Intern. Med.* **1997,** *157* (8), 913–919. https://doi.org/10.1001/archinte.1997.00440290099011.

Song, J. M.; Lee, K. H.; Seong, B. L. Antiviral Effect of Catechins in Green Tea on Influenza virus. *Antivir. Res.* **2005,** *68* (2), 66–74. https://doi.org/10.1016/j.antiviral.2005.06.010

Soto-Domínguez, A.; García-Garza, R.; Ramírez-Casas, Y.; Morán-Martínez, J.; Serrano-Gallardo, L. B. El Extracto Acuoso de Orégano (*L. graveolens* HBK) del Norte de México Tiene Actividad Antioxidante sin Mostrar un Efecto Tóxico In Vitro e In Vivo. *Int. J. Morphol.* **2012,** *30* (3), 937–944. https://doi. org/10.4067/S0717-95022012000300029

Towle, M. *The Ethnobotany of Pre-Columbian Peru.*Routledge: New York, **2017**. https://doi.org/10.4324/9781351303965

Tucuch-Pérez, M.; Arredondo-Valdés, R.; Hernández-Castillo, F. D. Antifungal Activity of Phytochemical Compounds of Extracts from Mexican Semi-Desert Plants Against *Fusarium oxysporum* from Tomato by Microdilution in Plate Method. *NovaScientia* **2020,** *12,* 1–19. https://doi.org/10.21640/ns.v12i25.2345

World Health Organization. Noncommunicable Diseases Progress Monitor 2020, 2020. https://www.who.int/news-room/fact-sheets/detail/noncommunicable-diseases

Xu, D.; Hu, M. J.; Wang, Y. Q.; Cui, Y. L. Antioxidant Activities of Quercetin and its Complexes for Medicinal Application. *Molecules* **2019,** *24* (6). https://doi.org/10.3390/molecules24061123

Yuan, K.; Zhu, Q.; Lu, Q.; Jiang, H.; Zhu, M.; Li, X.; Huang, G.; Xu, A. Quercetin Alleviates Rheumatoid Arthritis by Inhibiting Neutrophil Inflammatory Activities. *J. Nutrit. Biochem.* **2020,** *84,* 108454. https://doi.org/10.1016/j.jnutbio.2020.108454

Zhang, L.; Han, Y. J.; Zhang, X.; Wang, X.; Bao, B.; Qu, W.; Liu, J. Luteolin Reduces Obesity-Associated Insulin Resistance in Mice by Activating AMPKα1 Signalling in Adipose Tissue Macrophages. *Diabetologia* **2016,** *59* (10), 2219–2228. https://doi.org/10.1007/s00125-016-4039-8

CHAPTER 13

Biostimulation of Grains and Plant Seeds using Probiotic Microorganisms: Shaping the Production of Sustainable and Healthy Foods for the Future

STEPHANY NEFERTARI CHÁVEZ-GARCÍA, DORA ELISA CRUZ-CASAS, ERIK DE LEÓN-JUÁREZ, ÁLVARO JAVIER PASTRANA-PASTRANA, STEFFANNY SÁNCHEZ-PORTILLO, and ADRIANA CAROLINA FLORES-GALLEGOS*

Food Research Department, School of Chemistry, Saltillo, Coahuila, México

ABSTRACT

Plants are a food source that is increasingly in demand in the current society. Due to various factors such as their high nutritional value or sustainability. For this reason, we are currently seeking to produce these plants from their grains and seeds with green technology. Biostimulation is a technique with several advantages that promotes the effective development of the plant. It is already known that the use of microorganisms is a useful, cheap, and potential tool for plant growth. However, there are microorganisms, such as probiotics, which are known to have multiple beneficial applications for consumer health. However, there are only a few reports that discuss how probiotics improve plant growth, as well as improve the nutritional quality of the plant. And also, they have positive effects on the consumer. Therefore, it is important to know the role of the probiotic biostimulation method. Describe what this method consists of and compare it with others. The beneficial biostimulation effects of chitin on facilitating plant growth

Sustainable Engineering and Agro-Food Processing: Renewable and Nonrenewable Resources.
José Juan Buenrostro-Figueroa, Julio César Tafolla-Arellano, Mónica Chávez-González, Pedro Aguilar-Zárate, A. K. Haghi (Eds.)

and improving stress tolerance have been found. As a biostimulant, plant growth-promoting bacteria (PGPB) have been widely used in horticultural plants to promote their growth. PGPB can improve plant nutrient availability, promote the release of plant hormones, and enhance abiotic stress resistance by releasing carriers, organic acids, and enzymes. In addition to explaining the mechanism by which probiotics promote plant growth. Microorganisms, including bacteria, yeast, fungi, and microalgae have been proven to have biostimulation activity. Many plant-associated microbes can synthesize plant hormones such as auxin, ethylene, and cytokinins which have crucial multi-faceted roles in plants. Moreover, some bacteria can produce abscisic acid and gibberellins; inoculation with these microbes helps plants resist stresses. And how the use of these microorganisms affects the nutritional quality of plants.

13.1 INTRODUCTION

Sprouting is a bioprocessing technique used by the food industry where the nutrients contained in the seeds are used to give rise to the growth of a sprout (Zhang et al., 2020; Samec et al., 2018). The best use of nutrients comes from the enzymatic activity that begins from the soaking of the seed, unfolding the larger nutrients into simple molecules such as amino acids, simple sugars, and other nutritional compounds (Liu et al., 2019). In addition to reducing the content of antinutritional components present in grains and seeds, making these foods more digestible and with better sensory attributes (Ohanenye et al., 2020), considering them as a rich source of phenolic compounds, vitamins, and minerals (Mekky et al., 2020). The current trend in the consumption of healthy foods has led to an increase in the production of sprouts, being considered a functional food with beneficial properties for the consumer. Today's research is focused on the study of the effects of the germination process as part of biostimulation in plants, changes in nutritional composition, and phytochemical profile. Therefore, research has focused on studying the effects that germination has on the nutritional composition and phytochemical profile (Galieni et al., 2020).

The germination process consists of three stages: soaking, enzyme activation, and growth. During each of its phases there is a series of factors that influence the growth and nutritional content of the plant. Soaking, also known as imbibition, has been reported as one of the most important stages for the reduction of antinutritional components, in addition to the fact that it has been reported that the substance in which the seed is soaked can influence the metabolism of the seed. seed (Vötterl et al., 2019). One of the substances in which favorable changes have been seen in the development of the plant are probiotics; which are live microorganisms that when administered in the right amount can provide a health benefit to the host (Markowiak and Śliżewska, 2017). This strategy would allow probiotics to be considered as biostimulatory substances, in addition to developing a food of excellent nutritional quality and with beneficial effects on consumer health (Świeca et al., 2019). The objective of this work was to study the effect of sprouting on biostimulation y nutritional quality and the probiotic effect of sprouts.

13.2 PROBIOTICS

The most widely accepted definition of probiotics is the one suggested by the Food and Agriculture Organization of the United Nations/World Health Organization in 2002. It defines them as "live microorganisms which, when administered in appropriate dosage, provide health benefits to the host."

The beneficial effects that probiotics have been shown to have on the health of the consumer include immunological, antiviral, antidiabetic, and anticancer activity; it restore the microbial balance in the human intestine, prevent and treat diarrhea caused by various factors, intestinal diseases, irritable bowel syndrome, among others (Cruz Casas et al., 2021).

The most commercialized probiotic microorganisms are those belonging to the genera *Lactobacillus*, *Bifidobacteria*, *Lactococcus*, *Streptococcus*, and *Enterococcus*. In addition to some strains of *Bacillus* and *Saccharomyces* (Markowiak and Ślizewska, 2017). However, LAB are considered the most important probiotic species. They are Gram-positive, catalase-negative, aerotolerant, fastidious, strictly fermentative, nonspore-forming organisms, among others. They are mainly isolated from fermented dairy products (yogurt, kefir, and cheese) and silage (cabbage) (Axelsson and Ahrné, 2000). Probiotics are also characterized by producing compounds of high importance such as bacteriocins, hydrogen peroxide, diacetyl, lactic acid, acetic acid, aromatic compounds, exopolysaccharides, as well as proteolytic and lipolytic capacity (Vivek et al., 2023). Probiotics belong to a multibillion-dollar industry in the food area and are highly consumed worldwide as food supplements. In the food industry, to be considered a probiotic microorganism, it must be metabolically stable and active in the food, survive in the digestive tract and have beneficial effects in the host intestine and transfer them to the consumer (Chandramouli et al., 2004; Vivek et al., 2023).

In other areas, the importance of probiotics has also increased, such as the agricultural industry. These microorganisms have been identified as plant growth promoters. In recent years, the use of microorganisms to stimulate plant growth has become more involved. They are an environmentally friendly alternative (Sanjuán et al., 2023).

Probiotic microorganisms have multiple applications in addition to those beneficial to human health. It is essential to apply these beneficial microorganisms in other areas, such as agriculture, where sustainable production can be achieved and the consumption of these plants would be beneficial for the consumer.

13.3 APPLICATIONS

13.3.1 HEALTH BENEFITS

The probiotic bacteria promote some benefits on host health when is provided as live supplements, many of the effects of probiotics depend on the interaction with dendritic cells, this is possible due to the capacity of polarizing the adaptative immune response. The maturity of these cells and the cytosine production depend on the probiotic strain, some examples of this are the stimulation and synthesis of interleukin 10 (IL-10) and IL-12 cytosines, producing an immune response (Vizcaíno et al., 2016). Probiotics can also prevent the overgrowth of dangerous pathogenic bacteria improving the epithelial barrier function or increasing the resistance of the gut to invasions (FAO, 2002). Specific strains have specific effects on the host health, including neurological and endocrinological effects, antimicrobial activities, and production of specific bioactive substances, as long as the ones described previously in this text (Hill et al., 2014).

Some studies have shown differences in microbiota between healthy and allergic patients, and despite few studies have found positive results using probiotics to treat allergies recent research has shown that a combination of *Bifidobacterium breve* M-16V with other probiotics reduces significantly the Th2 cytosine production (Tang et al., 2009), and other studio claims that the administration of *Lactobacillus rhamnosus* GG during the last 4 weeks of pregnancy and the 3 months after birth improved the immune protection provided by breastmilk (Vizcaíno et al., 2016). Talking about breastmilk, contains an important quantity of nondigestible carbohydrates, which stains in the gastrointestinal tract as a substratum to microbiota like Bifidobacterium, these bacteria affect microbiota in the first year of life and have been observed the highest concentration in a healthy children in comparison to the overweight child (Vael and Desager, 2009; WHO, 2013).

13.3.2 AGRICULTURE

The use of probiotics on plants is not a human invention, plants are always accompanied by microorganisms known as plant probiotic microorganisms (PPM), and they have a symbiotic association with plants (Picard et al., 2008). These microbes support the vegetable host stimulating growing, increasing yield, reducing pathogen infections, and abiotic/biotic and salt stress (Vandenberghe et al., 2017). Some of the most studied are the microorganisms that establish an endosymbiotic relationship with the plant such as *Fusarium oxysporum*, *Alternaria* sp., *Phoma*, *Aspergillus* spp., *Penicillium oxalicum*, and *Penicillium simplicissimum*, which have demonstrated their growth-promoting capacity, this positions probiotics as an alternative to the use of chemical fertilizers (Moreno-Gavíra et al., 2020). However, not only fungi are used as probiotics, lactic acid bacteria are mostly utilized in fermented. They can be considered a biocontroller, thanks to their activity against pathogens and the generally recognized as safe (GRAS) status (Tsuda et al., 2016). Probiotics are located in soil too, and due to the microflora of soil mediating many processes in plants, the improve of soils with probiotics is an alternative for fertilization with no adverse effects (Bhardwaj et al., 2014).

13.4 BIOSTIMULANTS IN PLANTS

One of the most challenging of agriculture today is to increase crop production to provide food for the world's population. Chemicals such as pesticides and fertilizers increase crop yields. However, this practice has significant implications for production costs and raises environmental concerns about water, land, and air resources. For agricultural productivity to be achieved, new approaches that promote more sustainable production are needed. An alternative approach in agriculture is biostimulants (Pandey et al., 2022).

The term "biostimulant" is used in agriculture to describe a formulation of components, substances, or microbes that stimulate growth, yield, and stress tolerance in plants and crops (White et al., 2021). A biostimulant can be classified into five main groups: seaweed and plant extracts, humic substances, hydrolyzed proteins and nitrogen-containing compounds, microorganisms, and inorganic compounds with biostimulant action. Each of them with its specific effect on plants (Franzoni et al., 2022).

Biostimulants are substances capable of increasing crop growth or tolerance to adverse environmental conditions, reducing disease incidence, increasing soil nutrient availability, improving plant nutrient use efficiency, and promoting the degradation and humification of organic substances in soils. The use of biostimulants has been increasingly integrated into agricultural production systems such as seed treatment, plant growth, and harvesting to enhance plant productivity by altering physiological processes. It is difficult to identify the mechanisms of action or biochemical pathways involved in the physiological processes induced by biostimulants. In part, this is because the raw materials used in biostimulant production are heterogeneous and mixtures are complex. However, a hypothesis is that plant metabolism is affected by biostimulants, increasing growth potential, development, and stress response (Yakhin et al., 2017). On the contrary, a projected USD 4.14 billion by 2025 is the projected growth rate for the global biostimulants market. The use of these components in agriculture is trending and becoming increasingly important. However, biostimulants cannot solve all agricultural problems with a one-size-fits-all approach. Depending on the climatic conditions, the type of crops, and the dosage, the efficacy of these compounds can vary (Antón-Herrero et al., 2023).

13.4.1 HYDROPONIC METHOD

The growth of the world's population primarily occurs in urban areas, which means that agricultural soil will compete with the urban center for land and water, which will mean more food need to be produced on a smaller area of land in the future. Hydroponics is the science of growing crops without the need for soil, it seeks to generate an optimal environment for plant growth through an optimal ratio of nutrients and climate control. Hydroponic systems are efficient in the use of water and nutrients because they allow their reuse, which contributes to a reduction in the consumption of water and fertilization compared to conventional agriculture. In addition, hydroponic systems do not present any problems when it comes to diseases, pests, or weeds that originate in the soil (Khan et al., 2020).

Numerous crops can be grown hydroponically, including beans, corn, cucumbers, lettuce, tomatoes, carrots, rice, pepper, spinach, strawberries, and potato. Hydroponics has become more and more flexible due to the development of new materials and equipment. Examples are the wick system, drip irrigation, aeroponics, and nutrient film techniques. Hydroponic systems offer technological opportunities in regions with poor agronomic conditions such as deserts and infertile lands, with the advantage that hydroponics is highly productive and allows process automation. However, the high initial cost of hydroponic systems is the main disadvantage of this crop production model. Constant control is required because many pathogens can grow in hydroponic conditions due to nutrient concentrations, the most common microorganisms being *Fusarium*, *Phytophthora*, and *Pythium*. The use of probiotics as a biocontrol alternative has emerged in recent years. Traditional chemical and physical disinfection procedures can be replaced by this method. Moreover, probiotics could improve the crop's nutraceutical properties and productivity. But problems arising from the lack of understanding of the mechanism that prevents damage by plant pathogens need to be addressed (Lee and Lee, 2015; Sambo et al., 2019).

13.4.2 PROBIOTIC METHODS

In nature, there are specific microbial communities associated with plants, both externally and internally within their tissues, which are known as the plant microbiome. These communities include rhizosphere and endosphere microorganisms. PPM are microorganisms, usually bacteria, that promote plant growth through a symbiotic or free-living relationship. This approach shows promise as an alternative to chemical fertilizers and pesticides (Menendez and Garcia-Fraile, 2017).

PGPB is one of the most common groups of PPM in soil. PGPB are constituted by different microorganisms (bacteria, fungi, yeasts, and microalgae) that can grow in, on, or around plant tissues, which promotes growth, enhances resistance to abiotic stress, or increases plant nutrient availability through direct or indirect mechanisms. Direct mechanisms associated with nutrient enhancement include biological nitrogen fixation, phosphate solubilization, and hormone synthesis. Indirect mechanisms are related to plant defense mechanisms against phytopathogens. Although nitrogen accounts for approximately 80% of the earth's atmosphere, neither plants nor animals have an easy way to obtain sufficient nitrogen for their growth. This situation results because molecular nitrogen is chemically very stable and therefore unusable for most biological organisms. It must, therefore, be fixed before it can be assimilated. Microbial processes, therefore, play a crucial role in supplying the nutrients for growth. Only a few prokaryotes can carry out biological nitrogen fixation (BNF), among which are genera of *Azospirillum*, *Azotobacter*, *Bacillus*, and *nitrobacter*. Phosphorus is the second most relevant nutrient for plants after nitrogen, genera such as *Pseudomonas*, *Bacillus*, *Paenibacillus*, *Deftia*, *Azotobacter*, *Klebsiella*, and *Flavobacterium* have been reported as efficient phosphorus solubilizers. Phytohormone biosynthesis could be involved in plant pathogenesis. A wide spectrum of bacteria can produce phytohormones such as auxins, cytokinins, and gibberellins. Likewise, some microbial strains are capable of producing siderophores, organic compounds capable of sequestering ferric iron Fe^{3+}, which increases the availability of the nutrient to the plant. Some endophytic strains are *Micrococcus* and *Streptomyces*. On the contrary, lactic acid bacteria are extensively employed in probiotic and fermented products and can be regarded as biocontrollers as they possess the ability to combat pathogenic microorganisms Overall, understanding the roles of different bacteria in nutrient cycling and hormone production can lead to more sustainable and efficient agricultural practices (de Souza Vandenberghe et al., 2017; Menendez and Garcia-Fraile, 2017; Mishra et al., 2021).

13.4.3 OTHER SUSTAINABLE METHODS

Other techniques to improve soil quality, growth, and food production are bioremediation and composting.

Bioremediation: Soil contamination by heavy metals generates problematic related to damage to soil quality and fertility, leading to reduced production. An alternative to address this technology through biological means is the use of microorganisms. Microorganisms possess inherent mechanism to degrade a variety of contaminant component and remove it from the environment. Microorganisms have advantages compared to traditional methods of contaminant control, it is a simpler technique, easy to implement, economical, and environmentally friendly because it does not produce additional toxic waste (Raklami et al., 2022). Compost: Composting is a biochemical process based on the action of aerobic microorganisms with the ability to degrade organic matter into a more stable material. The product obtained seeks to improve physical (e.g., porosity and moisture retention capacity), chemical (e.g., pH and cation exchange capacity), and microbiological properties that favor a physicochemical and biological balance for proper plant-soil interaction (Sánchez et al., 2017).

13.5 PROBIOTIC MECHANISM IN THE GROWING OF PLANTS

The soil is one of the nonrenewable natural resources that is used as a medium for the development of food crops, being the sustenance of food for humans and animals. Its main function is to be the main source of substrates such as organic and mineral components, water retention, and the basis of the nutritional contribution for microbial growth to provide a suitable environment for plant development (Belorkar, 2021). The micro-organisms in the phytomicrobiome provide many benefits to the plant for the acquisition of nutrients, improvement in soil fertility, and release and modulation of metabolites such as hormones, antimicrobial compounds, and other compounds that suppress the development of phytopathogens. Bacteria, fungi, and other microorganisms present in the soil are responsible for the decomposition of organic matter for the transformation of chemical substances into nutrients used by the plant. Some of the microorganisms work through a symbiosis together with the plant, modifying its metabolism, and causing a greater availability of micronutrients in the soil made available for use by the plant. In addition to playing an important role in tolerance to drought, soil acidity and salinity, and control of the development of pathogenic microorganisms that limit plant growth (Shalini et al., 2023). There are a number of microorganisms that have been studied for their nitrogen-fixing capacity and for the transformation of phosphorus into its simplest form for plant use (they participate in P solubilization, mineralization, absorption, and transport processes). Today the application of microorganisms within sustainable agriculture as biostimulants is a new strategy offered by the agro-economy, with many advantages over previously known stimulants. Among the advantages offered using this technique is tolerance to biotic and abiotic stress for the production of excellent quality crops and meeting one of the objectives of the WHO 2030 agenda for food security. Biostimulants are a replacement for agrochemicals that have detrimental effects on the environment and on the health of humans and animals (Singh et al., 2022). Despite the advantages offered, there are certain limitations since agriculture and the green strategy that it contemplates are not fully established in the crops of each country and the regulations have not considered implementing this type of technology. On the contrary, another consortium of microorganisms used are probiotics that act as biostimulants and are applied to the microbiome of crops to improve the development and growth of the plant, in addition to increasing the immune response (Mojumdar et al., 2022). The development of endophyte-based bioformulations such as probiotics that will modulate different mechanisms involved in the probiotic–plant interaction, such as the acquisition of nutrients, secretion of secondary metabolites, siderophores, and production of phytohormones and an increase in defense mechanisms in response to abiotic and biotic stress (Aamir et al., 2020). One of the best methods to incorporate microorganisms into the plant system is from bioformulations based on inert carriers (gums, methylcellulose, starch), providing stability for the plant and its secondary metabolites. As aforementioned, current agriculture seeks to counteract the effect of phytopathogens from the use of biostimulants (Mishra et al., 2022).

13.6 FORWARD PERSPECTIVES

Currently, probiotics are used for the development of products that help with the treatment of different human health diseases (dysgestive, depressive, among others). However, the use of these products is expanding every day since it is known the benefits they can provide in other areas of application such as crops, apart from helping against the spread of pathogenic microorganisms that affect these plants, they are attributed to the ability to transform organic matter, thus benefiting the development of the plant with some properties mentioned in previous sections. Among the benefits reported so far, some plant growth-stimulating rhizobacteria stand out (Ramna et al., 2021; García-Fraile et al., 2017; Chabot et al., 1996). Due to the potential that has been found in research on the effects that these microorganisms have, research has been promoted in agricultural areas in products such as biofertilizers (Rana et al., 2020). It has also been shown that there are some bacteria that not only offer benefits in terms of production but also improve the nutritional composition, which provides for this field of application the use of probiotics with dual purpose, combat diseases both in health and in the plant, improve crop development through increased biomass, production, shorten production time in food production such as sprouted products and improve the characteristics of the final products in terms of the contribution of nutrition of consumers (Le et al., 2018; Shanta et al., 2016).

For this reason, it is important to study in depth the uses of probiotics, the different fields of application, whether in the food, pharmaceutical, agricultural, and other industries, as well as the technologies that can be used to promote their application improving production costs and more environmentally friendly, identify the bioavailability that exists of these naturally and/or isolated; This also helps plants to present resistance to extreme conditions, which allows the generation of new trends and thus with the use of probiotics from this area to generate agricultural holdings in this way a contribution to new trends in sustainability.

KEYWORDS

- **sustainable production**
- **bioprocess**
- **sprouting**
- **health benefits**
- **biostimulation**

REFERENCES

Aamir, M.; Kumar Rai, K.; Zehra, A.; Kumar Dubey, M.; Kumar, S.; Shukla, V.; Ram, S. 8—*Microbial Bioformulation-Based Plant Biostimulants: A Plausible Approach Toward Next Generation of Sustainable Agriculture*;. **2020;** pp 195-225, ISBN 9780128196540 https://doi.org/10.1016/B978-0-12-819654-0.00008-9.

Antón-Herrero, R.; García-Delgado, C.; Antón-Herrero, G.; Mayans, B.; Delgado-Moreno, L.; Eymar, E. Design of a Hydroponic test to Evaluate the Biostimulant Potential of New Organic and Organomineral Products. *Sci. Hortic.* **2023,** *310* (December 2022). https://doi.org/10.1016/j.scienta.2022.111753

Axelsson, L.; Ahrné, S. Food and Medicine. *Appl. Microb. Syst.* **2000,** *81,* 367–388. https://doi.org/10.1108/eb058860

Belorkar, S. A. *Biostimulants: Promising Probiotics for Plant Health, Biocontrol Agents and Secondary Metabolites*; Woodhead Publishing, 2021; 469–481, ISBN 9780128229194, https://doi.org/10.1016/B978-0-12-822919-4.00020-X.

Bhardwaj, D.; Ansari, M. W.; Sahoo, K.; et al. Biofertilizers Function as Key Player in Sustainable Agriculture by Improving Soil Fertility, Plant Tolerance and Crop Productivity. *Microb. Cell Fact.* **2014,** *13,* 1–10.

Chabot, R.; Antoun, H.; Cescas, M. P. Growth Promotion of Maize and Lettuce by Phosphate solubilizing *Rhizobium leguminosarum* Biovar Phaseoli. *Plant Soil* **1996,** *184,* 311–321

Chandramouli, V.; Kailasapathy, K.; Peiris, P.; Jones, M. An Improved Method of Microencapsulation and its Evaluation to Protect *Lactobacillus* spp. in Simulated Gastric Conditions. *J. Microbiol. Methods* **2004,** *56,* 27–35. https://doi.org/10.1016/j.mimet.2003.09.002

Cruz Casas, D. E.; Cázares Vásquez, M. L.; García Flores, L. A.; Lara Salas, M. A.; Aguilar, C. N.; Rodríguez Herrera, R.; Flores Gallegos, A. C. Probiotics as Functional Foods. In *Advances in Probiotics for Sustainable Food*; Springer Nature, 2021**;** pp 121–148. https://doi.org/10.1007/978-981-15-6795-7_6

de Souza Vandenberghe, L. P.; Blandon Garcia, L.; Rodrigues, C. M.; Cândido Camara, M.; Vinícius de Melo Pereira, G.; de Oliveira, J.; Soccol, C. R. Potential Applications of Plant Probiotic Microorganisms in Agriculture and Forestry. *AIMS Microbiol.* **2017,** *3* (3), 629–648. https://doi.org/10.3934/microbiol.2017.3.629

Food and Agricultural Organization of the United Nations and World Health Organization: Joint FAO/WHO Working Group Report on Drafting Guidelines for the Evaluation of Probiotics in Food. Food and Agricultural Organization of the United Nations, 2002.

Franzoni, G.; Cocetta, G.; Prinsi, B.; Ferrante, A.; Espen, L. Biostimulants on Crops: Their Impact Under Abiotic Stress Conditions. *Horticulturae* **2022,** *8* (3). https://doi.org/10.3390/horticulturae8030189

Hill, C.; Guarner, F., Reid, G.; Gibson, G. R.; Merenstein, D. J.; Pot, B.; Morelli, L.; Canani, R. B.; Flint, H. J.; Salminen, S.; Expert Consensus Document. The International Scientific Association for Probiotics and Prebiotics consensus statement on the scope and appropriate use of the term probiotic. *Nat. Rev. Gastroenterol. Hepatol.* **2014,** *11,* 506–514.

Jiménez-Gómez, A.; Menéndez, E.; Flores-Félix, J. D.; García-Fraile, P.; Mateos, P. F.; Rivas R. Effective Colonization of Spinach Root Surface by *Rhizobium.* In *Biological Nitrogen Fixation and Beneficial Plant-Microbe Interaction*; González-Andrés, F., James, E., Eds.; Springer: Cham, 2016; pp 109–122

Khan, S.; Purohit, A.; Vadsaria, N. Hydroponics: Current and Future State of the art in Farming. *J. Plant Nutrit.* **2020,** *44* (10), 1515–1538. https://doi.org/10.1080/01904167.2020.1860217

Le, T. A.; Pék, Z, Takács, S.; Neményi, A.; Daood, H. G.; Helyes, L. The Effect of Plant Growth Promoting Rhizobacteria on the Water-Yield Relationship and Carotenoid Production of Processing Tomatoes. *Hortscience* **2018,** *53,* 816–822. https://doi.org/10.21273/hortsci13048-18

Lee, S.; Lee, J. Beneficial Bacteria and Fungi in Hydroponic Systems: Types and Characteristics of Hydroponic Food Production Methods. *Sci. Hortic.* **2015,** *195*, 206–215. https://doi.org/10.1016/j.scienta.2015.09.011

Markowiak, P.; Ślizewska, K. Effects of Probiotics, Prebiotics, and Synbiotics on Human Health. *Nutrients* **2017,** *9*. https://doi.org/10.3390/nu9091021

Menendez, E.; Garcia-Fraile, P. Plant Probiotic Bacteria: Solutions to Feed the World. *AIMS Microbiol.* **2017,** *3* (3), 502–524. https://doi.org/10.3934/MICROBIOL.2017.3.502

Mishra, P.; K. Tiwari, S. K.; Singh, A. K.; Malik M. Z. Chapter 8 - Harnessing Endophytic Microbial Diversity for Stress Alleviation: Current Perspectives for Crop Improvement **2022,** 149–174, ISBN 9780323905688, https://doi.org/10.1016/B978-0-323-90568-8.00008-0.

Mishra, P.; Mishra, J.; Arora, N. K. Plant Growth Promoting Bacteria for Combating Salinity Stress in Plants – Recent Developments and Prospects: A Review. *Microbiol. Res.* **2021,** *252* (June), 126861. https://doi.org/10.1016/j.micres.2021.126861

Mojumdar, A.; Tanaya Behera, H.; Das, S.; Ray, L. CHAPTER 13 - Microbe-based Plant Biostimulants and Their Formulations for Growth Promotion and Stress Tolerance in Plants, In *New and Future Developments in Microbial Biotechnology and Bioengineering*; 2022; pp 213–230. ISBN 9780323851633, https://doi.org/10.1016/B978-0-323-85163-3.00004-1.

Moreno-Gavíra, A.; Diánez, F.; Sánchez-Montesinos, B.; Santos, M. *Paecilomyces variotii* as A Plant-Growth Promoter in Horticulture. *Agronomy* **2020,** *10* (4):597. https://doi.org/10.3390/agronomy10040597

Organización Mundial de la Salud. Estrategia Mundial para la Alimentación del Lactante y el Niño Pequeño. *Ginebra* [Online] 2002. Disponible en :http://www.who.int/nutrition/publications/gs_infant_feeding_text_spa.pdf.

Pandey, S. S.; Jain, R.; Bhardwaj, P.; Thakur, A.; Kumari, M.; Bhushan, S.; Kumar, S. Plant probiotics – Endophytes pivotal to plant health. *Microbiol. Res.* **2022,** *263* (April), 127148. https://doi.org/10.1016/j.micres.2022.127148

Picard, C.; Baruffa, E.; Bosco, M. Enrichment and Diversity of Plant-probiotic Microorganisms in the Rhizosphere of Hybrid Maize During four Growth Cycles. *Soil Biol. Biochem.* **2008,** *40,* 106–115.

Raklami, A.; Meddich, A.; Oufdou, K.; Baslam, M. Plants—Microorganisms-Based Bioremediation for Heavy Metal Cleanup: Recent Developments, Phytoremediation Techniques, Regulation Mechanisms, and Molecular Responses. *Int. J. Mol. Sci.* **2022,** *23* (9). https://doi.org/10.3390/ijms23095031

Ramna, Z.; Malik Nawaz, S.; Muhammad, A.; Muhammad Sohail, A. Sustainable Development and Biodiversity 27 Soil Microbiomes for Sustainable Agriculture Functional Annotation 2021. https://doi.org/https://doi.org/10.1007/978-3-030-73507-4_13

Rana, K. L.; Kour, D.; Kaur, T.; Devi, R.; Yadav, A. N.; Yadav N.; et al. Endophytic Microbes: Biodiversity, Plant Growth-Promoting Mechanisms and Potential Applications for Agricultural Sustainability. *Antonie Van Leeuwenhoek* **2020,** *113,* 1075–1107. https://doi.org/10.1007/s10482-020-01429-y

Sambo, P.; Nicoletto, C.; Giro, A.; Pii, Y.; Valentinuzzi, F.; Mimmo, T.; Lugli, P.; Orzes, G.; Mazzetto, F.; Astolfi, S.; Terzano, R.; Cesco, S. Hydroponic Solutions for Soilless Production Systems: Issues and Opportunities in a Smart Agriculture Perspective. *Front. Plant Sci.* **2019,** *10* (July). https://doi.org/10.3389/fpls.2019.00923

Sánchez, Ó. J.; Ospina, D. A.; Montoya, S. Compost Supplementation with Nutrients and Microorganisms in Composting Process. *Waste Manag.* **2017,** *69* (26), 136–153. https://doi.org/10.1016/j.wasman.2017.08.012

Sanjuán, J.; Nápoles, M. C.; Pérez-Mendoza, D.; Lorite, M. J.; Rodríguez-Navarro, D. N. Microbials for Agriculture: Why Do They Call Them Biostimulants When They Mean Probiotics? *Microorganisms* **2023,** *11,* 1–10. https://doi.org/10.3390/microorganisms11010153

Shalini, R.; Manoj, K. S.; Anjali, C. S.; Swarnmala, S. Chapter 17 - Microbial Endophytes as Probiotics for the Plant Health: An Overview **2023,** 269–281, ISBN 9780323906203, https://doi.org/10.1016/B978-0-323-90620-3.00011-8.

Shanta, N.; Schwinghamer, T.; Backer, R.; Allaire, S. E.; Teshler, I.; Vanasse, A.; Whalen, J.; Baril, B.; Lange, S.; MacKay, J.; Zhou, X.; Smith, D. L. Biochar and Plant Growth Promoting Rhizobacteria Effects on Switchgrass (*Panicum virgatum* cv. Cave-in-rock) for Biomass Production in Southern Québec Depend on Soil Type and Location. *Biomass Bioenergy* **2016,** *95,* 167–173. https://doi.org/10.1016/j.biombioe.

Singh, A. L.; Rani, K.; Jena, R.; Kona, P.; Reddy, K.; Gangadhara, K. CHAPTER 8 - Microbes-based Bio-stimulants Towards Sustainable Oilseeds Production: Nutrient Recycling and Genetics Involved **2022,** 111–130, ISBN 9780323851633, https://doi.org/10.1016/B978-0-323-85163-3.00013-2.

Tang, M. L. Probiotics and Prebiotics: Immunological and Clinical Effects in Allergic Disease. *Nestle. Nutr. Workshop Ser Pediatr Program* **2009,** *64,* 219–257

Tsuda, K.; Tsuji, G.; Higashiyama, M.; et al. Biological Control of Bacterial Soft rot in Chinese Cabbage by *Lactobacillus plantarum* Strain BY under field conditions. *Biol. Control.* **2016,** *100,* 63–69.

Vael, C.; Desager, K. The Importance of the Development of the Intestinal Microbiota in Infancy. *Curr. Op. Ped.* **2009,** 21 (6): 794–800.

Vandenberghe, L.; Garcia, L. M. B.; Rodrigues, C.; Camara, M. C.; Pereira, G.; De Oliveira, J.; Soccol, C. R. Potential Applications of Plant Probiotic Microorganisms in Agriculture and Forestry. *AIMS Microbiol.* **2017,** *3,* 629–648.

Vivek, K.; Mishra, S.; Pradhan, R. C.; Nagarajan, M.; Kumar, P. K.; Singh, S. S.; Manvi, D.; Gowda, N. N. A Comprehensive Review on Microencapsulation of Probiotics: Technology, Carriers and Current Trends. *Appl. Food Res.* **2023,** *3,* 100248. https://doi.org/10.1016/j.afres.2022.100248

Vizcaíno, R.; Macias-Tomei, C.; Márquez, S.; Julio, C.; Morales, A.; Torres, N. USOS CLÍNICOS DE LOS PROBIÓTICOS. *Archivos Venezolanos de Puericultura y Pediatría* **2016,** *79* (1), 029–040. Recuperado en 30 de abril de 2023, de http://ve.scielo.org/scielo.php?script=sci_arttext&pid=S0004-06492016000100007&lng=es&tlng=es.

White, J. F.; Chang, X.; Kingsley, K. L.; Zhang, Q.; Chiaranunt, P.; Micci, A.; Velazquez, F.; Elmore, M.; Crane, S.; Li, S.; Lu, J.; Molina, M. C.; González-Benítez, N.; Beltrán-García, M. J.; Kowalski, K. P. Endophytic Bacteria in Grass Crop Growth Promotion and Biostimulation. *Grass Res.* **2021,** *1* (1), 1–9. https://doi.org/10.48130/gr-2021-0005

Yakhin, O. I.; Lubyanov, A. A.; Yakhin, I. A.; Brown, P. H. Biostimulants in Plant Science: A Global Perspective. *Front. Plant Sci.* **2017,** *7* (January). https://doi.org/10.3389/fpls.2016.02049

Index

A

Abiotic stress tolerance, 156
  abscisic acid (ABA), 157
  1-aminocyclopropane-1-carboxylate deaminase (ACCD), 158
  ethylen (ET), 157
  hormone regulation, 157–158
  monodehydroascorbate reductase (MDAR), 158
  phenylalanine ammonia lyase (PAL), 158
  physiological adaptation, 159–160
Abscisic acid (ABA), 157
Agitated bioreactors (ABs), 237
Agrobacterium infection, 255
Agro-industrial waste, 7–8
  accumulation and poor disposal, 32–33
  biorefineries, 11
  classification, 27–28
  coffee bean, 28–29
  feedstock, 9–10
  generation of, 25–27
  grounds, 30–32
  integrated biorefinery systems, 10
  pulp, 30
  revalorization of, 33–36
Alkaloids, 37
1-Aminocyclopropane-1-carboxylate deaminase (ACCD), 158
Ancestral knowledge, 264
  ethnobotany, 265
  medicinal plants, 265
  secrets, 265
Anthocyanins, 40–41
Antioxidant activities, 103–106
Artocarpus, 116–117
Artocarpus altilis fosberg, 133
Artocarpus heterophyllus Lam., 131–132
Azotobacter chroococcum, 166

B

Bag bioreactors (BB), 232
Biocontrol agent
  antibiosis, 153–154
  competition, 151–152
  control capabilities, 150, 151
  mechanism of action (MOA), 151
  mycoparasitism, 152–153
  plant diseases, 154
  reduction of, 154–155
Biofertilizers
  defined, 160
  T. asperellum, 161
  T. koningiopsis, 161
  T. virens, 162
Biogas, 53–54
  landfills, 55–56
  modeling as tool, 57–59
  MSW decomposition, 56–57
Biotic stress
  DNA recombinant, 249
  genome edition techniques, 249
    clustered regularly interspaced short palindromic repeats (CRISPR), 252–254
    DNA sequences, 250
    double-stranded break (DSB), 250
    transcriptional activator-like effector nucleases (TALEN), 250–252
    zinc finger nucleases (ZFNs), 250–251
  tropical crops
    agrobacterium infection, 255
    Duncan grapefruit, 255
    genome edition in, 254–258
    Magnaporthe oryzae, 256
    MAX1 (Solyc08g062950), 258
    OsSEC3A gene, 257
    Phelipanche aegyptiaca, 257
    tomato yellow leaf, 257
    virus-infecting plants, 256
    Xanthomonas citri spp., 255
    ZYMV virus infection, 256
Blackberry juice, 78–79
Byrsonima crassifolia, 123–124, 132

C

Caffeine, 37
Case studies
  landfill, 62
  modeling results, 61–62
  San Luis Potosi, Mexico, 59–60
Chemical fertilizers, 164–165
Chihuahua desert endemic plants
  ancestral knowledge, 264
    ethnobotany, 265
    medicinal plants, 265
    secrets, 265
  importance of
    Mexican territorial surface, 266
    phytochemicals, 267
  natural compounds, bioactivities, 269–270
    Euphorbia antisyphilitica zucc (candelilla), 272–273
    Larrea tridentata (gobernadora), 273
    Lippia graveolens (oregano), 271–272
    Mesquite (*Prosopis* spp.), 273
    nutraceuticals effect, 274
    Opuntia ficus, 271
  phytochemicals
    secondary metabolism and production, 267–268
Circular economy (CE), 114
Clustered regularly interspaced short palindromic repeats (CRISPR), 252–254
Coffea arabica
  characteristics of, 24
Coffea canephora
  characteristics of, 24–25
Coffee industrialization
  agro-industrial waste
    accumulation and poor disposal, 32–33
    classification, 27–28
    coffee bean, 28–29
    generation of, 25–27
    grounds, 30–32
    pulp, 30
    revalorization of, 33–36
  diterpenes
    cafestol and kahweol, 38
    caffeic acid, 39
    chlorogenic acids, 39
    ellagic acid, 39
    ferulic acid, 39
    nonflavonoid polyphenols, 39
    polyphenolic compounds, 38
  flavonoid polyphenols, 39
    anthocyanins, 40–41
    flavanols, 40
  phytochemicals from, 36
    alkaloids, 37
    caffeine, 37
    methylxanthines, 37
    trigonelline, 37
  USDA reports, 25
Cucurbita ficifolia, 125–126, 131
Curcuma longa L., 92
  materials and methods
    DPPH free radical-scavenging, 97–98
    encapsulation efficiency, 97
    lipid oxidation inhibition (LOI) assay, 98
    matrix selection, 95
    microcapsules, 94–95
    protein saturation, 97
    scanning electron microscopy (SEM), 98–99
    solubility in water, 96–97
    Taguchi-based statistical method, 95–96
  microcapsules, 93
  polyphenols
    microencapsulation of, 94
  results and discussion
    antioxidant activities, 103–106
    encapsulation efficiency, 101
    matrix selection, 99–100
    microcapsules, 107
    microencapsulation, 99–100
    solubility, 100–101, 103–106
    Taguchi experimental design, 100
  spray-drying, 93

D

Diterpenes
  cafestol and kahweol, 38
  caffeic acid, 39
  chlorogenic acids, 39
  ellagic acid, 39
  ferulic acid, 39
  nonflavonoid polyphenols, 39
  polyphenolic compounds, 38

Domestic waste
methane production, 52–53
Double-stranded break (DSB), 250
DPPH free radical-scavenging, 97–98
Duncan grapefruit, 255

E

Engineer oleogels
challenges in, 216–217
gelator influences, 215–216
molecular self-assembly, 211–214
solvent, 215–216
Euphorbia antisyphilitica zucc (candelilla), 272–273

F

Flavonoid polyphenols, 39
anthocyanins, 40–41
flavanols, 40

G

Genome edition techniques, 249
clustered regularly interspaced short palindromic repeats (CRISPR), 252–254
DNA sequences, 250
double-stranded break (DSB), 250
transcriptional activator-like effector nucleases (TALEN), 250–252
zinc finger nucleases (ZFNs), 250–251
Glass transition temperature, 83–84
Guava juice, 81

H

Heavy metal decontamination, 169
biosorption, 170–171

J

Jenbacher-J320 ICM, 64

L

Larrea tridentata (gobernadora), 273
Life cycle assessment (LCA), 114
Lipid oxidation inhibition (LOI) assay, 98
Lippia graveolens (oregano), 271–272
Litchi chinensis (litchi), 3
agro industrial waste, 7–8
biorefineries, 11
feedstock, 9–10
integrated biorefinery systems, 10
production and consumption, 5–7
sustainable biorefineries
biomass feedstocks, 16
conversion technologies, 14–15, 17
environmental and social impacts, 15
feedstock availability and supply, 14
integration and scale-up, 15
policy and market support, 15–16
process integration, 17–18
waste valorization, 17–18
transformation
microwave-assisted extraction (MAE), 12
pectin extraction, 13
phenolic compounds, 11–13
polysaccharide extraction, 13
soxhlet extraction, 12
supercritical fluid extraction (SFE), 12–13
ultrasound-assisted extraction (UAE), 12
valuable components, 8–9
Lower Calorific Power (LCP), 63
Lulo pulp, 79

M

Magnaporthe oryzae, 256
MAX1 (Solyc08g062950), 258
Mesquite (*Prosopis* spp.), 273
Methane correction factor (MCF), 61
Methylxanthines, 37
Mexican territorial surface, 266
Mexico
neglected and underutilized species (NUS)
Annona purpurea, 124–125
Artocarpus, 116–117
Byrsonima crassifolia, 123–124
Cucurbita ficifolia, 125–126
Myrciaria floribunda, 119–120
Parmentiera aculeata, 117–118
Solanum betaceum, 120
Theobroma cacao L, 121–122
Microbial activities
Azotobacter chroococcum, 166
Phytopthtora capsici, 167
Serendipita indica, 167
soils and roots, 165
undesirables (pathogens), 166

Microorganisms biostimulants
  defined, 155
Microwave-assisted extraction (MAE), 12
Monodehydroascorbate reductase (MDAR), 158
Municipal solid waste (MSW), 50
Myrciaria floribunda, 119–120, 133

N

Neglected and underutilized species (NUS)
  bioactive compounds, 128–129
  biological activities, 126–127
  economic impact, 132
    Artocarpus altilis fosberg, 133
    multi and transdisciplinary strategy, 134
    Myrciaria floribunda, 133
    Parmentiera aculeate, 133
    Physalis peruviana, 133
    Saccharomyces cerevisiae, 133
    Saccharomyces pastorianus, 133
    Solanum betaceum, 133
    traditional Mexican fermented beverages (TMFB), 134, 135
  environmental impacts, 132
    Artocarpus altilis fosberg, 133
    multi and transdisciplinary strategy, 134
    Myrciaria floribunda, 133
    Parmentiera aculeate, 133
    Physalis peruviana, 133
    Saccharomyces cerevisiae, 133
    Saccharomyces pastorianus, 133
    Solanum betaceum, 133
    traditional Mexican fermented beverages (TMFB), 134, 135
  in Mexico
    Annona purpurea, 124–125
    Artocarpus, 116–117
    Byrsonima crassifolia, 123–124
    Cucurbita ficifolia, 125–126
    Myrciaria floribunda, 119–120
    Parmentiera aculeata, 117–118
    Solanum betaceum, 120
    Theobroma cacao L, 121–122
  potential applications
    Byrsonima crassifolia, 132
    Cucurbita ficifolia, 131
    jackfruit (*Artocarpus heterophyllus* Lam.), 131–132
    Teobroma bicolor, 130
  social impact, 132
    Artocarpus altilis fosberg, 133
    multi and transdisciplinary strategy, 134
    Myrciaria floribunda, 133
    Parmentiera aculeate, 133
    Physalis peruviana, 133
    Saccharomyces cerevisiae, 133
    Saccharomyces pastorianus, 133
    Solanum betaceum, 133
    traditional Mexican fermented beverages (TMFB), 134, 135

O

Oleogels
  defined, 210
  engineer oleogels
    challenges in, 216–217
    gelator influences, 215–216
    molecular self-assembly, 211–214
    solvent, 215–216
  functional properties, 217–220
Opuntia ficus, 271
Oregano
  applications
    consumption in, 196
    oregano essential oil (OEO), 197–198
  bioactive compounds
    extraction method of, 200–201
  botanical characteristics, 191–192
  chemical composition, 192, 194
  global importance, 195–196
  Mexico, 194
  producing countries, 196
  species, 194–195
  sustainable use
    harnessing works, 198–199
    residues, 198
    solid-state fermentation, 199, 201
Oregano essential oil (OEO), 197–198
OsSEC3A gene, 257

P

Packed column bioreactors (PCBs), 234, 235
Parmentiera aculeata, 117–118, 133
Pectin extraction, 13
Phelipanche aegyptiaca, 257
Phenylalanine ammonia lyase (PAL), 158

Physalis peruviana, 133, 166
Phytopthtora capsici, 167
Pineapple juice, 81–83
Plant growth-promoting bacteria (PGPB), 282
  applications
    agriculture, 285
    health benefits, 284–285
  biostimulants in plants, 285
    hydroponic method, 286–287
    probiotic methods, 287–288
    sustainable methods, 288
  probiotics
    beneficial effects, 283
    mechanism in, 289–290
    microorganisms, 284
  sprouting
    defined, 282
    germination process, 282, 283

R

Resource recovery from waste
  biogas, 53–54
    landfills, 55–56
    modeling as a tool, 57–59
    MSW decomposition, 56–57
  case studies
    landfill, 62
    modeling results, 61–62
    San Luis Potosi, Mexico, 59–60
  domestic waste
    methane production, 52–53
  electrical measurement
    Jenbacher-J320 ICM, 64
    Lower Calorific Power (LCP), 63
  methane correction factor (MCF), 61
  MSW in developing countries, 51
    municipalities, 52
  municipal solid waste (MSW), 50
  United States Environmental Protection Agency (USEPA), 62
  valorization, 51

S

Saccharomyces cerevisiae, 133
Saccharomyces pastorianus, 133
Scanning electron microscopy (SEM), 98–99
Secondary metabolism and production, 267–268
Serendipita indica, 167
Soil bioremediator
  degradation mechanisms, 168–169
  synthetic agrochemicals degrader, 167–168
Solanum betaceum, 120, 133
Solid-state culture bioreactors
  characteristics
    antifungals, 231
    fungal herbicides, 230–231
    fungal insecticides, 229–230
  propagules production, 231
    agitated bioreactors (ABs), 237
    bag bioreactors (BB), 232
    packed column bioreactors (PCBs), 234, 235
    tray bioreactors (TB), 233–234
Solid-state fermentation, 199, 201
Soxhlet extraction, 12
Soy protein isolate (SPI), 79
Spray drying, 72, 93
  application in the industry, 77
  atomization, 74
  co-current drying, 74
  drying process, 73
  nonsticky materials, 73
  physical properties, 74
    hygroscopicity, 76
    moisture content, 75
    solubility, 76
    water activity, 76
  pulverizing sugar-rich foods, 84–85
  seasonal fruits, 77
    blackberry juice, 78–79
    glass transition temperature, 83–84
    guava juice, 81
    lulo pulp, 79
    pineapple juice, 81–83
    tamarind juice, 79–80
    watermelon juice, 80–81
  soy protein isolate (SPI), 79
  stages in, 74
  sticky materials, 73
  water evaporation, 74
Sprouting
  defined, 282
  germination process, 282, 283

Supercritical fluid extraction (SFE), 12–13
Sustainable biorefineries
  biomass feedstocks, 16
  conversion technologies, 14–15, 17
  environmental and social impacts, 15
  feedstock availability and supply, 14
  integration and scale-up, 15
  policy and market support, 15–16
  process integration, 17–18
  waste valorization, 17–18
Sustainable Development Goal (SDG), 114

T

Taguchi-based statistical method, 95–96
Tamarind juice, 79–80
Teobroma bicolor, 130
Theobroma cacao L, 121–122
Three of the nonexpressor of pathogenesis-related 3 *(TcNPR3),* 254, 255
Tomato yellow leaf, 257
Traditional Mexican fermented beverages (TMFB), 115, 134, 135
Transcriptional activator-like effector nucleases (TALEN), 250, 251–252
Transformation
  microwave-assisted extraction (MAE), 12
  pectin extraction, 13
  phenolic compounds, 11–13
  polysaccharide extraction, 13
  soxhlet extraction, 12
  supercritical fluid extraction (SFE), 12–13
  ultrasound-assisted extraction (UAE), 12
Tray bioreactors (TB), 233–234
Trichoderma
  abiotic stress tolerance, 156
    abscisic acid (ABA), 157
    1-aminocyclopropane-1-carboxylate deaminase (ACCD), 158
    ethylen (ET), 157
    hormone regulation, 157–158
    monodehydroascorbate reductase (MDAR), 158
    phenylalanine ammonia lyase (PAL), 158
    physiological adaptation, 159–160
  biocontrol agent
    antibiosis, 153–154
    competition, 151–152
    control capabilities, 150, 151
    mechanism of action (MOA), 151
    mycoparasitism, 152–153
    plant diseases, 154
    reduction of, 154–155
  biofertilizer
    defined, 160
    T. asperellum, 161
    T. koningiopsis, 161
    T. virens, 162
  chemical fertilizers, 164–165
  climate change, 149
  endophytic fungi, 156
  heavy metal decontamination, 169
    biosorption, 170–171
  microbial activities
    Azotobacter chroococcum, 166
    Phytopthtora capsici, 167
    Serendipita indica, 167
    soils and roots, 165
    undesirables (pathogens), 166
  microorganisms biostimulants
    defined, 155
  modern agriculture, 148
  nanoparticles, 171
    biosynthesis of, 172–174
    NPS, effect, 174–175
  plant biostimulants
    defined, 155
  plant growth
    effects, 156
  soil bioremediator
    degradation mechanisms, 168–169
    synthetic agrochemicals degrader, 167–168
  soil inhabitant and root colonizer, 150
Trigonelline, 37
Tropical crops
  agrobacterium infection, 255
  Duncan grapefruit, 255
  genome edition in, 254–258
  Magnaporthe oryzae, 256
  MAX1 (Solyc08g062950), 258
  OsSEC3A gene, 257
  Phelipanche aegyptiaca, 257
  tomato yellow leaf, 257
  virus-infecting plants, 256
  Xanthomonas citri spp., 255
  ZYMV virus infection, 256

U

Ultrasound-assisted extraction (UAE), 12
Undesirables (pathogens), 166
UNESCO Universal Declaration on Cultural Diversity of 2001, 264
United Nations Conference on Environment and Development (UNCED), 228
United States Environmental Protection Agency (USEPA), 62

V

Valorization, 51
Virus-infecting plants, 256

W

Waste valorization, 17–18
Water evaporation, 74
Watermelon juice, 80–81

X

Xanthomonas citri spp., 255

Z

Zinc finger nucleases (ZFNs), 250–251
ZYMV virus infection, 256